Geology of the Elsinore Fault Zone San Diego Region

GEOLOGIC TIME SCALE

ERA	PERIOD	EPOCH	Millions of Years Ago
CENOZOIC	QUATERNARY	Holocene	0.011
		Pleistocene	1.8
	TERTIARY NEOGENE	Pliocene	5.3
		Miocene	23.8
	PALEOGENE	Oligocene	33.7
		Eocene	55
		Paleocene	65
MESOZOIC	CRETACEOUS	Late	99
		Early	142
	JURASSIC	Late Middle Early	205.7
	TRIASSIC	Late Middle Early	250
PALEOZOIC	PERMIAN	Late Early	290
	CARBONIFEROUS PENNSYLVANIAN	Late	323
	MISSISSIPPIAN	Early	354
	DEVONIAN	Late Middle Early	417
	SILURIAN	Late Early	443
	ORDOVICIAN	Late Middle Early	495
	CAMBRIAN	Late Middle Early	545
	PRECAMBRIAN		
ORIGIN OF EARTH			4,550

From Abbott, Patrick L. 1999. *The Rise and Fall of San Diego: 150 Million Years of History Recorded in Sedimentary Rocks.* San Diego: Sunbelt Publications

Geology of the Elsinore Fault Zone San Diego Region

SDAG/SCGS Volume Number 31- 2003

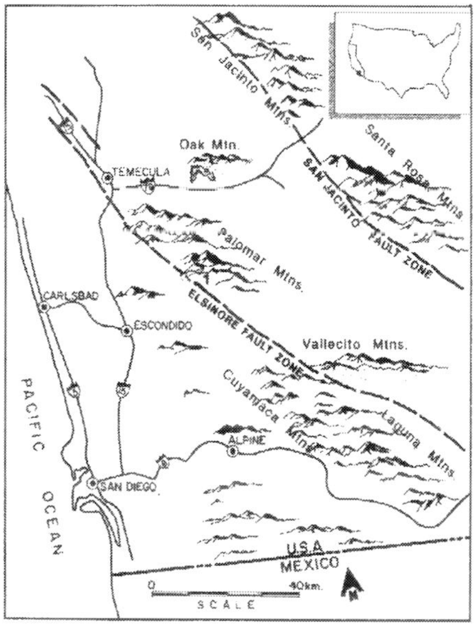

SAN DIEGO ASSOCIATION OF GEOLOGISTS

SOUTH COAST GEOLOGICAL SOCIETY

Co-editors: Monte L. Murbach Michael W. Hart

Published in Conjunction with the 2003 Field Trip:
"Hot Springs and Tourmalines of Eastern San Diego County"

Distributed by Sunbelt Publications
www.sunbeltbooks.com

Geology of the Elsinore Fault Zone, San Diego Region

 First edition 2003, sixth printing 2021

Edited by Monte L. Murbach and Michael W. Hart
Book design and composition by W.G. Hample & Associates
Project management by Lowell Lindsay
Printed in the United States of America

Please direct comments and inquiries to:

SDAG
P.O. Box 191126
San Diego, California 92159-1126
w.w.w.sandiegogeologists.org

SCGS
P.O. Box 10244
Santa Ana, California 92711-0244
www.southcoastgeo.org

or the distributor:

Sunbelt Publications
1250 Fayette St.
El Cajon, California 92020
(619) 258-4911, fax: (619) 258-4916
www.sunbeltbooks.com
mail@sunbeltpub.com

24 23 22 21 9 8 7 6

Library of Congress Cataloguing-in-Publication Data

Geology of the Elsinore Fault Zone, San Diego Region / co-editors, Monte L. Murbach, Michael W. Hart.-- 1st ed.
p. cm.
"SDAG/SCGS volume number 31, 2003."
"Published in conjunction with the 2003 field trips: Hot springs and tourmalines of eastern San Diego County."
Includes bibliographical references.
ISBN 0-916251-65-9
1. Geology-California-San Diego County-Guidebooks. 2. Geology-California-Imperial County-Guidebooks. 3. Faults (Geology)--California-San Diego County-Guidebooks. 4. Faultgs (Geology)--California-Imperial County-Guidebooks. 5. Elsinore Fault (Calif.)--Guidebooks. 6. San Diego County (Calif.)--Guidebooks. 7. Imperial County (Calif.)--Guidebooks. I. Murbach, Monte L. II. Hart, Michael W.
QE90.S116G46 2003
557.94'98--dc22
20032022828

Cover photographs courtesy of:
Front — Philip T. Farquharson, aerial view from Julian to the east down Banner Canyon to Granite Mountain. Frank Morris, Pilot.
Front inside — Portion of Geologic Map of the Santa Ysabel Quadrangel, R. Merriam, 1958.
Back — P. Farquharson, blue-gray beryl and quartz crystals on Clevelandite extracted from Cryo-Genie mine, 2002.
Back inside — Charles E. Houser, blue cap pink tourmaline from Pala District.

DEDICATION

DARYL WALKER STREIFF

February 11, 1943 - August 1, 2003

Daryl Streiff, a well-respected engineering geologist, passed away during compilation of this field trip guidebook. This volume is dedicated to him, and fittingly so. Daryl worked at the southern end of the Elsinore fault for his senior thesis at San Diego State University (see abstract below). He also worked on other projects related to the field trip area, more recently including the Diamond Valley Lake (aka Domenigoni or Eastside) reservoir project near Hemet (see photo below). Daryl was also a member and past president of the San Diego Association of Geologists and a member of the South Coast Geological Society. SDAG and SCGS are the joint organizers of this field trip.

Daryl graduated from San Diego State University with a degree in geology in 1972. During college, he began his career with Woodward-Clyde Consultants as a Field/Laboratory Technician. Upon graduation, he maintained employment with the firm, and eventually became Chief Geologist of their San Diego office. During his professional career, Daryl worked primarily in southern California. However, he also provided his geologic expertise on projects in Alaska, Utah, Nevada, Iran, Thailand, Malaysia, Western Australia, Kosrae Island (Micronesia), American Samoa, and the US Virgin Islands. Most recently, Daryl had worked during semi-retirement as an independent engineering geologist. He became well known for his adeptness at reviewing rock core for major pipeline, tunnel and dam projects in the western United States. Daryl was notably a multi-year president of the SDSU Geology Alumni Association and the recipient of their Honored Alumnus award in 1999. He was also a local Chairman for the 1991 Geological Society of America annual meeting held in San Diego.

Daryl died in his Pagosa Springs, Colorado home with his wife, Karen, by his side. He had endured a long, brave battle with pancreatic cancer. He loved his family, friends, God, fishing, hiking, skiing, NASCAR racing, and....... geology. He is missed!

Geology of the Yuha Basin Southwestern Imperial County, California

Daryl Streiff, Spring 1971

The Yuha Basin, in the southwestern Imperial Valley of Southern California, consists of 1,300 feet of marine mudstones, siltstones and sandstones of the Imperial Formation, which form a transitional contact with the overlying 980 feet of terrestial sandstones, claystones and mudstones of the Palm Spring Formation. The formations are of late Pliocene and early Pleistocene age respectively. Terraces that surround the basin form an unconformable contact with the Imperial and Palm Spring Formations and isolate the basin from surrounding geologic features.

The sediments are exposed along a northeast-southwest, doubly-plunging anticline that was formed by the compressive components of north-south regional shear stress. The anticline has been offset on the southwest nose by a right-lateral oblique fault that is regionally correlative to the Laguna Salada and the Elsinore fault zones. All of the faults present in the mapped area are reverse faults, and were caused by regional shear stress.

Rock-core review for Diamond Valley Lake project, Hemet

2003 Officers

SDAG

President	Patrick Brooks brooksgp@efdsw.navfac.navy.mil
Vice President	Monte Murbach mmurbach@petra-inc.com
Secretary	Margaret Eggers meggers@eggersenv.com
Treasurer	Philip Farquharson geoguy@cq-squared.com
Publications Manager	Lowell Lindsay llindsay@sunbeltpub.com
Webmaster	Carolyn Glockhoff carolyn@caro-lion.com

SCGS

President	Mike Hart mwhart@aol.com
Vice President	Mark Legg mrlegg@attglobal.net
Secretary	Bob Lemmer rlemmer@leightongeo.com
Treasurer	John Bonsangue jbonsangue@ocwd.com
Publications Manager	Ed Steiner esteiner@leightongeo.com

2003 Corporate Sponsors

SDAG

Anteon International Corporation
Dr. Patrick L. Abbott
Dr. Richard Berry
CG-Squared Productions
Christian Wheeler Engineering
Joe Corones
ENV America, Inc.
Dr. Margaret R. Eggers, RG, CHG
William J. Elliott
Eldon Gath, Earth Consultants, Inc.
Katherine Freese
Carolyn Glockhoff
Dr. Sarah Gray
Hargis + Associates, Inc.
Rob Hawk
John Hoobs, Geocon, Inc.
Kleinfelder, Inc.
Lowell Lindsay, Sunbelt Publications
Dr. Monte Marshall
Pacific Soils Engineering, Inc.
Prosonic Corp., Sean Roy
Robert Walters Equipment Rentals
Bob Smillie, TerraCosta Consulting Group
Southland Geotechnical Consultants
David and Jan Steller
Dr. Anne Sturz
Carole L. Ziegler

SCGS

Camp, Dresser, and McKee, Inc.
Cotton, Shires & Associates, Inc.
Earth Consultants International, Inc.
William J. Elliott, Engineering Geologist
Geomatrix
Goffman, McCormick & Urban, Inc.
Gary Guacci, R.G.
Gossco Manufacturing, Inc.
Layne Christensen Company
Leighton and Associates, Inc.
Pacific Surveys
Roscoe Moss Company
Stony-Miller Consultants, Inc.
WDC Exploration and Wells
Water Well Renovations, Inc.
Marina D. West, R.G.

The generosity of these corporate sponsors allows our organizations to continue to keep membership dues low, sponsor reduced rates for students, produce the monthly newsletter, purchase supplies and host the annual field trip, produce our geologic guidebooks, and provide scholarships to local geology students. To contact any of the above sponsors, please visit our websites at www.sandiegogeologists.org and www.southcoastgeo.org.

Contents

Introduction and Acknowledgments

This publication covers the geology of the Elsinore fault zone, San Diego region, and includes the 2003 field trips for the San Diego Association of Geologists (SDAG) and South Coast Geological Society (SCGS). This book is useful for anyone interested in geology, mines, minerals, hot springs, history, and Holocene faulting along the southern Elsinore fault zone (EFZ) in eastern San Diego County and western Imperial County.

The region discussed in this book and covered by each 2-day field trip consists of the area along the Elsinore fault zone generally between Mount Palomar at the north and the Coyote Mountains to the south. The trip will parallel and cross the Elsinore fault zone through the mountains, valleys and desert landscapes of southern California. The beautiful mountain community of Julian, with its rich gold mining history, is in the northern central portion of the route. San Diego's main hydrothermal features lie along the EFZ field trip route: Warner Springs, Agua Caliente County Park, and Jacumba Hot Springs. Related areas are in Riverside County (Murrieta and Glen Ivy) and southwest of Mexicali (Guadalupe Hot Springs).

This publication is unique as it is the first time that SDAG and SCGS have joined to produce a cooperative guidebook for two field trips. The field trips will follow nearly identical routes but will occur on different dates. The SDAG field trip is scheduled for October 25 and 26, while the SCGS trip follows two weeks later on November 8 and 9, 2003.

This book has three main sections: road log, new publications, and reprints of associated geologic papers. The road log follows the route of both trips and provides instructions and useful information for self-guided driving tours through many points of geologic and historic interest in the region. Attached to the road log is a resource list and a summary of major geologic works performed along the field trip route. The new publications section includes a new paper by the world-renowned California geologist, Thomas Dibblee, Jr., and excellent new publications by other local geologists and scientists. The reprinted papers section represents a collection of selected and previously published geology-related papers in this region.

The SDAG and SCGS field trips begin with Julian Pie. Since we are in the famous Julian apple district, it would not be appropriate to start the trip without Julian Pie for breakfast! We will look at some recently collected tourmalines right after the continental breakfast with a tour at the Cryo-Genie mine. After a look at offsets along the Elsinore fault we will enjoy lunch at Menghini Winery. Camping for the evening will occur at the group camp section of the wonderful San Diego County Park, Agua Caliente Hot Springs., where people can soak their tired feet. Chuck Houser, tourmaline enthusiast and engineering geologist, will provide a slide presentation of pegmatites and tourmalines of San Diego County. Sunday's highlight will include lunch at the Pack Rat mine while enjoying incredible mountain and desert views near Jacumba. Please refer to the road log for additional details and field trip stops.

In planning the field trip, we would like to thank the following people:

Lowell Lindsay for tremendous editorial and publication support, and most of all, his enthusiasm for this project.

Diana Lindsay for providing editorial detail including significant historical and geological notes to the road log.

Phil Farquharson for providing his very graphic support for many photographs and maps in the guidebook, in particular the cover photograph and the road log guide maps.

Hample and Associates for design and composition of the guidebook.

Diane Murbach for handling the food and beverage logistics and helping with registration, and editing the road log appendix and introduction.

Toni and Mike Menghini Menghini Winery for use of their facilities and catering the food and fine vinos. A great resting area for anyone in the Julian area!

Dana Gochenour and Jim Means for allowing us access to the Cryo-Genie and Pack Rat mines, respectively.

Field Trip Stop Leaders: Jeffrey Patterson, Dave Schug, Mike Hart, Monte Murbach, George Copenhaver, Tom Rockwell, Chuck Houser, Mike Walawender, and anyone else coerced at the last minute.

Chuck Houser and Phil Farquharson for the Saturday night tourmaline presentation.

Monte Murbach, Vice President,
San Diego Association of Geologists

Mike Hart, President,
South Coast Geological Society

List of Current Publications — 2003

San Diego Association of Geologists www.sandiegogeologists.org

2003 — *Geology of the Elsinore Fault Zone, San Diego Region*
"Hot Springs and Tourmalines of Eastern San Diego County"
2001 — *Coastal Processes and Engineering Geology of San Diego, California*
2000 — *Geology and Enology of the Temecula Valley, Riverside County*
1999 — *Water for Southern California: Water Resources Development*
1998 — *Geology and Geothermal Resources of the Imperial and Mexicali Valleys*
1997 — *Santa Cruz Island Geology*
1995 — *Paleontology and Geology of Anza-Borrego Desert State Park*
1994 — *Geology and Natural History, USMC Camp Pendleton*
1992 — *Natural History of the Coronado Islands, Baja California, Revisited*
1991 — *Environmental Perils, San Diego Region*
1990 — *Geotechnical Engineering Case Histories in San Diego County*
1989 — *Seismic Risk in the San Diego Region*
1988 — *Landslides in Crystalline Basement Terrain — San Diego and Riverside Counties*

South Coast Geological Society www.southcoastgeo.org

2003 — *Geology of the Elsinore Fault Zone, San Diego Region*
"Hot Springs and Tourmalines of Eastern San Diego County"
2002 — *Structural Geomorphology, Petroleum and Water Resources, Northwestern Transverse Ranges*
2001 — *Geology of the Owens Valley and Inyo Mountains*
2001 — *Geologic Guidebook to the Long Valley-Mono Craters Region of Eastern California*
2000 — *Geology of the Las Vegas Area*
1999 — *The Long Valley Caldera, Mammoth Lakes, and Owens Valley Region, Mono County, California*
1998 — *Black Canyon of the Colorado River, Arizona/Nevada and Castle Mountains Gold Mine, San Bernardino County, California*
1997 — *Southern San Andreas Fault — Whitewater to Bombay Beach — Salton Trough*
1995 — *The Garlock Fault and El Paso Mountains*
1992 — *The Regressive Pleistocene Shoreline*
1991 — *Southern Coast Ranges*
1990 — *Lower Colorado River Extensional Terrane and Whipple Mountains*
1988 — *Geology of the Death Valley Region*
1982 — *Geology and Mineral Wealth of the California Transverse Ranges*

Related Publications www.sunbeltbooks.com

Patrick L. Abbott (1999) *The Rise and Fall of San Diego: 150 Million Years of History Recorded in Sedimentary Rocks*

Michael L. Cassiliano (2002) *The Stratigraphic Nomenclature of the Plio-Pleistocene Palm Spring Group, Anza-Borrego Desert*

Leland Fetzer (2002) *A Good Camp: Gold Mines of Julian and the Cuyamacas*

John Minch (1998) *Roadside Geology and Biology of Baja California*

Paul Remeika and Lowell Lindsay (1992) *Geology of Anza-Borrego: Edge of Creation*

Michael J. Walawender (2000) *The Peninsular Ranges: A Geological Guide to San Diego's Back Country*

II.

2003 Joint SDAG/SCGS Field Trip Road Log:

"Hot Springs and Tourmalines of Eastern San Diego County"

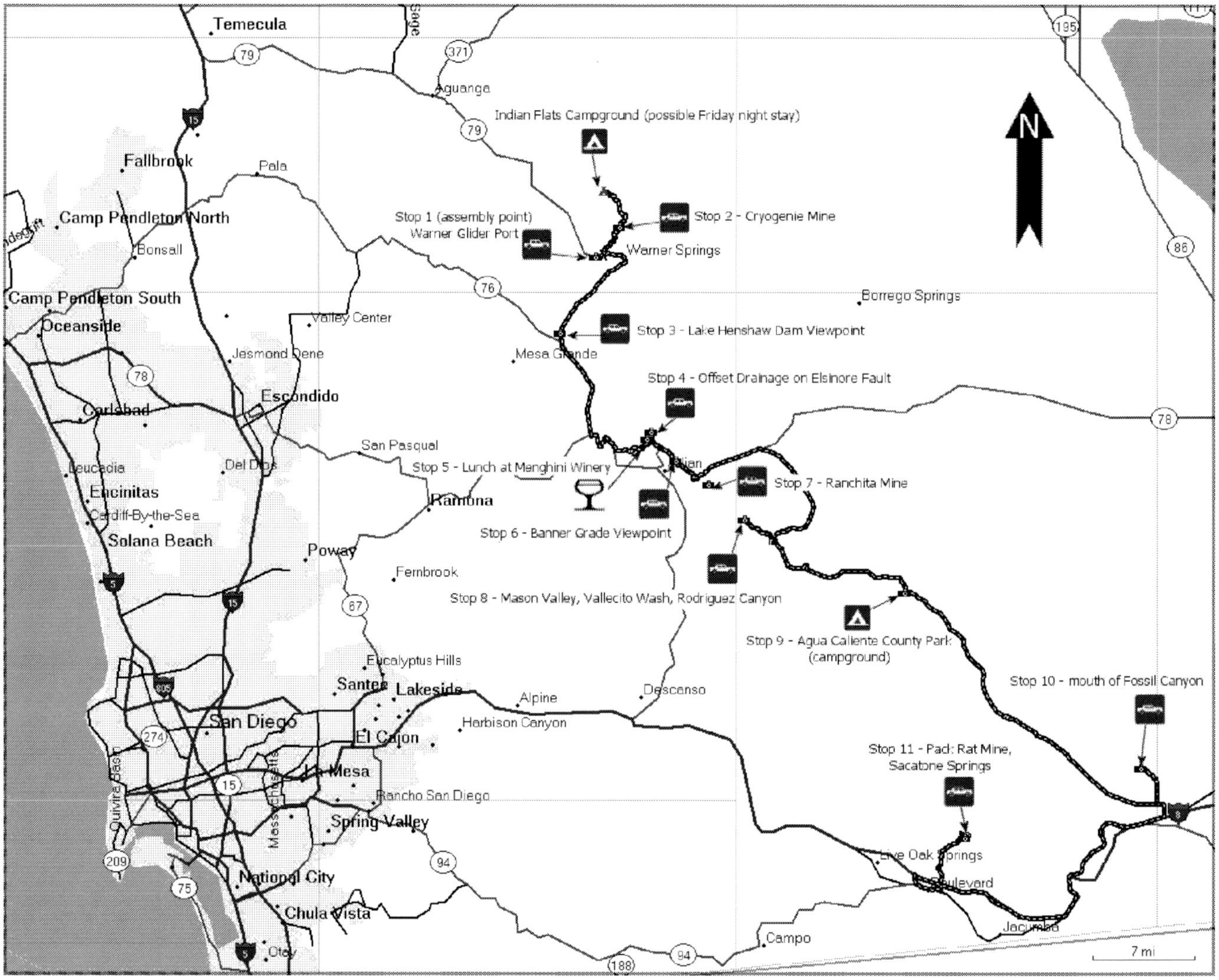

Map of San Diego County showing Stops 1-11 on 2003 SDAG/SCGS Field Trip.

Datum: WGS 84			
Name	Description	Lat/Lon hddd°mm.mmm' Position	UTM Position
Stop One	Warner Airport Terminal	N33° 17.099' W116° 40.105'	11 S 530875 3682931
Lost Valley Road Turnoff	Turnoff To Cryogenie Mine	N33° 17.191' W116° 39.617'	11 S 531632 3683102
Stop Two	Cryogenie Parking	N33° 18.813' W116° 38.417'	11 S 533484 3686106
Indian Flats Campground	Indian Flats Campground	N33° 20.949' W116° 39.591'	11 S 531650 3690046
Stop Three	Call Box 79-281	N33° 12.586' W116° 42.660'	11 S 526933 3674579
Stop Four	Elsinore Fault Offset Stream	N33° 06.743' W116° 36.127'	11 S 537122 3663817
Stop Five	Menghini Winery	N33° 06.298' W116° 36.415'	11 S 536677 3662992
Stop Six	Banner Grade Viewpoint	N33° 05.023' W116° 34.377'	11 S 539856 3660649
Stop Seven	Ranchita Mine	N33° 03.633' W116° 31.992'	11 S 543577 3658096
Stop Eight Turnoff	Oriflamme Canyon Road	N33° 00.463' W116° 27.381'	11 S 550781 3652273
Stop Eight	Rodriguez Canyon Mouth	N33° 01.573' W116° 29.493'	11 S 547483 3654308
Stop Nine	Agua Caliente County Park Entrance	N32° 57.303' W116° 18.145'	11 S 565199 3646521
Stop Ten	Fossil Canyon Mouth	N32° 46.986' W116° 01.200'	11 S 591774 3627666
Stop Eleven	Pack Rat Mine Overlook	N32° 42.827' W116° 13.807'	11 S 572152 3619820
Pack Rat Mine	Pack Rat Mine Upper Dike	N32° 42.968' W116° 13.867'	11 S 572056 3620079

2003 Joint SDAG/SCGS Field Trip Road Log
“Hot Springs and Tourmalines of Eastern San Diego County”

Diana Lindsay, Monte Murbach, Mike Hart

Warner Springs to Julian via Highway 79 through Santa Ysabel;
Julian to Scissors Crossing via Highway 78 down Banner Grade;
Scissors Crossing to Ocotillo via Highway S-2 through Anza-Borrego Desert State Park and Coyote Mountains;
Ocotillo to McCain Valley/Sacatone Overlook via Interstate 8 to Boulevard.

ODOMETER (TRIP) MILEAGE

Trip starts just east of Milepost (MP) 37.5 on Highway 79 at the airport.

0.0 Stop 1. Warner Springs Ranch Airport east of Warner Springs Resort on Highway 79. This is the field trip check-in spot, and Julian Pie and Coffee Fill-Up. This small airport is home to many glider and sailplane enthusiasts. We meet here, consolidate cars, head east on Highway 79 a short distance, and turn left onto Indian Flats Road, also called Lost Valley Road on some maps.

> MP 37.0. Sign to Lost Valley Road and Indian Flats Campground.

0.4 A brown sign at the road entrance reads: Indian Flats Campground 7 Miles. Over 100 years ago local Cupeño Indian bands called this area home. The Cupeño are related to the Cahuilla and are Shoshonean in culture. They were physically removed from their ancestral home in 1903 and placed on the Pala Indian Reservation by the owner of Warner Ranch, a former California governor.

3.0 From this mileage, start parking at pullouts.

3.2 Stop 2. Cryo-Genie Mine parking area. The Cryo-Genie is currently an active tourmaline mine. Jeffrey Patterson and the mine owner will give us a tour of the mine workings. <u>Access to this site is by permission only</u>.

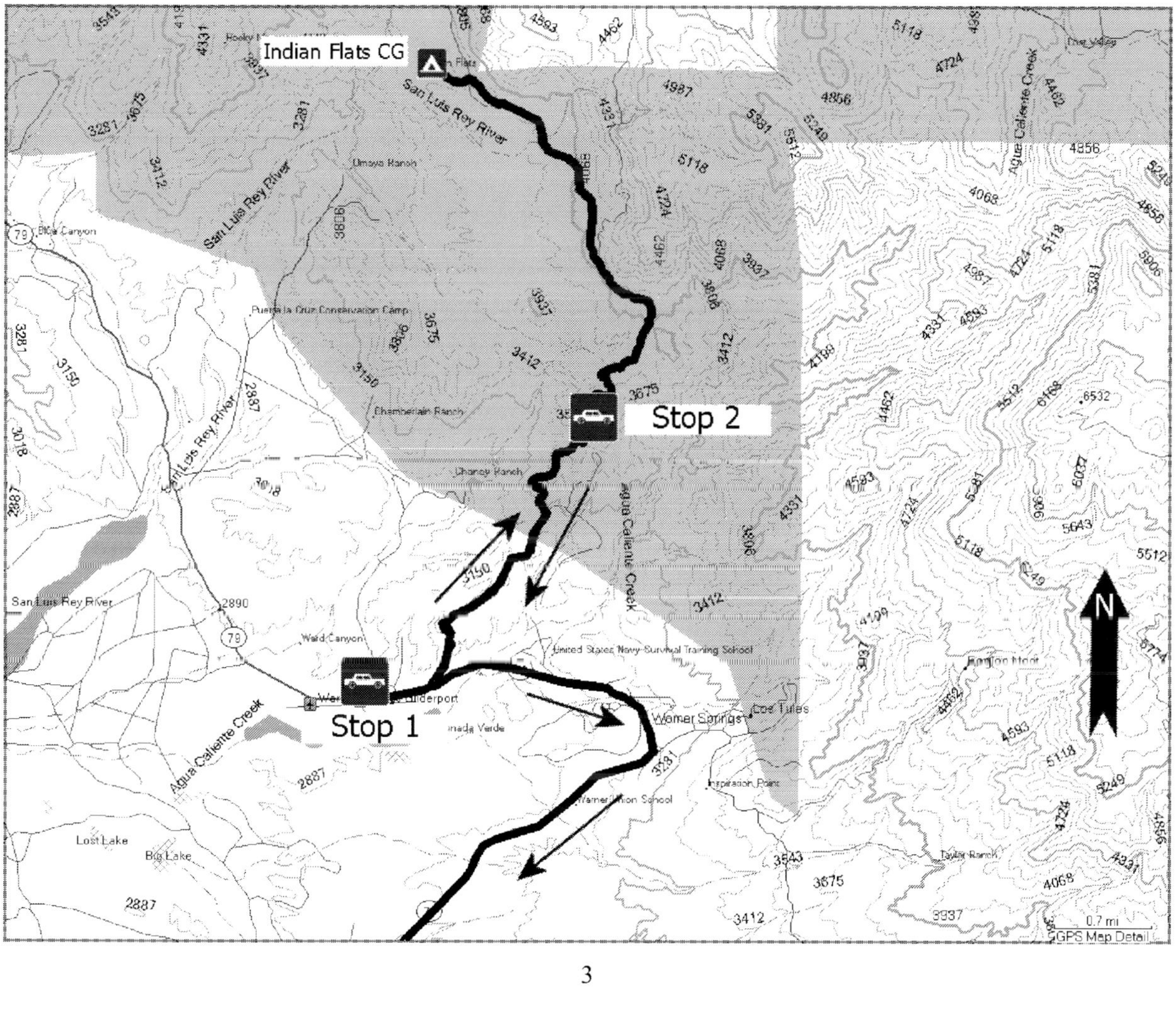

Stops 1 and 2.

The word tourmaline has its roots in India and Ceylon where traders first encountered this gem. "Turmali," as it was called in a Sinhalese dialect, means rainbow and refers to the variety of different colors found in a spectrum. This name reflects the many different colors of this gem — pink, red, lavender, blue, green, and bi-color, which has more variety than any other gem. Gem-quality pink tourmaline is five times as rare as gem diamond and more than 10 times as valuable as pure gold. It is harder than quartz (Mohs 7.253) and highly reflective when cut. Tourmaline was first discovered in San Diego County in the late 1860s, but serious production did not begin until 1890 when the Dowager Empress of China, Tzu Hsi, began purchasing all of the pink tourmaline that was available for export to China. Tiffany & Co. of New York handled orders. Over $800,000 of pink tourmaline was shipped to China from the Mesa Grande Mine alone between 1902 and 1910. The tourmaline boom ended with the death of the Dowager Empress in 1908 and the end of the Ch'ing Dynasty. Serious tourmaline mining did not resume again until the 1970s. The Cryo-Genie mine is an example of one of the newer operations. It was formerly known as the Lost Valley Truck Trail prospect, first located in 1904. In 1974 the property was claimed as the Cryo-Genie mine by Bart Cannon of Seattle, Washington, who worked it occasionally. Since 1994, Dana Gochenour of Tustin, California, has held the claim.

3.2 Additional parking on sides of road.

3.2 *Walk* to mine. Drive back down the road to the main highway.

3.5 Good overview of Lake Hodges Dam, valley and Mesa Grande to the east. During the Cupeño's "Trail of Tears" forced march to Pala, some Indians ran and hid in the surrounding hills or fled to Mesa Grande to seek refuge with local Indians who lived there. The Mesa Grande Indians are Kumeyaay, which is Yuman in culture and a different language than that spoken by the Cupeño.

5.9 Back at Highway 79. Turn Left. Head east, then south. Immediately on left, note the road cut exposing Temecula Arkose. Pass the Chapel of Saint Francis built in 1830 as an outlying church of Mission San Luis Rey to serve the Indians of the Warner Springs area.

MP 35.5 on Highway 79.

7.5 Warner Springs Ranch. Gasoline, small store, tennis courts, golf. Save it for another trip! Pala Indians who are descendants of the original Cupeño rented their ancestors' homes at the resort one weekend in 2003 on the 100th anniversary of the expulsion. Ceremonies held that weekend at the resort honored their grandparents and their sacred hot springs.

9.2 Mount Palomar Observatory, visible on skyline to right, is the white dome in the distance. It took over a decade before Caltech's 200-inch Hale telescope became operational in 1948. Today the observatory also has several smaller telescopes on the grounds.

MP 32.0 on Highway 79.

11.1 Pass Borrego Springs Junction. Less than a mile from this junction on Highway S-2 is the historic Butterfield Overland Mail Warner Springs Stage Station, located just below the original Warner Ranch established by Jonathan Trumbull Warner, an American who came to California during the Mexican period and received a large land grant from the Mexican government. Proceed on Highway 79 to Santa Ysabel (nearing Pie Country!)

14.6 Stop 3. Highway 79 Call Box 79-281. Park at the nice wide pullout. Here Dave Schug will give us an overview of the Lake Henshaw dam construction. This is a 15 minute stop. Back on the highway continue to Santa Ysabel. The hill near Lake Henshaw is called Monkey Hill, reputedly named for a monkey that had escaped from a circus caravan nearby.

15.9 Pass intersection with Highway 76. Stay on Highway 79.

MP 24 on Highway 79.

19.0 Elsinore fault zone, on left, traverses up through the valley on Indian reservation land. Also note the Santa Ysabel Mission built in 1818, an "asistencia" of the San Diego Mission, to serve the Kumeyaay Indians of the Santa Ysabel area. The Kumeyaay were the southern neighbors of the Cupeño of Warner Hot Springs. Their territory extends down into Baja, California. The Elsinore fault takes its name from the type locality — the Elsinore Mountains and town of Elsinore (now called Lake Elsinore), which, according to the Elsinore *News*, was a name selected from Shakespeare's works.

The Cupeño, and culturally related Cahuilla, explain earthquakes as part of their creation story. Two brothers, Mukat and Tamaioit, created the earth and sky, and each created their own people. Mukat's people are those who inhabit the earth today, but Tamaioit's people were ugly. After an argument, Tamaioit gathered his creations and retreated into the center of the earth, where today he still bellows in anger in the form of earthquakes.

MP 20.5 on Highway 79.

22.2 Santa Ysabel is the junction with Highway 78. The Santa Ysabel General Store at this historic crossroads was built in the 1870s during the Julian gold rush days. The junction is now famous for Dudley's Bakery and its many variety of breads sold throughout the county. Turn left onto Highway 78. Proceed toward the town of Julian.

MP 54.0 on Highway 78 (also Highway 79 into Julian).

25.1 Pass Orfila Wine Tasting Room. Great wines. Stop NEXT time…

25.6 Turn left onto Wynola Road. Wynola was an early community developed in the 1870s. It is the area where the famous Julian apple industry began. Note the Spencer Valley School on the east side of Highway 78 before you make your turn. The school is a one-school district, still in operation. Hang onto your hat as you proceed down the road. This road is quite the winding one. Note the dead pine trees, which died because of the pine beetle.

28.9 Left turn on Farmer Road (intersection with Wynola Road). Proceed north.

29.6 Stop 4. Gate on left, "Badlands Provisions" sign. Go through the gate. We are entering private property with permission. We will be looking at the Elsinore fault exposed in a road cut and will look at an offset alluvial fan and stream channel.

Back to the cars for a short drive to lunch. Drive back up the hill (southerly) direction.

30.2 At the stop sign, turn right on Wynola Road.

30.3 Immediately turn right again on Julian Orchards Drive. Head to Menghini Winery.

30.5 Stop 5. Lunch Stop at Menghini Winery. Here we enjoy a catered lunch and enjoy the winery's newly landscaped grounds.

After lunch, head back to Wynola Road. You are traveling through the Julian Mining District, made famous some 20 years after the discovery of gold in California. A.E. Coleman discovered gold on what is now called Coleman Creek, and a rush to the Julian area began in late 1869. The town of Julian was established by ex-Confederate soldiers Drury, Frank, and James Bailey and their cousins Mike and Webb Julian. On January 20, 1870, "Drue" discovered a quartz ledge, which he named Warrior's Rest, and on the very next day, other miners discovered a rich ledge that became the famous Washington Mine. Sometime later, Drue discovered his largest strike in Chariot Canyon, which he named the Ready Relief. Almost immediately after Drue's first discovery, the Julian Mining District was formed, and Mike Julian was named recorder. Drue began laying out the town, which he named Julian City for his cousin Mike. In later years, Drue said he named the city for his cousin because he was "handsomer than himself." Today, Julian remains a hub of the mountain community, steeped in rich mining history.

31.0 Left turn onto Wynola Road. Head down toward Banner Grade. Elsinore fault is on the left — in the trees!

31.1 The Elsinore fault is at the base of hill. Note the faceted spurs.

32.1 Junction at Highway 78. Turn left. Go down hill, along Banner Grade. In the 1870s, Banner Grade was considered to be the most dangerous road in the county. L.H. Wilcox built it in 1871 as a toll road to the Banner mines. Banner Creek runs through the canyon of the same name. Devastating storms that caused the creek to overflow in 1873-74, 1916, and again in 1926, wiped out the town of Banner at the bottom of the grade each time. The town was not rebuilt after the last washout. At one time, Banner City had a population of 600-1000 with 40 buildings that included three stores, one hotel, boarding houses, and four, perhaps even seven, saloons. Proceed down the grade. There is a fault bench 100 feet up from the bottom. Look for lineaments and offset stream channels, vegetation lineaments, and sidehill benches.

33.5 Stop 6. Pull over on the right side pullout. Maximum 20 cars. This is a great view of the Elsinore fault in Banner Grade.

34.1 Photo opportunity and pullout for single cars. Continue down the grade. Note recent burn on the hillsides from the Pines Fire in 2002.

36.1 Near the bottom of the grade, turn right onto a dirt road. This road is located just before the Banner Store. This road is for HIGH CLEARANCE VEHICLES ONLY.

36.3 At the gate. Proceed to the Ranchito Mine. The Ranchito Mine and a large five-stamp mill are among the more significant ruins found in Rodriguez Canyon. Ownership of this mine in 1957 was with the family of Cave J. Couts, an Army officer who figured prominently in San Diego

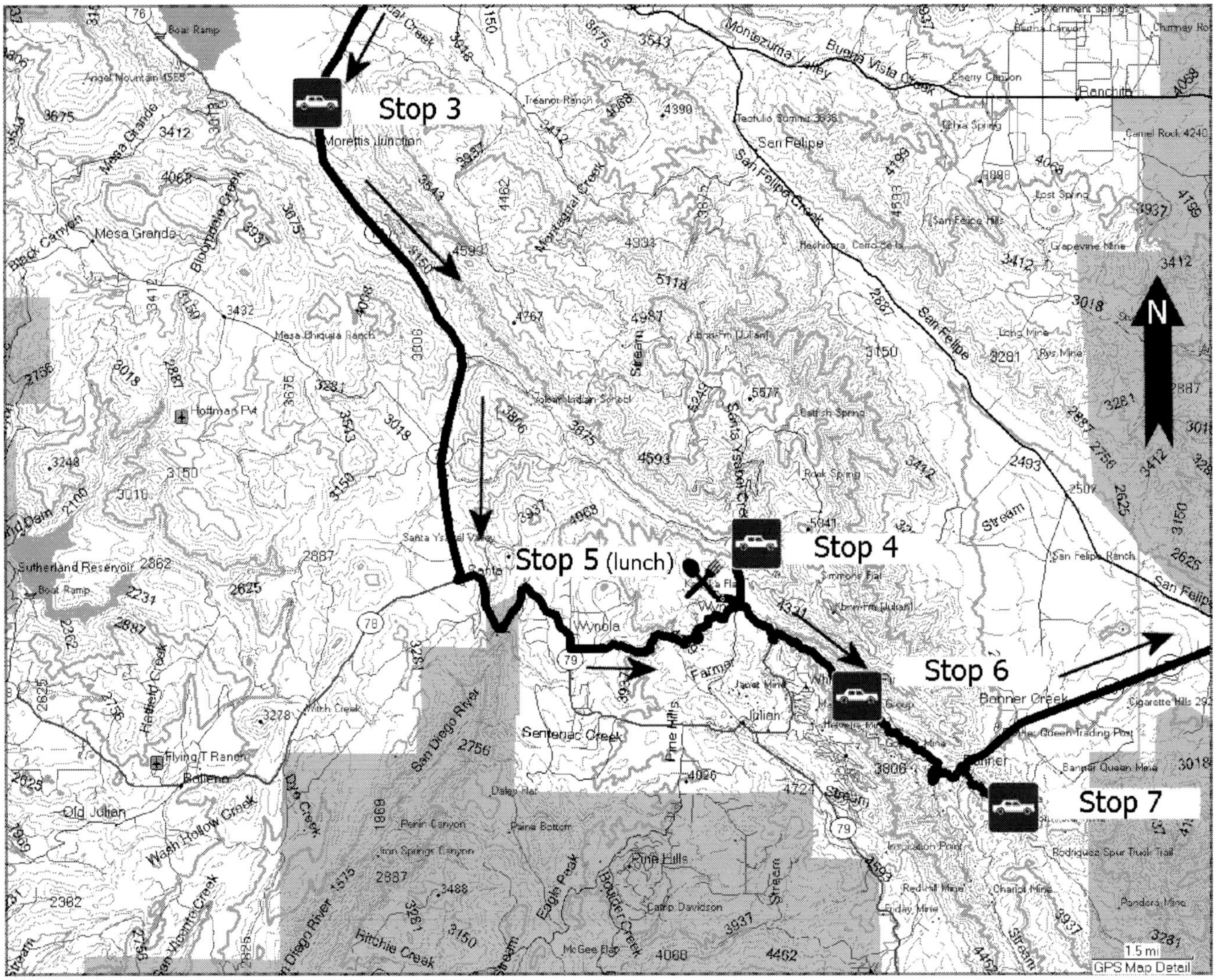

Stops 3 through 7.

history during and immediately after the Mexican War. Couts' son, Cave J. Couts Jr., purchased the Ranchito Mine in 1895 for $5500 cash. The Ranchito produced about 2500 ounces of pure gold and about 100 ounces of silver between 1895 and 1948. Considering the price of gold in 2000, that would approach a million dollars.

37.4 At the fork in the road, turn left, go up hill; swing left again on a well-traveled road. Looking north up Banner Grade, note the Elsinore fault on the right. The mountains on west are all Julian Schist.

38.2 Stop 7. Ranchito Mine works. We are at the gate to Right Fender Ranch. Note the right fender sign — it *is* a right fender! Park and walk a short distance. George Copenhaver will give

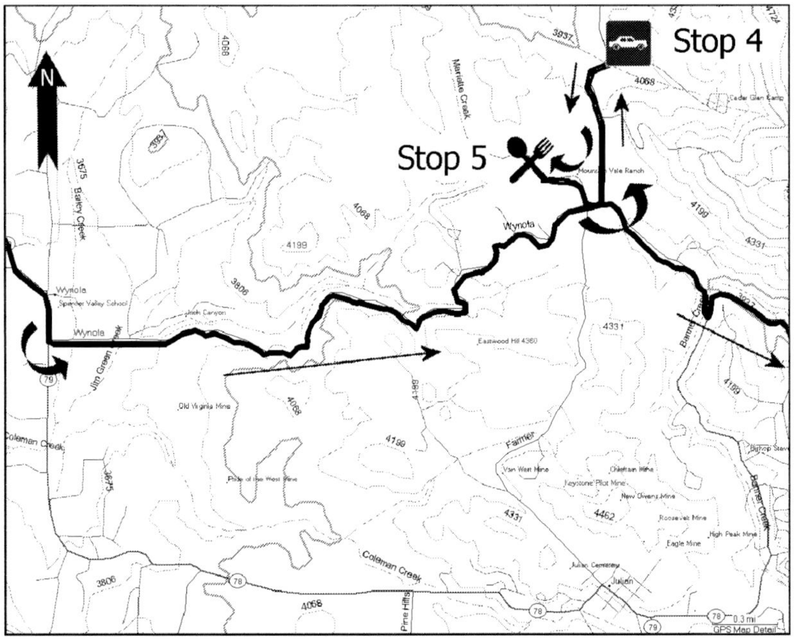

Stops 4 and 5.

us a short presentation on the history of gold mining in the Julian area.

Turn around and head back to the highway.

39.9 At the Y in the dirt road, turn right, and head down hill.

40.2 At Highway 78, turn right; proceed down hill past the Banner Recreation Ranch, once the center of Banner City. Banner was named for a small flag that was used to mark the site of a gold discovery. Some sources say it was an American flag that marked a discovery in Chariot Canyon by Louis B. Redman, while others contest it was a Confederate flag marking Drue and James Bailey's Ready Relief Mine, across the creek from Redman's mine.

MP 69.0 on Highway 78.

45.0 Highway S-2 junction (west), known as Scissors Crossing. Turn right onto S-2, heading towards Ocotillo along this historic byway. This route was part of the Southern Emigrant Trail to California first used after the discovery of gold in California in 1848. At this junction is a monument commemorating the Butterfield Overland Mail. Just north of this monument and west of the San Felipe Road is the site of the San Felipe Stage Station, which was located 18 miles from Vallecito and 10 miles from Warners Ranch. The station was active as a Butterfield stop from 1858 to 1861 when stagecoaches regularly ran between Tipton, Missouri, and San Francisco. Later, the station was used by Banning stages and by the military during the Civil War. It was the only wooden station on the western length of the Butterfield Mail. S-2 was erroneously named and signed as the "Great Southern Overland Stage Route of 1849" (wrong name and wrong dates!) by the county road department. As you travel, note the landslide in granitic rock on the right side. We are entering Shelter Valley, previously named Earthquake Valley. Developers in the valley felt that the name "Shelter" would be more conducive to real estate sales.

MP 21 on Highway S-2.

48.9 Pass Stagecoach Trail RV Park and Campground.

MP 22 on Highway S-2.

49.5 On the right side, note the landslide in metamorphic rock at the base of Granite Mountain near Cool Canyon (see photo on p. 14). This 5633-foot mountain is a granitic peak with many pegmatite dikes that were the focus of interest for many early gold prospectors. Granite Mountain Mine is found on its southwest slope. The Kumeyaay Indians who lived in this area have an answer as to how pegmatite dikes were created. In their creation story there was a giant monster, much like a centipede, that crawled over the land. Where he went, he left a slime path, which today we see as pegmatite dikes.

50.7 Blair Valley turnoff on left. Stay on the highway. Blair Valley, Little Blair Valley, and Earthquake (Shelter) Valley would form a 10-mile-long trough west of the Pinyon Mountains if it were not for a semicircular ridge of granite that separates the three into distinct areas. The granite is laced with pegmatite dikes that have long been the subject of prospecting. Blair Valley and Little Blair Valley are nestled between Granite Mountain and Whale Peak. The two valleys have dry playas that were interconnected lakes during the ice ages. Runoff from heavy rain floods Blair Dry Lake and the lower part of Blair Valley and can remain for several weeks.

MP 25.0 on Highway S-2.

52.2 Note the recumbent fold exposed in Julian Schist on hill to the right (see photo on next page). Julian Schist is one of the oldest rock bodies found in San Diego County. It is well exposed in the Julian area and takes its name from the mining town. Julian Schist dates back to the early Triassic or late Paleozoic time, around 250 Ma (mega-annum, million years). It is a metamorphic rock containing some minor gneiss and quartzite and some limestone and dolomite. It also contains gold, found chiefly in the quartzite beds, tungsten, manganese, and graphite and was one of the rock bodies sought by early-day prospectors.

53.4 Pass by the pullout for information on Box Canyon, the Mormon Battalion, and the route of the Butterfield Overland Mail (stage route). It was through this narrow canyon that the Mormon Battalion drove the first wagons to California in 1847. A 30-foot dry waterfall sits on the canyon floor. To avoid the waterfall, the road had to be routed up and around, which necessitated cutting the steep canyon walls by pick axe and hand tools until a passage was made wide enough for the wagons to roll through. In later years, the road was widened to accommodate the Butterfield stages.

MP 26.5 on Highway S-2.

Recumbent fold northwest of Milepost 25 on Highway S-2 between Blair Valley and Box Canyon, Anza-Borrego Desert State Park. Photo by Michael Hart.

54.5 At the mailbox and a SMALL brown sign that reads "Oriflamme Canyon," turn right onto the dirt road. High clearance vehicles are required. The name Oriflamme is French for "gold flame," and may refer to the gold mines found in the Oriflamme Mountain area or to the gold fever that swept the county, but in all likelihood, the origin of the name comes from a side-wheel steamship named *Oriflamme* that regularly stopped at San Diego Harbor as part of its route between San Francisco and the Atlantic Ocean. The first miners to arrive in 1870 during the Julian gold rush came on the *Oriflamme* and brought a two-stamp mill with them from San Francisco. Oriflamme was the name first given to a mine that was worked in the area, and the canyon and mountain name soon followed.

54.8 At the fork, turn right, go northwest and north.

56.2 Fault benches are on the right. This is a main strand of the Elsinore fault.

56.5 At the fork, bear right.

57.5 Stop 8. Take a short walk up to the canyon on the east. The fault zone is several hundred feet wide in metamorphic rocks exposed on the canyon wall. If there is time, we will walk a short distance to the south along the fault to view offset drainages. The Oriflamme Mine found in this canyon was the only gold producing mine in the present Anza-Borrego Desert State Park area. It was originally two separate claims, Lincoln Ledge and Warpath, but E.W. Morse renamed the claims Oriflamme when patented in 1882. The mine was located on the west side of Oriflamme Creek Canyon. Its peak production was probably between 1870 and 1885, although sporadic work continued until 1905. It is estimated that less than $25,000 came from the Oriflamme Mine. An attempt to get it into operation in the mid-1920s proved unsuccessful.

Go back to the vehicles. Turn around and head back downstream.

57.7 On the left (east), note the fault scarps and breaks-in-slopes on the hillside.

59.7 Turn left at the fork; go up hill (easterly direction) and back to the highway.

60.4 At the highway, turn right. The Elsinore fault is on the left (east) side of road.

62.7 The large road-clearing east (left) has an interpretive panel on the southeast side near the road (pullout on the left.) The panel commemorates the San Antonio-San Diego Mail Line. Congress had passed legislation in 1856 creating a mail and passenger service between San Francisco and the Mississippi River. Authority for the selection of both the route and the contractor was given to the Postmaster General, Aaron V. Brown, but Congress specified that an interim service should begin at once. John Butterfield of New York and James E. Birch of California submitted bids. Birch was awarded the interim service while the regular mail contract was awarded to John Butterfield. Birch's San Antonio-San Diego Mail Line employed mules to climb up Oriflamme Canyon to the Cuyamacas, and was popularly known as the "Jackass Mail." It was the first regularly scheduled overland mail and passenger stage line with connections to the Atlantic.

MP 33.0 on Highway S-2.

67.1 There is a landslide on the left at one o'clock in granitic rocks.

MP 38.0 on Highway S-2.

71.7 Stop 9. Turn right on Agua Caliente Hot Springs Road to the Group Camp section. No dogs are allowed at this campground; if you must have your dog, which must be leashed, you may camp almost anywhere in the state park as long as it is not in an established campground. No hard alcohol is allowed; beer, wine, and ice teas are okay. If you visit this campground on your own, keep in mind that it is closed between June 1 and September 1. The hot springs are at the foot of the Tierra Blanca Mountains at the east end of Vallecito Valley. The springs are part of the 910-acre Agua Caliente Regional Park, owned and operated by the San Diego County Department of Parks and Recreation. The springs were originally home to the Haawii ("water in rock"), a band of Kumeyaay known as Kwaaymii.

71.9 At ACHS store. Proceed to group campsite.

Next Day. Sunday Morning. After a hearty breakfast, we plan to head out at 8:30 AM.

72.1 Back at the highway, turn right, and go south past airstrip on the left.

MP 42.0 on Highway S-2. Badlands are found on the left in Palm Spring Formation.

77.8 Pass by a sign on the right: Canebrake Canyon. Canebrake literally refers to "a place overgrown with cane" or "a dense thicket of cane." The canyon, wash, and community that have grown at the mouth of the canyon take their name from the thick growth of reeds found in the canyon. Today this area is occupied by small homeowners who obtained their property after 1946 when the BLM surveyed the land and established five-acre plots available for a five-year lease "for such purposes as home, cabin, health and recreational sites." After leaseholders made certain improvements, they were able to purchase the property for a nominal fee. This land acquisition became known as the "Jack Rabbit Homestead Movement."

MP 46.0 on Highway S-2.

79.5 Pass Indian Canyon dirt road turn off on the right. There is a large alluvial fan up to the canyon. There are great campsites up the canyon (for future reference). The narrow gorge leads into Indian Valley where the Kumeyaay Indians had encampments, both in North Indian Valley and in South Indian Valley.

MP 47.0 on Highway S-2.

82.0 The road to Bow Willow Campground is on right. This excellent campground has water, fire pits, ramadas, and pit toilets. It was also a popular camping area used by the Kumeyaay Indians, as can be noted by the many mortero grinding holes found in the rocks. A short hike towards Mountain Palm Springs leads to various palm groves.

MP 47.5 on Highway S-2.

82.3 Carrizo Creek Wash. Upcanyon leads to Carrizo Gorge and the tracks of the San Diego & Arizona Eastern

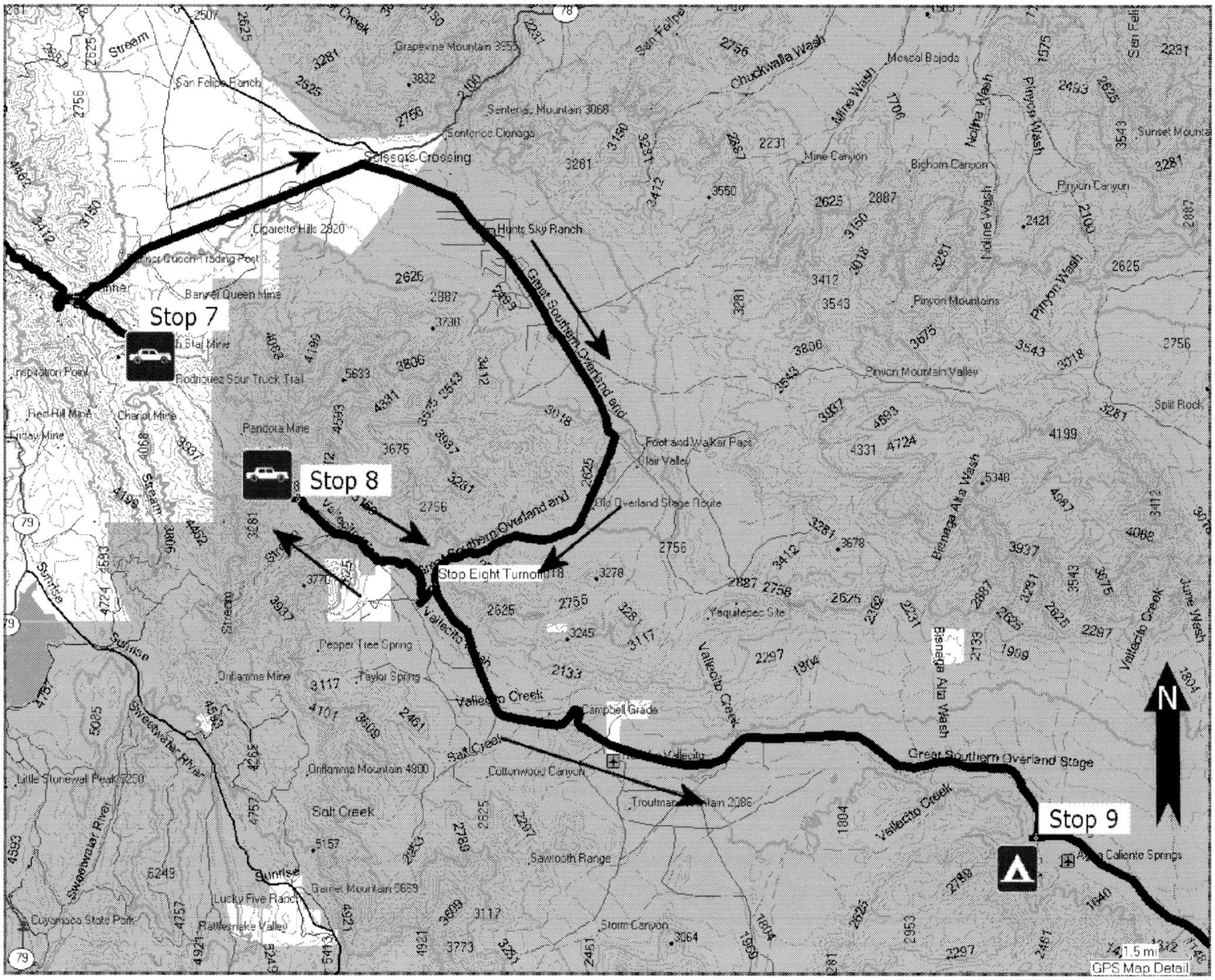

Stops 7, 8, and 9.

Railway (SD&AE), the "impossible railroad" built by John D. Spreckels who raised $18 million in order to link San Diego with the East. Overcoming numerous calamities, the railway was finally opened in 1919, 12 years after the groundbreaking ceremony. After virtual abandonment for 20 years due to landslides and tunnel fires, it is scheduled to reopen to through railroad traffic in 2004.

84.2 Plio-Pleistocene Palm Spring Formation is exposed in the road cut at this switchback.

85.0 Pass by the pullout and entrance sign to Canyon Sin Nombre, "Canyon Without Name." This is the beginning of a 4-wheel route to follow the "journey through time" down Diablo Drop-off into Fish Creek and Split Mountain (for a future trip!).

MP 51.5.

85.2 Pass by Carrizo Badlands Overlook. The monument notes that "Marine reefs, lake deposits, and silt from the Colorado River delta have been cast into layers over millions of years. These deposits, once horizontal, have been tilted up along the trace of the Elsinore fault. The elements of erosion have sculpted the desolate badlands. Paleontologists studying the fossils of the Carrizo Badlands have discovered this is one of the richest deposits of its kind on Earth. The time period of one-half million to five million years ago is represented by the hardened bones of mastodons, llamas, horses and tapirs. Grasslands and scattered forests of the area supported large animal herds." The journey through time travels some 20 miles and 10 million years northeast from here.

MP 53.5 on Highway S-2.

87.1 On the left side is the West Dolomite Mine Trail.

The largest and most accessible deposits of dolomite limestone in San Diego County are found in a low group of hills near the Dos Cabezas siding and in the southwest flank of the Coyote Mountains. Also called dolomite marble or magnesium calcium carbonate, it was used mainly as sources of crushed and broken stone for use as roofing granules, poultry grit, and decorative stone. The Coyote Mountains deposit is the one referred to as the Dolomite Mine. It was located the furthest from the rail siding at Dos Cabezas and the dolomite had to be trucked eight rugged miles to the siding. The deposit was first worked in 1924. By 1957 only a small shack and a loading bin remained on the property.

89.0 Volcanic Hills to the right. This area, with names like Red Hill and Lava Flow Wash, offers the most accessible examples of volcanic activity in the Anza-Borrego region. Variously called Jacumba basalt and Alverson andesite, approximate dating on these flows ranges from 15 to 22 Ma. The separation of compositionally similar Table Mountain near Jacumba, some 10 miles west and 3000 feet higher, is an indication of the major regional dip-slip and strike-slip faulting along the Elsinore fault zone.

There are two notably red volcanic hills in the Anza-Borrego area — the Red Hill of this area and a hill in Red Rock Canyon in the closed Carrizo Impact Area. The Cahuilla creation story explains why they are red. When Mukat the creator died, his people built a fire to cremate him. They were worried about the coyote, who might try to steal the body and eat it. So, they sent coyote off a long distance on an errand and quickly burned the body. Coyote saw the funeral fire and rushed back, jumping over the heads of the people who had formed a tight circle around the funeral pyre. He quickly snatched the heart of Mukat, which was all that was left, and ran miles away to a hill where he ate the heart of Mukat. Wherever blood had dripped as he ran with the heart and on the hill where he consumed the heart, you will find red pigment.

MP 56.0 on Highway S-2.

90.0 San Diego/Imperial County Line (elevation 914'). At one time the plaque on the monument on the left side of the road commemorated the completion of the highway in 1933. The plaque was stolen in the 1970s, but a new aluminum cast plaque was installed sometime in the late 1990s. This new plaque whimsically reads: "This is the Desert. There is nothing out here. Nothing" — signed "N. Karavasiles."

96.6 At the stop sign on the outskirts of Ocotillo, turn left on Shell Canyon Road, and go north to the base of the mountains. Shell Canyon and Alverson Canyon are old names for Fossil Canyon. The road to Fossil Canyon still retains the old name. The canyon has been famous for its coral and shells since 1890. At the south end of the canyon are fossilized coral reefs, which rise atop the narrow sandstone and mudstone walls.

98.8 Go straight on the dirt road before the gravel pit and then bear left to the west.

99.6 Stop 10. Park at the mouth to Fossil Canyon (Shell Canyon). Get out and walk north along the Elsinore fault. At this location, Tom Rockwell, or his designated representatives, will give us a tour of the fault exposures and point out evidence of recent fault activity. Fossil Canyon is part of the BLM Coyote Mountains Wilderness Area. The Coyote Mountains are a complex suite of almost every major rock type and geologic structure found throughout the Anza-Borrego region. The historic outline is briefly as follows, with dating approximate:

(1) 450 Ma — marine sediments in a shallow sea and organic limestone later metamorphosed into the marble stripes seen today high on mountain faces with large fragments found in washes, eroded from veins and dikes found throughout the mountains.

(2) 100 Ma — an intrusion of granitic magma, which uplifted the marine sediments, metamorphosed the overlying sediments into schist, gneiss, marble, and dolomite. These metasediments are seen today as remnant roof pendants.

(3) 15-20 Ma — basaltic lava flows, related to initial rifting of the Gulf of California. This flow is known as Alverson andesite or Alverson Formation.

(4) 4-6 Ma — Pliocene-Miocene marine advance and retreat from the area, leaving the near-shore sands, offshore yellow muds, and delta-fan oyster-shell hash "reefs" of the marine Imperial Formation draped over earlier bedrock like frosting on a cake.

100.4 Back on the pavement, head south.

102.6 At the stop sign, turn left onto Imperial Highway S-2. Go east, then south, heading toward Interstate 8 freeway and the community of Ocotillo. Ocotillo was developed in 1933 when the dirt road version of the Imperial Highway met paved Highway 80 from San Diego. Prior to that time, Coyote Wells to the east had been the focal point of development. Ocotillo has gas stations, stores, a motel, the Lazy Lizard Saloon, and freeway access.

103.9 Turn right, onto the freeway, I-8, and go west.

Call Box (CB) 8-41; MP 4.0 on I-8.

111.7 Cross over Devil's Canyon Bridge. At one time Devil's Canyon was a notorious section of the wagon road leading to San Diego. Washouts, sand, and rocks made this route very difficult for wagons. You can see part of the route as you look down from the bridge.

MP 77.7 on I-8. San Diego/Imperial County Line. Table Mountain is the flat-topped Miocene-age volcanic highland on the north (right). These Jacumba volcanics correlate to the Alverson andesite seen above at MP 56 or mile 89 on Highway S-2. The loading ramp of the abandoned Mica Gem Mine soon comes into view about a mile north of the freeway. Muscovite mica books may still be seen here.

MP 73.0 on I-8.

124.7 Miocene volcanic neck and ash deposits are visible.

127.2 Exit I-8 freeway at Campo, Boulevard, and Manzanita exit. At the stop sign, go south (left) under the freeway.

127.4 At the stop sign, turn east (left) onto old Highway 80.

129.7 Turn north (left) on McCain Valley Road.

132.1 The paved road turns to dirt at the sign: "Entering McCain Valley Recreation Area." McCain Valley was named for George Washington McCain and sons who first arrived there in 1868 and began grazing cattle. One son, Robert, who was the first to be born in McCain Valley, bought up surrounding parcels and leased others so that he eventually controlled 40,000 acres of rangeland in eastern San Diego County and was called the "dean of San Diego County cattlemen." McCain Valley is now a resource conservation area of 38,692 acres managed by the BLM.

132.5 At the brown mileage sign, turn right; follow the arrow to Sacatone Springs and Overlook (2 miles). The roadhead overlooks spectacular Carrizo Gorge, just east of McCain Valley.

134.3 Take a left at the fork and left again up to the mine parking area.

134.4 Stop 11. Pack Rat Mine. Park everywhere. We will tour this lower scale mine works. Chuck Houser, stop leader of the pack, will give a talk on the history of the mine and will show us a collection of tourmalines and other minerals collected from the Pack Rat and other pegmatite mines in San Diego. We will tour the mine area, enjoy the incredible views, and have lunch. Also at this location we will have a short talk on the history of the Carrizo Gorge railway, visible to the east, by George Copenhaver, as well as a talk on the local geology by Mike Walawender.

"Within a 5 km radius of the town of Jacumba, California, lies the Jacumba pegmatite group. The pegmatites intrude Jurassic gneiss, schist, marble, and quartzite; Cretaceous granitoid bodies and dikes; and Tertiary andesitic pyroclastic tuffs, and agglomerates," according to the *Journal of Pegmatology*.

End of the Field Trip. Return to I-8.

141.2 Turn right on Ribbonwood, cross under the freeway.

141.4 Turn left onto the freeway on-ramp, go west on I-8. Great field trip!

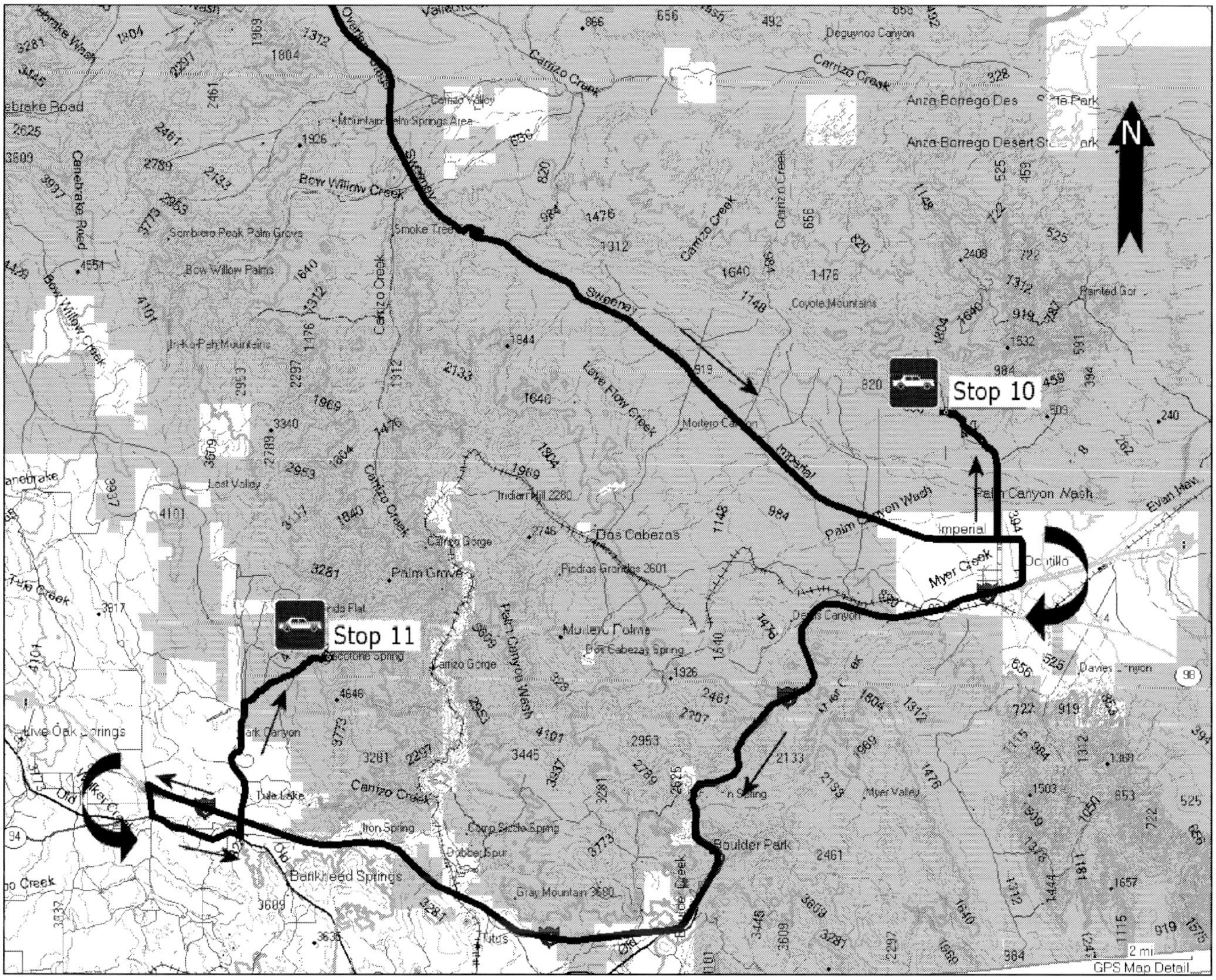

Stops 10 and 11.

HISTORICAL AND NATURAL HISTORY NOTES from Diana Lindsay's
ANZA-BORREGO A TO Z: PEOPLE, PLACES, AND THINGS,
ANZA-BORREGO DESERT REGION (with Lowell Lindsay),
JACKPOT TRAIL: INDIAN GAMING IN SOUTHERN CALIFORNIA,

and the following websites:
http://www.palagems.com/california_gem_mining.htm
http:// www.mmmgems.com/stewart/stewhist.htm
http://pegmatology.com/cbrowndoc1.htm
http://home.earthlink.net/-goke/lostvalleynews.htm
http://royalty.nu/asia/china/tzuhsi.htm

Sources for Cahuilla and Kumeyaay creation stories include:
Lora Cline, 1984, *Just Before Sunset*
Melicent Lee, 1989, *Indians of the Oaks*
Chief Francisco Patencio, 1943, *Stories and Legends of the Palm Springs Indians*

These were presented in the Indigenous Earth Sciences Project (IESP) at UCR in 1999 under the direction of Eric M. Riggs, now in the Department of Geological Sciences at SDSU.

Landslide...west of Hwy S-2. Landslide in metasedimentary units west of Milepost 22 on Highway S-2, southwest edge of Earthquake Valley, Anza-Borrego Desert State Park, Cool Canyon on right, Mason Valley in distance, upper left center. Photo by Michael Hart.

Central-Southern Elsinore Fault Zone: Summary and Resource Information for the SDAG/SCGS Field Trip, San Diego County, California

***Monte L. Murbach**, Petra Geotechnical, Inc.*
***Diane Murbach**, City of San Diego*

INTRODUCTION

This paper is a resource for the central-southern Elsinore fault zone (EFZ) as part of the annual San Diego Association of Geologists (SDAG) field trip, October 25 and 26, and the annual South Coast Geological Society (SCGS) field trip, November 7-9, 2003. The theme of each trip is *Hot Springs & Tourmalines of Eastern San Diego County*. Each trip starts at the Warner Springs Airport and continues south along the Elsinore fault zone. The trip spends a night at Agua Caliente hot springs, then heads farther south along the Elsinore fault zone through the Coyote Mountains, ending near Interstate 8 east of San Diego, California. As you travel along the central-southern Elsinore fault zone, you see laterally offset drainages, closed depressions and faceted topography marking the recent traces of this fault, supporting Holocene activity for this active fault.

ELSINORE FAULT ZONE

The northwest-trending Elsinore fault zone is one of the largest active faults in southern California, though in historical times has been considered one of the quietest. The Elsinore fault is part of the San Andreas system of faults. The type of faulting is right-lateral strike slip. To the east are the San Jacinto and San Andreas fault zones. To the west, the Newport Inglewood/Rose Canyon fault zone.

The Elsinore fault zone splays into two segments along some zones, and several of the fault zones possess their own names. The northern end of the Elsinore fault zone includes the Chino and Whittier faults near Whittier and Ontario, California. The southern end of the Elsinore fault continues as the Laguna Salada fault into Mexico. The central-southern Elsinore faults include the double-stranded Aguanga, Agua Tibia, Earthquake Valley, and the Agua Caliente faults, from Mount Palomar southward to Lake Henshaw, Julian, and to the Agua Caliente campground. The Elsinore strand is also known as the Julian and Coyote Mountains segments in this area.

According to the Southern California Earthquake Data Center website (SCECDC, 2003), historic surface rupture occurred during an earthquake in 1892 on the Laguna Salada fault zone to the south (M7). In addition, a magnitude 6.0 earthquake most likely occurred on the Elsinore fault in 1910. This earthquake was probably centered northwest of Lake Elsinore, produced no known surface rupture and did little damage.

Numerous studies have been performed on the central-southern Elsinore fault zone. This paper presents a summary of major works performed, including a resource list for the portion of the fault zone between Palomar Mountain on the north and the Coyote Mountains to the south. The earliest study referenced that discusses this fault is an 1893 publication by H.W. Fairbanks. Smith (1979) summarized that this fault was first shown on a published map utilizing the name "Elsinore" by Lawson and others in 1908. Wood (1916) tabulated the historical earthquakes of California and suggested that a number of earthquakes may have occurred along the Elsinore fault. Merriam (1958) and Weber (1963) prepared studies that detailed this fault. Newer detailed mapping for the Coyote Mountains portion of the Elsinore fault by Rockwell (1990) is presented. See Figures 1A, 1B, 1C.

SUMMARY OF WORKS

The following works are presented in chronological order. Not everything related to the southern Elsinore fault is presented. However, this is a good compilation for those interested in studying this portion of this fault. All of the papers discussed are referenced at the end of the text. The current availability is also indicated.

Bulletin 159 (1951), Crystalline Rocks of Southern California, discusses field mapping performed predominantly in the 1930's in the Peninsular Ranges. The bulletin includes a discussion on faults (including the Elsinore fault) in the Peninsular Ranges. The bulletin is out of print.

Elsinore Fault / Coyote Mountains Neotectonics

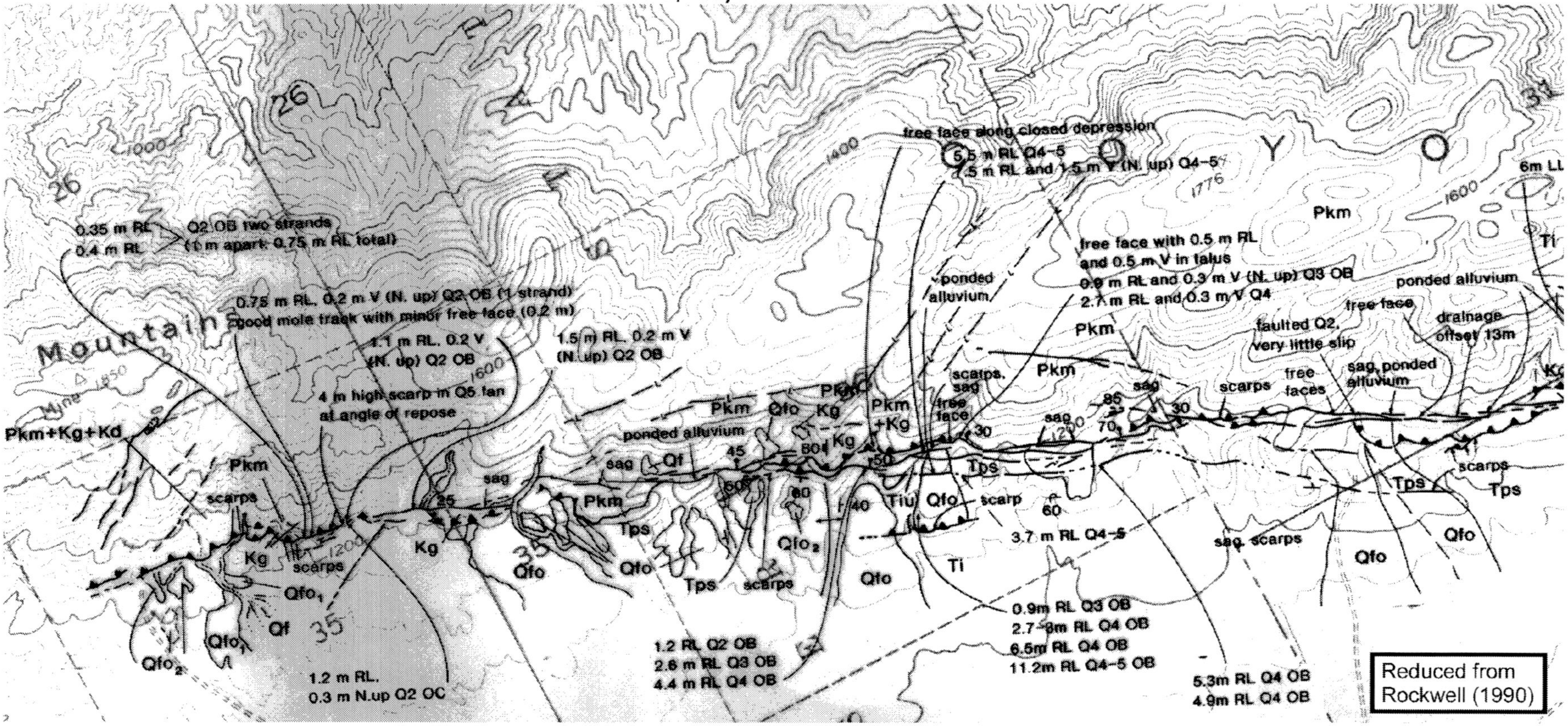

(continued on next figure)

EXPLANATION

Tiu	Upper Imperial Formation	Qfo	Undif. Pleistocene Alluvial Fan	OB --	Offset Alluvial Bar
Ti	Imperial Formation (undif.)	Qal	Active Channel Alluvium	OC --	Offset Channel
Tibs	Imperial Formation (basal sandstone unit)	Kg	Cretaceous Granitic Rocks (undif.)	OCW --	Offset Channel Wall
Tibg	Imperial Formation (basal conglomerate unit)	Pkm	Pre-Cretaceous Metamorphic Rocks	OR --	Offset Rill
Tps	Palm Spring Formation	peg	Pegmatite Dike		
		Ta	Alverson Andesite		

Scale: 1 mile

Figure 1a. Coyote Mountains portion of the Elsinore fault. Courtesy Thomas Rockwell

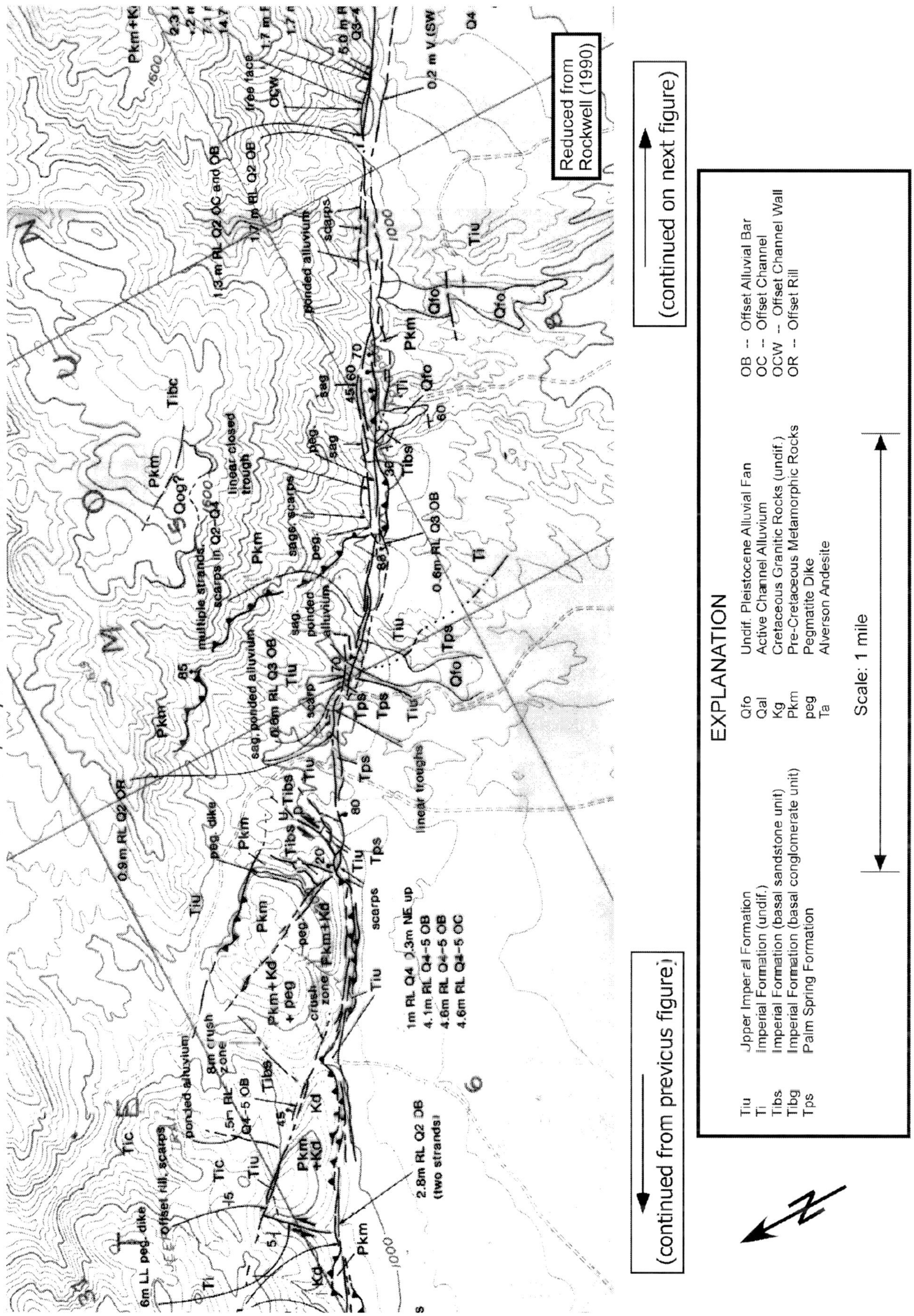

Figure 1b. Coyote Mountains portion of the Elsinore fault. Courtesy Thomas Rockwell

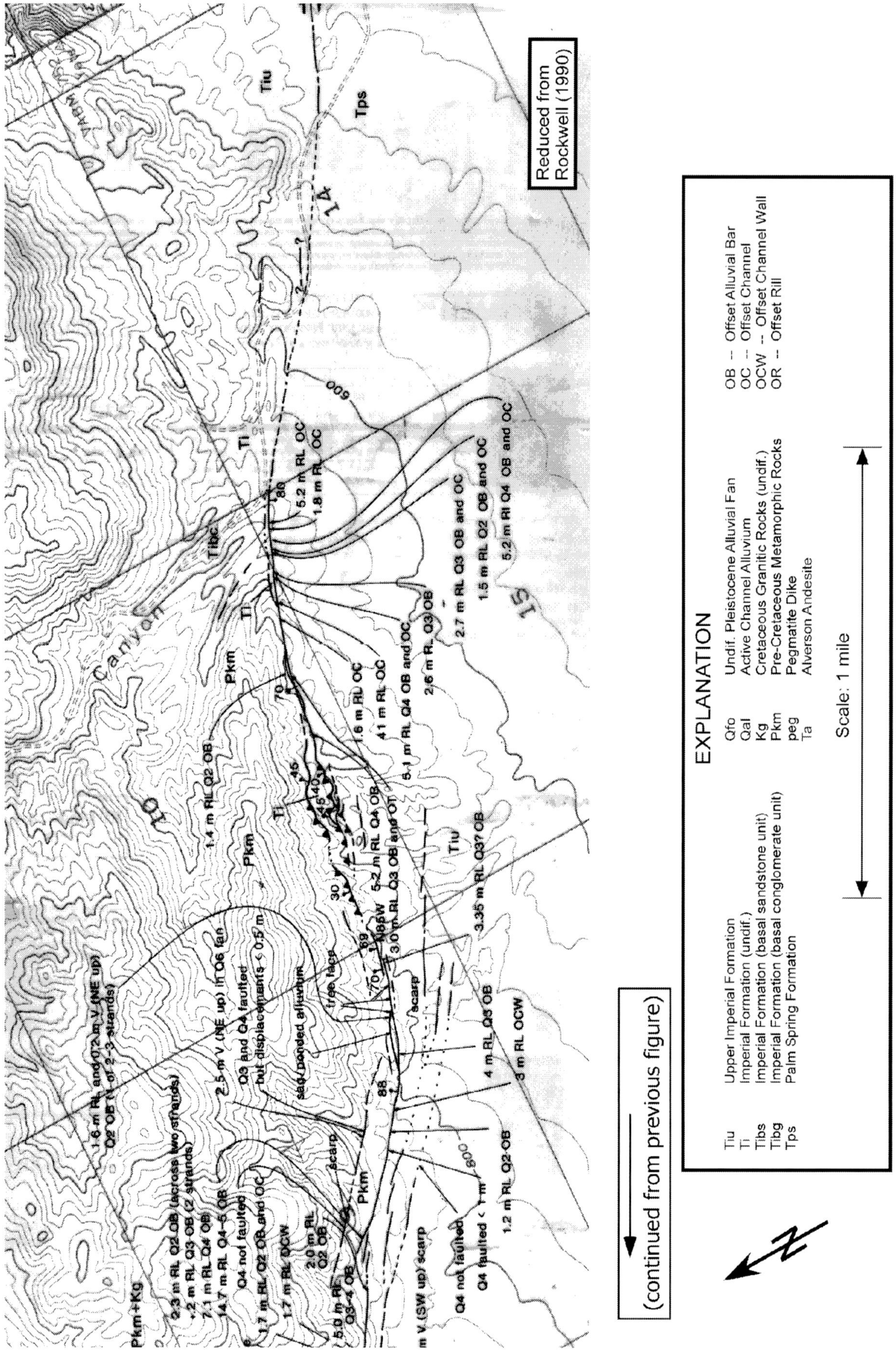

Figure 1c. Coyote Mountains portion of the Elsinore fault. Courtesy Thomas Rockwell

Bulletin 177, Geology and Mineral Resources of Santa Ysabel Quadrangle, has a detailed geology map (Merriam, 1958) depicting the trace of the Elsinore fault from Lake Henshaw to the north and the Banner Grade area to the south. There is no age discussion of the fault in this paper. The map is in color, with a scale of 1:62,500 allowing for fairly detailed mapping. A portion of this map is presented on the inside front cover of this volume. Bulletin 177 is a good reference for the geology of the area, but is out of print.

The San Diego-El Centro Sheet geologic map (CDMG, 1962) is a high quality color map at a scale of 1:250,000. The map depicts the southern Elsinore fault zone at the eastern portion of the map. The map presents detailed geology, however, the base map is small-scaled. The map is currently available for purchase through the California Geological Survey.

The California Division of Mines and Geology (DMG), County Report 3 (Weber, 1963), includes a large, color geologic map of San Diego County. The scale of the map is 1 inch equals 2 miles. The southern Elsinore fault trace is clearly depicted on this map. The map includes mine localities at the time of publication. This publication is still a great resource, in particular for eastern San Diego County. The bibliography is extensive listing older publications on the geology of southern California. This report is no longer available through the State. It is still available at some university libraries and private collections.

Allison (1974), representing Standard Oil Company of California at that time, prepared a reconnaissance and geophysical study on the Elsinore and Chariot Canyon fault systems. This earlier paper discusses previous mapping performed along the southern Elsinore fault zone in the early 1900s, and is well worth reading. This guidebook is out of print.

Todd (1977), compiled a geologic and fault map of the Aqua Caliente Springs quadrangle, San Diego County, California. The base map is large-scaled and the geologic detail on this map is good. Geologic notes are hand printed. Availability of this map is unknown.

Todd (1978), compiled a geologic and fault map of the Monument Peak quadrangle, San Diego County, California. The geologic mapping is large-scaled, with good geologic detail. Availability of this map is unknown.

Smith (1979) performed detailed mapping of the southern Elsinore fault zone as part of the Fault Evaluation Report (FER # 84) CDMG Open-File Report 90-13. The map depicts fault features including hillside valleys, vegetation lineaments, hillside benches, springs and scarps. Unfortunately, the report is a microfiche copy, so is somewhat difficult to read. The report is available through the CDMG, (renamed the California Geological Survey).

Freckleton (1981) prepared a paper with a groundwater/geologic map of the Santa Ysabel and Mesa Grande Indian Reservations. Although the emphasis of this paper relates to groundwater in the area, detailed geology and faulting which includes a portion of the southern Elsinore fault, are presented on a clean and clear, black and white, large-scale map. This publication is out of print.

Clark, M.M. (1982), provides detailed mapping along the southern Elsinore fault between Lake Henshaw and Mexico. The map plates are scaled at 1:24,000, and are excellent quality with detailed mapping. The maps show recently active breaks along the Elsinore and associated faults. This map is currently available through the California Geological Survey. Note: Rockwell (1990) provides additional detail of the Elsinore fault zone along the Coyote Mountains segment.

DMG Open-File Report 84-59 (1984) includes a detailed geologic and fault map of southern California and northern Baja California. The map is a black and white with a scale of 1:250,000. The small-scale map has good details, however, some of the small print may require reading glasses. This publication is currently available through the California Geological Survey.

Rockwell and others (1990) studied the Holocene soils offsets in the Coyote Mountains and determined that the Elsinore fault in this segment has ruptured several times in the Holocene. Rockwell's 1990 study focused on resolving displacement per event for the Coyote Mountain segment of the Elsinore fault. The entire length of the fault along the Coyote Mountains was mapped and displacements of late Holocene deposits were measured according to their relative age. Measurements of slip between the youngest and oldest mapped Holocene units ranged between one and several meters. Based on vertical and horizontal components, a value of Ms 6.8 earthquake for the Coyote Mountains segment was estimated. Rockwell (personal communications, 2003) indicated that carbon dating of collected charcoal samples in the alluvial soils offset by the fault are currently being performed. In addition, a small fault trench exposing the fault segment may still be open, though recent rain storms may have destroyed the trench exposures. A reduction of Rockwell's 1990 geologic map is included in this paper (Figures 1A, 1B, 1C).

DMG (1992), Preliminary Fault Activity Map, compiled by Charles W. Jennings, presents an oversized plate of the State of California. The scale is 1:750,000. Though the map is small-scaled, the mammoth size of the plate allows for detailed plotting of fault strands. The map indicates activity of the fault strands based on a color code, and provides reference numbers for each significant segment. Notes and numbers are hand written. The referenced numbers correlate to fault studies at that particular segment. This is a good reference. The map was finalized in 1994.

Magistrale and Rockwell (1996) prepared an updated paper on the central and southern Elsinore fault zone, southern California. Their abstract presents an excellent updated summary and should be read. See Figure 2. Quotes from their paper include: *We examine hypocenter patterns, geomorphic features, total slip determinations, and slip-rate information to define the active elements of the southern and central portions of the right-lateral Elsinore fault zone in southern California. In addition to the Elsinore fault, the seismicity patterns define a continuous fault between about 33 degrees N and 33.5 degrees N, 7 to 12 km east of the Elsinore fault, that corresponds to the Aguanga, Agua Tibia, and Earthquake Valley faults. This part of the*

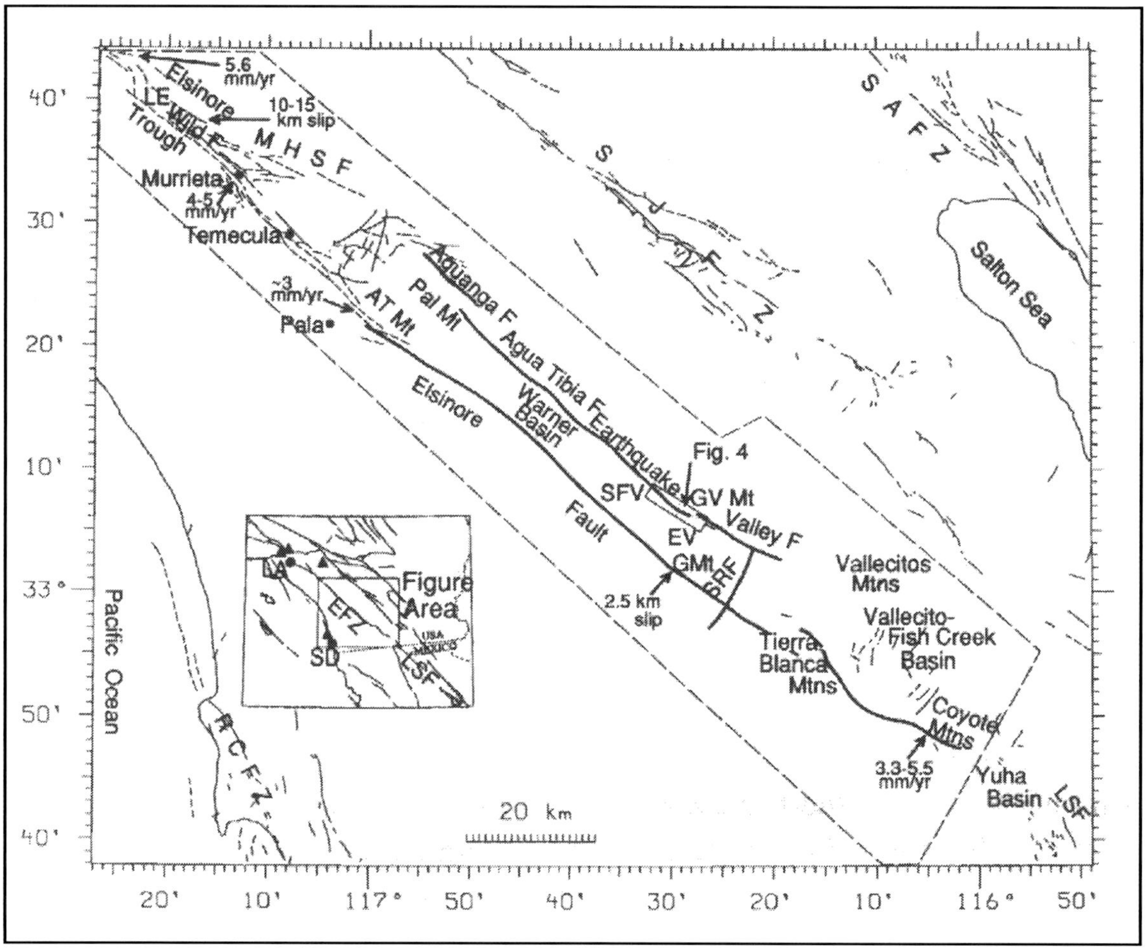

Figure 2. Regional map, showing faults and features named in text. Dashed line encloses area of relocated seismicity. slip rate and total slip determinations are indicated. Triangles (inset) indicate seismometers used to relocate 1940 earthquakes. Abbreviations: ATMt, Agua Tibia mountain; EFZ, Elsinore fault zone; EV, Earthquake Valley; F, fault; GMt, Granite mountain; GVMt, Grapevine mountain; LA, Los Angeles (inset); LE, Lake Elsinore; LSF, Laguna Salada fault; MHSF, Murrieta Hot Springs fault; PalMt, Palomar mountain; RCFZ, Rose Canyon fault zone; SAFZ, San Andreas fault zone; SD, San Diego (inset); SFV, San Felipe Valley; SJFZ, San Jacinto fault zone; SRF, Sawtooth Ridge fault; Wild F, Wildomar fault. *From: Magistrale and Rockwell, 1996.*

fault zone is double stranded. Total slip and current slip rate are apparently distributed between the two strands. Slip transfer between the strands occurs at the ends of the eastern strand over two steps with topographic signatures reflecting the step geometries in a right-lateral fault zone: a contractional left step occupied by Palomar Mountain, and an extensional right step within the Vallecito-Fish Creek basin. We suggest that the two strands of the double-stranded portion of the fault zone have different ages, and the proportion of slip carries by each strand may have changed through time. Hypocenter patterns define several northeast-striking faults within the fault zone; we attribute a 1940 ML 5.1 earthquake to one of them.

Walawender (2000) authored a geological guide to San Diego's back country. The book discusses the development of the Peninsular ranges and the age relationships of the various rock types. There is also a chapter on gems and gold. There are nine field excursions with geological history of 30 sites. Numerous black and white illustrations are included. This great resource is still available.

ACKNOWLEDGMENTS

Thanks to Drs. Miles Kenny and Thomas Rockwell for assisting by providing some of the resources, and thanks to Philip Farquharson, 2003 SDAG Treasurer, for providing graphic support and the aerial view of the Coyote Mountains section of the EFZ (Figure 3). Thanks also to Joe Cota for locating and providing reference materials for older publications.

WEBSITE RESOURCES

www.consrv.ca.gov (California Geological Survey)

http://scamp.wr.usgs.gov

www.geology.sdsu.edu

www.seismo-watch.com

www.scecdc.scec.org/clsfault.html

www.usgs.gov

REFERENCES

Allison, M.L., 1974, Geologic and geophysical reconnaissance of the Elsinore-Chariot Canyon fault systems, in Recent Geological and Hydrologic Studies, Eastern San Diego County and Adjacent Areas, A Guidebook Prepared for the December 14 and 15, 1974 Field Trip of the San Diego Association of Geologists (fourth annual field trip of the association), Hart, M.W., and Dowlen, R.J., editors.

California Department of Natural Resources, Division of Mines, 1951, Crystalline Rocks of Southwestern California, Bulletin 159.

California Department of Conservation, Division of Mines and Geology, 1962, Geologic Map of California: San Diego-El Centro Sheet, fourth printing 1992.

California Department of Conservation, Division of Mines and Geology, 1980, Special Studies Zones — Earthquake Valley Quadrangle Official Map.

California Department of Conservation, Division of Mines and Geology, 1980, Special Studies Zones — Julian Quadrangle Official Map.

California Department of Conservation, Division of Mines and Geology, 1984, Preliminary Geologic Map of the California-Baja California Border Region, DMG Open File Report 84-59.

California Department of Conservation, Division of Mines and Geology, 1992, Preliminary fault activity map the California, DMG Open-File Report 92-03.

Clark, M.M., 1982, Map showing recently active breaks along the Elsinore and associated faults, California, between Lake Henshaw and Mexico, Map I-1329, Miscellaneous Investigation Series, U.S. Geological Survey, reprinted 1986.

Freckleton, J.R., 1981, Water resources of the Santa Ysabel and Mesa Grande Indian Reservations, San Diego County, California, United States Geological Survey, Water Resources Investigations, Open-File Report 81-342.

Goodmacher, J., and Rockwell, T.K., 1990, Properties and inferred ages of soils developed in alluvial deposits in the southwestern Coyote Mountains, Imperial County, California, in Rockwell, T.K., et al, 1990, Friends of the Pleistocene, 1990, winter field trip, western Salton Trough — soils and neotectonics.

Grant, L.B., and Rockwell, T.K., 2002, A northward-propagating earthquake sequence in coastal southern California?, in Seismological Research Letters, July/August 2002, Seismological Society of America, Volume 73, Number 4.

Figure 3. View north over Coyote Mountains (lower right), Fish Creek Mountains (upper right). Elsinore Fault runs diagonally from left center edge to lower right corner. Highway S-2 roughly parallels fault trace in lower left quadrant. Photo by Phillip T. Farquharson, pilot Frank Morris (2003)

Kenney, M.D., 2002, Geology Field Trip Handout, Southern Elsinore fault zone, southern Anza-Borrego State Park & Western Salton Trough region: A rock record of 500 Million Years of Geologic Events.

Magistrale, H., and Rockwell, T.K, 1996, The central and southern Elsinore fault zone, southern California, in Bulletin of the Seismological Society of America, Volume 86, No. 6, pp. 1793-1803, December.

Merriam, R., 1958, Geologic Map of the Santa Ysabel Quadrangle, in California Department of Natural Resources, Division of Mines, 1958, Geology and Mineral Resources of Santa Ysabel Quadrangle, San Diego County, California.

Rockwell, T.K., and Pinault, C.T., 1986, Holocene slope events on the southern Elsinore fault, Coyote Mountains, southern California: in Guidebook and volume on neotectonics and faulting in southern California (P. Ehlig, Ed.), Cordilleran Section, Geological Society of America, pp. 193-196.

Rockwell, T.K., 1989, Behavior of individual fault segments along the Elsinore-Laguna Salada fault zone, southern California and northern Baja California: Implications for the characteristic earthquake model: U.S. Geological Survey Redbook on Fault Segmentation and the Earthquake Generation Process: U.G. Geological Survey Open File Report.

Rockwell, T.K., 1990, Holocene Activity of the Elsinore fault in the Coyote Mountains, southern California, in Rockwell, T.K., et al, 1990, Friends of the Pleistocene, 1990, winter field trip, western Salton Trough — soils and neotectonics.

Rockwell, T.K., 2003, personal communications regarding the southern Elsinore fault and penultimate French wines.

San Diego State University, 2002, Geology of the western Salton Trough, Anza-Borrego Field Trip Guide.

Smith, D., Elsinore fault, 1979, Fault Evaluation Report (FER) number 84, in California Department of Conservation, Division of Mines and Geology, 1990, Microfiche copies of fault evaluation reports for the Peninsular Ranges, Open-File Report 90-13.

Southern California Earthquake Data Center (SCECDC), 2003, website sponsored by SCEC, USGS, and California Institute of Technology, Seismological Laboratory.

Thorup, Kimberly M., 1997, Paleoseismology of the Central Elsinore Fault in Southern California: Results from three trench sites, Masters Thesis San Diego State University.

Todd, V.R., 1977, Geologic map of the Aqua Caliente Springs quadrangle, San Diego County, California, U.S. Geological Survey, Open-File Report 77-742

Todd, V.R., 1978, Geologic map of the monument peak quadrangle, San Diego County, California, U.S. Geological Survey, Open-File Report 78-697.

Walawender, M.J.,2000, The Peninsular ranges, a geological guide to San Diego's back country, San Diego State University.

Weber, Jr., F.H., 1963, Geology and Mineral Resources of San Diego County, California, California Division of Mines and Geology, County Report 3.

III.

New Papers

Garnet-Bearing Granitoids and Pegmatite Dikes in the Indian Hill-Tule Mountain Roof Pendant: Implications for the Origin of Gem-Bearing Pegmatites in Southern California

Michael J. Walawender, *Department of Geological Sciences*
San Diego State University, San Diego, California 92182

ABSTRACT

The Indian Hill-Tule Mountain roof pendant is an elongate body of sillimanite-grade metasedimentary rocks located in the eastern zone of the Peninsular Ranges Batholith. Surrounded by the 94-Ma La Posta pluton, it contains several small to moderate sized bodies of undeformed garnetiferous two-mica granites and numerous granitic dikes with pegmatitic textures and structures. The dikes are consistently terminated at their contact with the granites but have compelling geochemical affinities for the finer grained plutonic units. Linear trace-element patterns suggest a mixing/unmixing relationship but their individual REE patterns are distinctive. The pegmatitic dikes and the small bodies of granite were derived, at least partly, from source rocks enriched in ^{18}O and ^{87}Sr.

Experimental studies on partial melting of metagraywacke show a sequential breakdown from muscovite (+ plagioclase + quartz + biotite) at lower temperatures to biotite (+ feldspar + quartz + an aluminosilicate) at somewhat higher temperatures. The granitic melts produced by the water-absent melting reactions equilibrated with plagioclase (+ quartz + biotite + aluminosilicate) and garnet (+ quartz + aluminosilicate + feldspar), respectively. The measured REE patterns from the two rock units match the residual mineralogy predicted from the melting experiments. This implies that the pegmatitic dikes represent the initial water-undersaturated melts produced by muscovite breakdown in the prebatholithic metasedimentary complexes.

The garnet-bearing monzogranites are the product of biotite-controlled fusion at slightly higher temperatures in the same source. The heat energy for this event was the latent heat within the encompassing La Posta pluton. The gem-bearing pegmatitic dikes throughout southern California may thus owe their origin to the association of large La Posta-type plutons and the prebatholithic metasedimentary packages.

INTRODUCTION

The eastern and medial portions of the Peninsular Ranges Batholith are host to abundant arrays of granitic pegmatite dikes that are world famous for producing rough gem material as well as superb crystal specimens. From 1898 to 1914, the Himalaya Mine near Mesa Grande, California, was the world's leading supplier of colored tourmaline. Most of this material went to Imperial China where it was used for carving until the fall of the Manchu Dynasty. Today, natural history museums and private mineral collections around the world contain spectacular crystal specimens produced from southern California locales that include the Pala, Mesa Grande, and Ramona districts (Figure 1).

Although the pegmatite dikes commonly intrude a wide variety of rocks within the medial portion of the batholith, they are also spatially associated with small bodies of undeformed two-mica granite. This relationship is well exposed within the Indian Hill-Tule Mountain roof pendant in eastern San Diego County. Their spatial, temporal, and geochemical association suggest that these two units have a common origin that is tied in with the evolution of the eastern zone of the Peninsular Ranges Batholith (PRB). The purpose of this paper is three-fold:

1. To document the structural, geochemical, and temporal character of these units.
2. To place them into the structural and tectonic framework of the batholith.
3. To place constraints upon their source materials and origin.

GEOLOGIC SETTING

The PRB is a complex mixture of igneous and metamorphic rock that averages about 100 km in width and extends southward about 1000 km from Riverside County to the southern tip of the Baja California peninsula. The

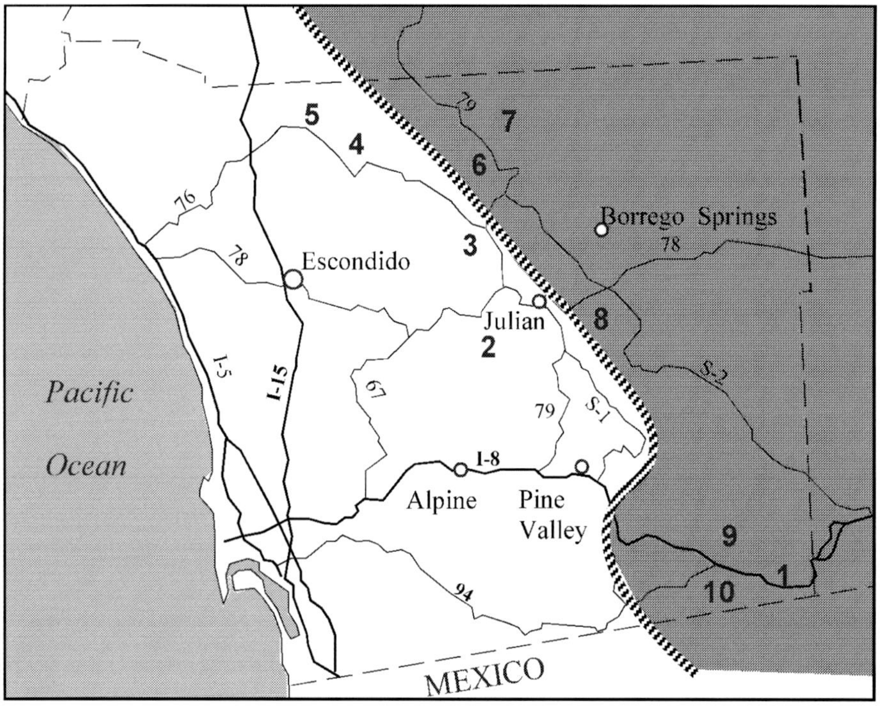

Figure 1. Index map of the pegmatite mining districts in San Diego County. The hatchured line represents the approximate boundary between the western and eastern zones of the PRB (after Walawender, 2000). 1 Jacumba, 2 Ramona, 3 Mesa Grande, 4 Rincon, 5 Pala, 6 Aguanga, 7 Chihuahua Valley, 8 Banner, 9 Tule Mountain, 10 Boulevard.

batholith, unlike its Mesozoic counterparts around the Pacific Basin, is calcic rather than calc-alkaline in nature and exhibits a well documented lithologic and geochemical asymmetry across its north-northwest structural grain. It has been subdivided into western and eastern zones, each with a distinctive combination of lithology, age, isotopic character, and geochemical composition (Gastil, 1975; Silver and others, 1979; DePaolo, 1980, 1981; Silver and Chappell, 1988). The western zone of the batholith contains scattered upper greenschist- to lower amphibolite-grade metasedimentary assemblages (Berggreen and Walawender, 1977; Gastil, 1975) that include quartzite, biotite schist, amphibolite, and minor metaconglomerate. Pelitic assemblages are scarce. Sparse and highly deformed Jurassic granitic rocks with zircon U-Pb ages as old as 231 Ma (Thompson and Girty, 1994; Girty and others, 1998) are intrusive into the metasedimentary package.

Cretaceous plutonic rocks in the western zone have been subdivided into three lithologic groups: gabbro, quartz diorite to tonalite, and granodiorite to monzogranite (Silver and others, 1979; Silver and Chappell, 1988; Walawender and others, 1991). The gabbroic plutons, based mainly on field relationships, appear to be the oldest and form multiple intrusive complexes that range in composition from peridotite through two-pyroxene gabbronorite to amphibole gabbro (Walawender, 1976; Walawender and Smith, 1980; Smith and others, 1983). The quartz diorite-tonalite plutons comprise the bulk of the western zone (Larsen, 1948; Gastil, 1975) and exhibit internal variations that range from pyroxene-amphibole quartz diorite to amphibole-biotite tonalite (Lombardi, 1992). Zircon U-Pb emplacement ages reported for these lithologies vary from 105 to 108 Ma (Walawender and others, 1991). The granodiorite-monzogranite plutons, which make up about 30% of the western zone, were commonly cited as the youngest western plutonic series. Although scattered zircon U-Pb radiometric ages indicate emplacement ages between 118 and 115 Ma, field relationships in several areas (Carollo and Walawender, 1993) still argue that at least some of the undated rock bodies are younger than the quartz diorite-tonalite plutons.

Metamorphic rocks in the eastern zone of the PRB consist of sillimanite-bearing mica schists, migmatitic gneisses, and lesser amounts of quartzite, marble and associated calc-silicate assemblages, and amphibolite (Germanario, 1982; Siem, 1988; Todd and others, 1988). Age constraints on these rocks are not well established but they may be as old as lower Paleozoic. The plutonic rocks of the eastern zone are limited in composition. They occur as large, undeformed, concentrically zoned (La Posta-type) plutons with tonalite margins and monzogranite cores. Contacts between internal units in these plutons are typically gradational over distances of several tens to a few hundreds of meters (Gastil and others, 1990; Walawender and others, 1990). A distinctive gneissic tonalite occurs locally within at least three of the La Posta-type plutons. It has sharp intrusive contacts with the undeformed interior zones and major- and trace-element chemistry consistent with the main La Posta series (Walawender and others, 1990). A zircon U-Pb age of 93 Ma at one locale (Pimental, 1998) places it into the La Posta timeframe.

The pegmatite dikes in southern California are intrusive into a variety of western PRB plutonic rocks that include gabbro (Pala, Mesa Grande), and mafic tonalite (Ramona). In the eastern zone of the batholith, host rocks include metasedimentary screens (Chihuahua Valley, Sacatone) and the early gneissic phase of the La Posta-type plutons (Chihuahua Valley, Table Mountain, Blair Valley). Most of the dikes that cut plutonic rocks, however, have metasedimentary screens in close proximity. Within the undeformed zones of the large La Posta-type plutons, pegmatitic dikes are scarce or absent.

THE INDIAN HILL-TULE MOUNTAIN ROOF PENDANT

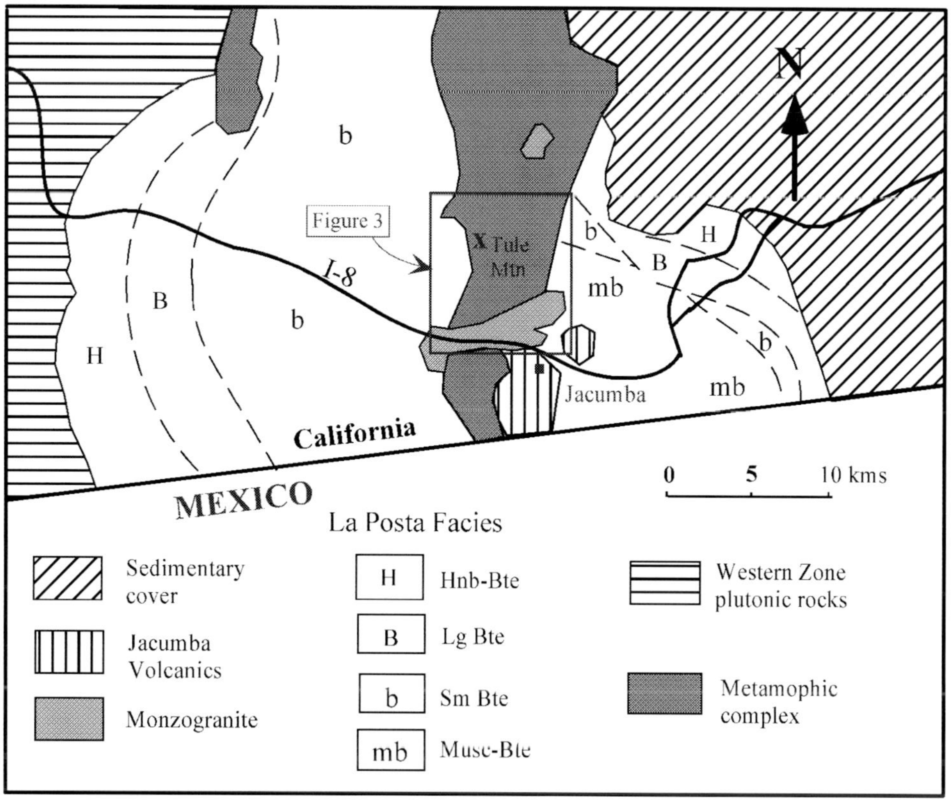

Figure 2. Generalized geologic map of the La Posta pluton and the Indian Hill-Tule Mountain roof pendant in eastern San Diego County (after Walawender and others, 1990). The unpatterned areas represent the pluton and its four facies. The smaller monzogranite body is the Indian Hill pluton. The larger one bisected by I-8 is the Boulevard pluton.

The Indian Hill-Tule Mountain roof pendant is located, in part, within the Anza-Borrego Desert State Park about 75 km east of the city of San Diego. It is surrounded by the 1400 km^2 La Posta pluton and lies within the eastern zone of the PRB (Figure 2). To the north, the roof pendant abuts against and may be offset by the Elsinore fault. It continues southward from the study area into Baja California for another 40 km (Gastil, 1975). The bulk of the roof pendant is made up of sillimanite-bearing migmatite gneiss that contains large isolated blocks of quartzite, amphibolite and marble. It is also pervasively intruded by small- to moderate-sized bodies of garnetiferous two-mica granite and by pegmatitic dikes up to a kilometer in length and ten meters in width.

The Indian Hill pluton is located in the northern part of the roof pendant (Figure 2). It consists of two leucocratic muscovite-biotite granitoids (Parrish and others, 1986; Parrish, 1990) that are distinguished in the field by differences in grain size, and the presence or absence of garnet and large muscovite flakes. The coarser grained (1-3 mm) garnet-bearing granitoid is the main phase and makes up approximately 90% of the exposed area. This main phase is monzogranite to granodiorite in composition and contains up to 5% muscovite. The garnet-free granitoid is finer grained (<1 mm) and granodiorite in composition. Muscovite makes up less than 1% of the rock (Table 1).

A second, unnamed granitoid body cuts across the western side of the metamorphic complex just north of the international border along Interstate 8 (Figure 2). This unit, hereafter referred to as the Boulevard pluton, is a homogeneous garnet-biotite-muscovite monzogranite. It is intrusive into the Small Biotite facies of the La Posta pluton and contains meter-sized inclusions of the early hornblende-bearing gneissic phase of the La Posta series.

The pegmatitic dikes within the metamorphic complex are similar in modal composition to the garnet-bearing monzogranite. Tourmaline (schorl) and garnet are common accessory minerals. Part of the dike system just north of Tule Mountain (Figure 3) has produced gem spodumene, aquamarine, and spessartine. The thicker dikes are internally complex but can be traced along strike for up to one kilometer. Thickness vary from less than 1 m to more than 10 m. Many dikes have developed a crude concentric zonation similar to that described by Jahns and Burnham (1969), Jahns (1979), and Walawender (2000) for other southern California pegmatite dikes. Others exhibit distinct layers of pegmatite and finer grained granite that are parallel to the margins of the dike. Locally, the dikes form subparallel arrays that, when viewed on the scale of the

TABLE 1. MODAL DATA.

Sample No.	IH-1	IH-2	IH-4	IH-7	IH-9a	IH-27	IH-34	IH-39
Unit	GMB	GMB	GMB	MB	GMB	MB	GMB	MB
Quartz	38.4	40.6	37.1	43.2	37.9	43.9	35.2	40.7
Plagioclase	38.8	33.7	36.4	39.6	42.2	45.5	33.5	42.5
Alk. Fs	13.2	17.2	17.0	7.1	11.2	4.1	21.9	10.3
Biotite	4.1	4.0	3.8	7.1	2.1	3.6	3.5	4.0
Muscovite	1.4	2.5	0.8	0.9	3.6	0.8	2.0	1.4
Garnet	tr	0.2	0.5	0.0	0.7	0.0	0.7	0.0
Sericite	3.7	1.7	3.8	1.1	2.0	1.7	3.0	1.1
Myrmekite	tr	0.6	0.2	tr	tr	0.2	tr	tr
Chlorite	0.3	0.4	0.2	tr	0.3	0.2	tr	tr

Sample No.	IH-40	IH-51	IH-61	IH-93	LP-17	LP-28	LP-45	LP-46
Unit	GMB	MB	dike	dike	GMB	GMB	GMB	GMB
Quartz	35.6	36.5	33.8	36.9	33.2	32.6	27.3	32.9
Plagioclase	32.0	35.8	42.6	29.8	48.4	42.6	39.3	42.9
Alk. Fs	23.0	16.4	17.0	21.3	12.7	18.8	24.4	20.9
Biotite	2.3	4.1	0.5	2.9	4.1	2.7	4.2	0.6
Muscovite	5.5	0.4	1.7	4.8	1.3	2.0	4.2	1.4
Garnet	tr	0.0	tr	tr	tr	tr	tr	0.6
Sericite	1.6	6.5	4.4	4.0	0.1	tr	tr	tr
Myrmekite	tr	0.3	tr	0.3	tr	0.6	0.6	tr
Chlorite	tr	tr	0.0	0.0	tr	1.3	tr	0.8

Sample No.	T-2	T-4	T-5	T-8	T-10	T-12	T-7
Unit	dike	dike	dike	dike	dike	dike	GMB
Quartz	38.8	40.4	30.0	34.0	22.4	32.4	33.2
Plagioclase	35.6	39.2	33.2	33.6	44.5	38.2	38.8
Alk. Fs	17.2	14.2	21.6	26.4	14.7	24.6	21.2
Biotite	tr	0.0	0.0	tr	0.0	0.0	0.0
Muscovite	2.8	4.4	9.6	4.8	6.8	2.8	1.2
Garnet	0.0	1.8	2.0	1.2	11.6	1.6	tr
Sericite	3.6	tr	tr	tr	tr	tr	0.6
Myrmekite	0.0	0.0	0.0	0.0	0.0	0.0	0.0
Chlorite	tr	0.0	tr	tr	tr	tr	0.4
Tourmaline	0.0	0.0	3.6	tr	tr	0.4	0.0

Note: GMB = garnet-bearing granitoids, MB = garnet-free granitoids, dike = aplitic portion. All data in volume perce

roof pendant, form an arcuate pattern that wraps around the central portion of the roof pendant (Figure 3).

AGE RELATIONSHIPS

A U-Pb concordia diagram for the Indian Hill pluton (Walawender and others, 1990) shows clear discordance with a lower intercept age of 84.4 +/- 6 Ma and an upper intercept age of 1162 +/- 430 Ma. The upper intercept age, within error limits, is consistent with inherited zircons found in other La Posta-type plutons and implies that all of the plutons in the eastern zone of the PRB have a lower crustal component. A Rb-Sr isochron age for the same sample of the Indian Hill pluton yielded an age of 89.6 +/- 2.6 Ma with an $^{87/86}Sr_i$ of 0.70628. Based on the assigned error limits, Walawender and others (1990) argued for a preferred emplacement age of 89 Ma. The garnetiferous monzogranite of the Boulevard pluton is intrusive into the metamorphic roof pendant and the interior facies of the La Posta pluton. It yielded a concordant zircon U-Pb age of 93 +/- 1 Ma (Clinkenbeard, 1987). Multiple zircon fractions from three locations within the La Posta pluton produced consistent ages of 94 +/-1 Ma (Clinkenbeard, 1987; Walawender and others, 1990), and suggest that both plutons are related to a single thermal event.

Relative age relationships are compatible with the radiometric data. The Indian Hill pluton has intruded only the metamorphic complex, whereas the Boulevard pluton is intrusive into both the screen and the La Posta pluton. The pegmatitic dikes spatially associated with both the Indian Hill (Parrish, 1990), and Boulevard plutons (Henninger, 1996) are cut off at the pluton margins and are thus older than either body. Reid (2000) and Augustine (1998) demonstrated that the pegmatitic dikes cutting the western side of the Indian Hill-Tule Mountain screen are also cut by and thus older than the enveloping Small Biotite facies of the La Posta pluton. These dikes have produced substantial quantities of gem materials and appear to be the same rare-element type (Cerny, 1991) found in the major gem-producing districts of southern California (Figure 1). These districts have reported Ar-Ar ages between 95 and 100 Ma (Foord and others, 1991) that are only slightly older than the La Posta and associated La Posta-type plutons in the eastern zone of the PRB. Thus, the age data for all plutonic units point to a single thermal event centered about the emplacement of the 94 Ma La Posta pluton.

GEOCHEMISTRY

Major- and trace-element data for the granitoid plutons and the pegmatitic dikes are reported in Table 2. Major and selected trace-element analyses were performed at the analytical facilities at San Diego State University and are described by Parrish (1990) and Clinkenbeard (1987). All REE analyses were done at the Analytical Laboratory, Washington State University using ICP/MS. Procedures are described in Carollo (1993). Samples of the dike system were collected from segments that were either narrow, fine-grained irregular dikelets cutting through the main body of the pegmatite or discontinuous fine-grained segments entirely enclosed within the pegmatite. Similar fine-grained granitic rocks have been described elsewhere as aplite and recognized as more rapidly (pressure-?) quenched portions of the silicate melt responsible for the dike (Jahns and Burnham, 1969; Jahns, 1979).

The granitic units that comprise both the Indian Hill and the Boulevard plutons are peraluminous with A/CNK values generally between 1.1 and 1.2. Silica contents are clustered between about 74 and 77% (Table 2). Trace-element variations, however, suggest a petrogenetic link between the Indian Hill and Boulevard plutons and the spatially associated pegmatitic dikes. Figures 4a-c plot Zr against compatible elements Rb, Sr, and Ba. Both the Zr-Rb and Zr-Sr plots are broadly linear with the garnet-bearing and garnet-free granitoids falling into separate fields. Also shown are the compositional fields of the enclosing La Posta pluton and its most evolved unit (Clinkenbeard, 1987). The pegmatitic dikes fall at the low Zr end of the linear pattern, suggesting that a mixing/unmixing relationship exists between all units.

The Ba-Zr diagram (Figure 4c) is also consistent with a mixing/unmixing origin but indicates two separate processes. The small garnet-free unit in the Indian Hill pluton falls between the garnet-bearing granitoid analyses and the La Posta rocks. This pattern could result from mixing of the more evolved portion of the La Posta pluton and the residual liquids within the contemporaneous Indian Hill pluton. A mixing/unmixing relationship between the dike system and garnet-bearing granitoids satisfies their geochemical pattern but without two obvious melt end members, these two units are more likely related through progressive anatexis of a common source region, i.e. an unmixing origin.

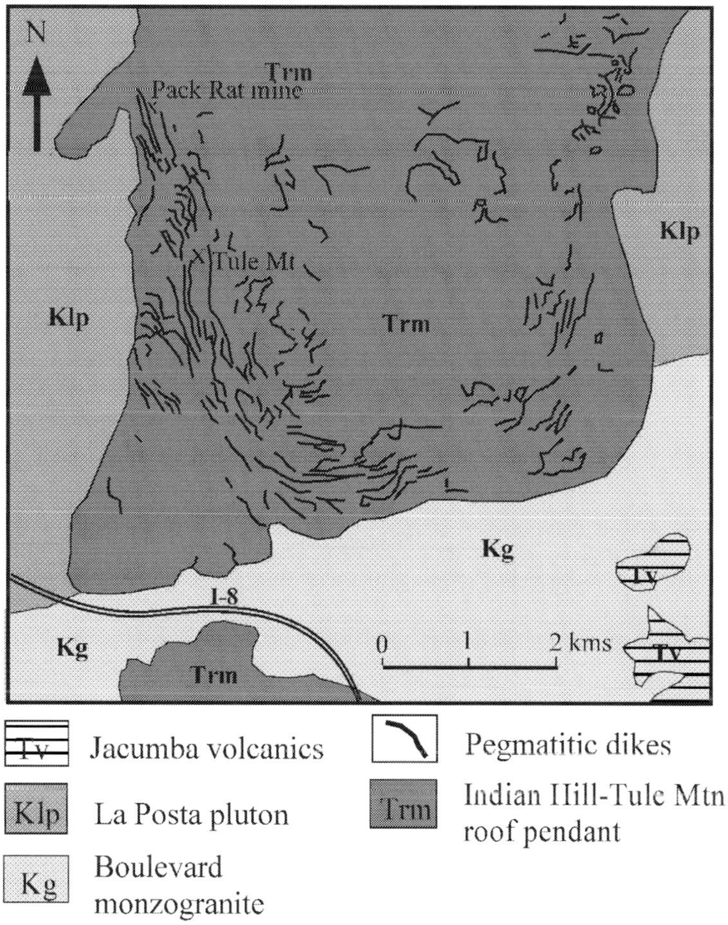

Figure 3. Generalized geologic map showing the distribution of pegmatite dikes in the central portion of the Indian Hill-Tule Mountain roof pendant (after Henninger, 1996).

REE patterns for the garnet-bearing granitoids and the pegmatitic dikes in the Indian Hill-Tule mountain roof pendant are shown in Figures 5a and b, respectively. The granitoid rocks (Figure 5a) have a steep slope with minimal to absent Eu anomalies, whereas the pegmatite dike system (Figure 5b) has overall lower concentrations of the LREE's, a pronounced negative Eu anomaly, and a flat HREE pattern. The granitoid pattern implies the presence of garnet (Nash and Crecraft, 1985), whereas the dike pattern, in terms of its pronounced negative Eu anomaly and flat HREE, imitates feldspar (plagioclase) REE partition coefficient patterns (Schnetzler and Philpotts, 1970; Dunn and Sen, 1994). Since both units contain feldspar and garnet in roughly equal amounts (Table 1), an origin through fractional crystallization seems unlikely. Alternately, these patterns suggest different source regions with garnet (+/- amphibole) controlling the melt chemistry for the granitoid plutons, and feldspar (plagioclase) controlling the dike chemistry.

The pegmatitic dikes and garnet-bearing granitoids within the Indian Hill-Tule Mountain roof pendant give some indication that older continental crust was involved in their generation. Whole rock del^{18}O for both the Indian Hill and the Boulevard plutons falls between 11.5 and 12.1‰, whereas this parameter for quartz separates falls between 12.0 and 13.2‰ (Walawender and others, 1990). These values suggest that the source region was enriched in heavy oxygen, a characteristic feature of source rocks that have undergone a weathering cycle at or near Earth's surface. Parrish (1990) and Walawender and others (1990) reported a $^{87/86}Sr_i$ for the garnet-bearing phase of

TABLE 2. MAJOR- AND TRACE-ELEMENT ANALYSES.

Sample	IH-1	IH-2	IH-4	IH-5	IH-7	IH-8	IH-9A	IH-11	IH-12	IH-20	IH-22	IH-27
unit	GMB	GMB	GMB	GMB	MB	GMB	GMB	GMB	GMB	GMB	dike	MB
SiO_2	74.42	73.93	75.64	76.08	74.52	75.44	76.45	75.69	75.13	75.55	76.31	74.03
TiO_2	0.10	0.12	0.11	0.14	0.22	0.07	0.06	0.05	0.12	0.08	0.03	0.19
Al_2O_3	13.32	13.57	13.86	14.08	14.41	13.98	13.86	13.81	13.71	12.91	13.98	14.59
Fe_2O_3	1.23	1.35	1.26	1.505	1.78	1.01	0.99	0.90	1.26	0.93	0.48	1.64
FeO	1.11	1.21	1.13	1.35	1.60	0.91	0.89	0.81	1.13	0.84	0.47	1.48
MnO	0.04	0.05	0.04	0.05	0.03	0.03	0.09	0.05	0.03	0.03	0.09	0.03
MgO	0.40	0.20	0.62	0.46	0.75	0.39	0.17	0.38	0.30	0.05	0.08	0.66
CaO	1.54	1.52	1.55	1.64	2.34	1.44	1.12	1.30	1.62	1.27	0.94	2.25
Na_2O	3.21	3.21	3.37	3.91	3.52	3.20	3.33	3.17	3.23	3.26	3.60	3.31
K_2O	3.46	3.50	3.47	3.52	2.37	3.75	3.66	3.95	3.68	3.91	4.29	2.40
P_2O_5	0.03	0.04	0.03	0.04	0.05	0.04	0.04	0.02	0.03	0.35	0.03	0.06
LOI	0.42	n.d.	n.d.	n.d.	n.d.	n.d.	0.45	n.d.	n.d.	n.d.	n.d.	0.25
Rb	115	117	118	111	96	135	143	151	123	132	169	87
Ba	1817	1488	1494	1536	1485	1335	1086	1118	1607	1248	631	1379
Sr	292	287	290	282	519	258	210	206	302	216	144	493
Zr	91	97	91	93	166	81	65	58	94	54	40	171
Y	6.64	n.d.	n.d.	n.d.	n.d.	n.d.	11	n.d.	n.d.	n.d.	n.d.	5
La	22.46	n.d.	n.d.	n.d.	n.d.	n.d.	17.76	n.d.	n.d.	n.d.	n.d.	19.65
Ce	41.33	n.d.	n.d.	n.d.	n.d.	n.d.	33.77	n.d.	n.d.	n.d.	n.d.	37.33
Pr	4.58	n.d.	n.d.	n.d.	n.d.	n.d.	3.78	n.d.	n.d.	n.d.	n.d.	4.16
Nd	17.44	n.d.	n.d.	n.d.	n.d.	n.d.	14.18	n.d.	n.d.	n.d.	n.d.	15.86
Sm	3.61	n.d.	n.d.	n.d.	n.d.	n.d.	3.09	n.d.	n.d.	n.d.	n.d.	2.97
Eu	0.81	n.d.	n.d.	n.d.	n.d.	n.d.	0.45	n.d.	n.d.	n.d.	n.d.	0.90
Gd	2.34	n.d.	n.d.	n.d.	n.d.	n.d.	2.38	n.d.	n.d.	n.d.	n.d.	1.81
Tb	0.32	n.d.	n.d.	n.d.	n.d.	n.d.	0.39	n.d.	n.d.	n.d.	n.d.	0.25
Dy	1.42	n.d.	n.d.	n.d.	n.d.	n.d.	2.23	n.d.	n.d.	n.d.	n.d.	0.99
Ho	0.23	n.d.	n.d.	n.d.	n.d.	n.d.	0.38	n.d.	n.d.	n.d.	n.d.	0.15
Er	0.58	n.d.	n.d.	n.d.	n.d.	n.d.	1.00	n.d.	n.d.	n.d.	n.d.	0.32
Tm	0.07	n.d.	n.d.	n.d.	n.d.	n.d.	0.15	n.d.	n.d.	n.d.	n.d.	0.04
Yb	0.45	n.d.	n.d.	n.d.	n.d.	n.d.	0.98	n.d.	n.d.	n.d.	n.d.	0.24
Lu	0.07	n.d.	n.d.	n.d.	n.d.	n.d.	0.13	n.d.	n.d.	n.d.	n.d.	0.04

Sample	IH-34	IH-39	IH-40	IH-43	IH-44	IH-45	IH-49	IH-51	IH-54	IH-61	IH-66	IH-93
unit	GMB	MB	GMB	GMB	MB	GMB	GMB	MB	GMB	dike	dike	dike
SiO_2	74.03	73.82	75.06	74.97	74.01	75.52	73.98	74.75	74.06	74.80	76.01	75.11
TiO_2	0.07	0.16	0.06	0.08	0.18	0.10	0.11	0.13	0.10	0.07	0.03	0.14
Al_2O_3	13.42	14.04	13.52	12.69	13.68	13.58	13.01	14.30	13.51	13.25	13.67	13.22
Fe_2O_3	1.01	1.50	0.92	0.98	1.43	1.19	1.27	1.43	1.16	0.69	0.42	0.97
FeO	0.91	1.35	0.83	0.88	1.29	1.07	1.14	1.29	1.05	0.62	0.38	0.87
MnO	0.04	0.03	0.03	0.05	0.03	0.04	0.04	0.03	0.03	0.17	0.04	0.07
MgO	0.07	0.47	0.16	0.07	0.37	0.29	0.18	0.54	0.40	0.06	0.22	0.17
CaO	1.28	1.84	1.28	1.26	1.88	1.60	1.57	1.66	1.80	1.18	0.96	1.27
Na_2O	3.02	3.64	3.29	3.37	3.57	3.16	3.19	3.88	3.21	2.71	3.90	2.89
K_2O	4.06	2.97	3.57	3.74	3.00	3.64	3.28	3.38	3.20	4.53	3.66	3.97
P_2O_5	0.04	0.04	0.03	0.03	0.06	0.03	0.03	0.04	0.08	0.05	0.07	0.08
LOI	n.d.	0.41	n.d.	n.d.	n.d.	n.d.	n.d.	n.d.	n.d.	0.40	n.d.	0.46
Rb	122	115	126	118	118	108	112	106	96	164	140	157
Ba	1244	1365	1114	1383	1392	1535	1526	1716	1724	702	462	703
Sr	223	386	216	236	407	289	312	346	372	183	135	170
Zr	60	131	63	63	117	101	83	129	101	34	10	41
Y	<1	7	<1	<1	<1	<1	<1	<1	<1	29	<1	28
La	n.d.	17.11	n.d.	n.d.	n.d.	n.d.	n.d.	n.d.	n.d.	9.57	n.d.	10.40
Ce	n.d.	31.85	n.d.	n.d.	n.d.	n.d.	n.d.	n.d.	n.d.	19.22	n.d.	20.70
Pr	n.d.	3.70	n.d.	n.d.	n.d.	n.d.	n.d.	n.d.	n.d.	2.18	n.d.	2.39
Nd	n.d.	14.19	n.d.	n.d.	n.d.	n.d.	n.d.	n.d.	n.d.	8.77	n.d.	9.44
Sm	n.d.	2.92	n.d.	n.d.	n.d.	n.d.	n.d.	n.d.	n.d.	2.72	n.d.	2.94
Eu	n.d.	0.79	n.d.	n.d.	n.d.	n.d.	n.d.	n.d.	n.d.	0.47	n.d.	0.47
Gd	n.d.	2.01	n.d.	n.d.	n.d.	n.d.	n.d.	n.d.	n.d.	3.00	n.d.	3.11
Tb	n.d.	0.28	n.d.	n.d.	n.d.	n.d.	n.d.	n.d.	n.d.	0.68	n.d.	0.68
Dy	n.d.	1.45	n.d.	n.d.	n.d.	n.d.	n.d.	n.d.	n.d.	4.58	n.d.	4.57
Ho	n.d.	0.24	n.d.	n.d.	n.d.	n.d.	n.d.	n.d.	n.d.	0.99	n.d.	0.93
Er	n.d.	0.59	n.d.	n.d.	n.d.	n.d.	n.d.	n.d.	n.d.	3.25	n.d.	2.90
Tm	n.d.	0.08	n.d.	n.d.	n.d.	n.d.	n.d.	n.d.	n.d.	0.54	n.d.	0.42
Yb	n.d.	0.48	n.d.	n.d.	n.d.	n.d.	n.d.	n.d.	n.d.	4.04	n.d.	2.79
Lu	n.d.	0.07	n.d.	n.d.	n.d.	n.d.	n.d.	n.d.	n.d.	0.64	n.d.	0.43

Note: n.d. = not determined, MB = garnet-free granitoids, GMB = garnet-bearing granitoids, dike = aplitic portion. Major elements in wt. %, trace elements in ppm.

TABLE 2 (Cont.)

Sample	T2	T4	T5	T7	T8	T10	T11	T12	LP-59	LP-62	LP17	LP46
unit	aplite	aplite	aplite	GMB	dike	dike	GMB	dike	GMB	GMB	GMB	GMB
SiO_2	78.94	76.24	75.05	74.82	76.8	71.90	74.09	76.32	76.29	74.56	72.90	73.48
TiO_2	0.03	0.03	0.02	0.18	0.01	0.01	0.20	0.02	.07	0.08	0.22	0.09
Al_2O_3	12.67	14.17	14.64	13.47	13.72	16.09	14.96	13.75	14.16	13.51	15.23	15.67
Fe_2O_3	n.d.	n.d.	n.d.	n.d.	n.d.	n.d.	n.d.	n.d.	0.98	1.18	n.d.	n.d.
FeO	0.24	0.53	0.74	1.11	0.44	1.28	1.29	0.37	n.d.	n.d.	1.55	0.66
MnO	0.01	0.14	0.31	0.02	0.08	1.13	0.05	0.12	0.04	0.04	0.04	0.04
MgO	0.16	0.15	0.22	0.41	0.15	0.11	0.55	0.12	0.39	0.24	0.58	0.23
CaO	0.50	0.81	0.66	1.51	0.65	0.72	2.74	0.79	1.32	1.54	2.10	1.75
Na_2O	4.27	4.14	4.20	3.30	3.75	6.74	4.48	5.18	2.93	3.60	3.71	4.02
K_2O	3.37	4.11	3.82	4.08	4.6	1.43	1.68	2.69	3.58	3.26	2.94	3.56
P_2O_5	0.05	0.05	0.12	0.06	0.05	0.26	0.06	0.06	0.04	0.03	0.07	0.08
LOI	n.d.	n.d.	n.d.	n.d.	n.d.	n.d.	n.d.	n.d.	0.53	0.54	0.28	0.50
Rb	112	145	144	88	121	112	63	101	125	123	107	130
Ba	422	610	21	1618	604	7	842	68	1264	1331	1217	733
Sr	117	109	23	247	85	7	413	41	241	248	349	264
Zr	41	37	30	89	31	59	112	24	58	53	64	65
Y	9	16	21	15	14	22	9	16	12	12	<1	<1
La	5.19	10.06	4.31	33.01	6.03	3.78	16.83	6.53	17.92	19.93	n.d.	n.d.
Ce	10.09	19.97	9.29	61.27	12.31	8.73	32.51	13.23	34.40	38.59	n.d.	n.d.
Pr	1.15	2.30	1.09	6.50	1.45	1.10	3.69	1.61	3.91	4.38	n.d.	n.d.
Nd	4.50	9.02	4.23	24.76	5.77	4.18	14.80	6.60	15.36	16.81	n.d.	n.d.
Sm	1.28	2.17	1.76	4.87	1.65	2.62	3.13	2.09	3.51	3.87	n.d.	n.d.
Eu	0.24	0.29	0.05	0.76	0.24	0.02	1.03	0.11	0.60	0.66	n.d.	n.d.
Gd	1.25	2.09	1.87	3.77	1.80	3.00	2.41	2.00	2.67	2.94	n.d.	n.d.
Tb	0.23	0.40	0.47	0.53	0.38	0.78	0.34	0.39	0.43	0.45	n.d.	n.d.
Dy	1.37	2.59	3.19	2.84	2.53	4.32	1.85	2.45	2.39	2.39	n.d.	n.d.
Ho	0.26	0.48	0.58	0.48	0.45	0.60	0.32	0.48	0.41	0.39	n.d.	n.d.
Er	0.00	1.10	1.72	1.16	1.10	1.54	0.78	1.31	1.02	0.97	n.d.	n.d.
Tm	0.09	0.17	0.33	0.17	0.18	0.28	0.11	0.21	0.13	0.13	n.d.	n.d.
Yb	0.59	1.04	2.56	0.98	1.14	2.11	0.72	1.30	0.81	0.78	n.d.	n.d.
Lu	0.09	0.15	0.40	0.15	0.16	0.31	0.12	0.19	0.11	0.11	n.d.	n.d.

Sample	LP28	LP61	LP62	LP45	LP59	LP60
unit	GMB	GMB	GMB	GMB	GMB	GMB
SiO_2	73.91	73.93	74.56	74.91	75.04	76.29
TiO_2	0.08	0.12	0.08	0.11	0.09	0.07
Al_2O_3	14.6	14.43	13.51	14.52	14.04	14.16
Fe_2O_3	n.d.	n.d.	n.d.	n.d.	n.d.	n.d.
FeO	1.04	0.98	1.06	0.93	1.05	0.88
MnO	0.04	0.06	0.04	0.05	0.04	0.04
MgO	0.18	0.34	0.24	0.16	0.31	0.39
CaO	1.50	1.81	1.54	1.39	1.51	1.32
Na_2O	3.63	3.35	3.60	3.11	3.50	2.93
K_2O	3.56	3.24	3.26	3.76	3.46	3.58
P_2O_5	0.08	0.06	0.03	0.06	0.03	0.04
LOI	0.42	0.60	0.54	0.79	0.45	0.53
Rb	122	107	123	157	125	137
Ba	1185	848	1331	806	1264	772
Sr	217	248	248	181	241	200
Zr	58	54	53	42	58	40
Y	<1	<1	<1	<1	<1	<1
La	n.d.	n.d.	n.d.	n.d.	n.d.	n.d.
Ce	n.d.	n.d.	n.d.	n.d.	n.d.	n.d.
Pr	n.d.	n.d.	n.d.	n.d.	n.d.	n.d.
Nd	n.d.	n.d.	n.d.	n.d.	n.d.	n.d.
Sm	n.d.	n.d.	n.d.	n.d.	n.d.	n.d.
Eu	n.d.	n.d.	n.d.	n.d.	n.d.	n.d.
Gd	n.d.	n.d.	n.d.	n.d.	n.d.	n.d.
Tb	n.d.	n.d.	n.d.	n.d.	n.d.	n.d.
Dy	n.d.	n.d.	n.d.	n.d.	n.d.	n.d.
Ho	n.d.	n.d.	n.d.	n.d.	n.d.	n.d.
Er	n.d.	n.d.	n.d.	n.d.	n.d.	n.d.
Tm	n.d.	n.d.	n.d.	n.d.	n.d.	n.d.
Yb	n.d.	n.d.	n.d.	n.d.	n.d.	n.d.
Lu	n.d.	n.d.	n.d.	n.d.	n.d.	n.d.

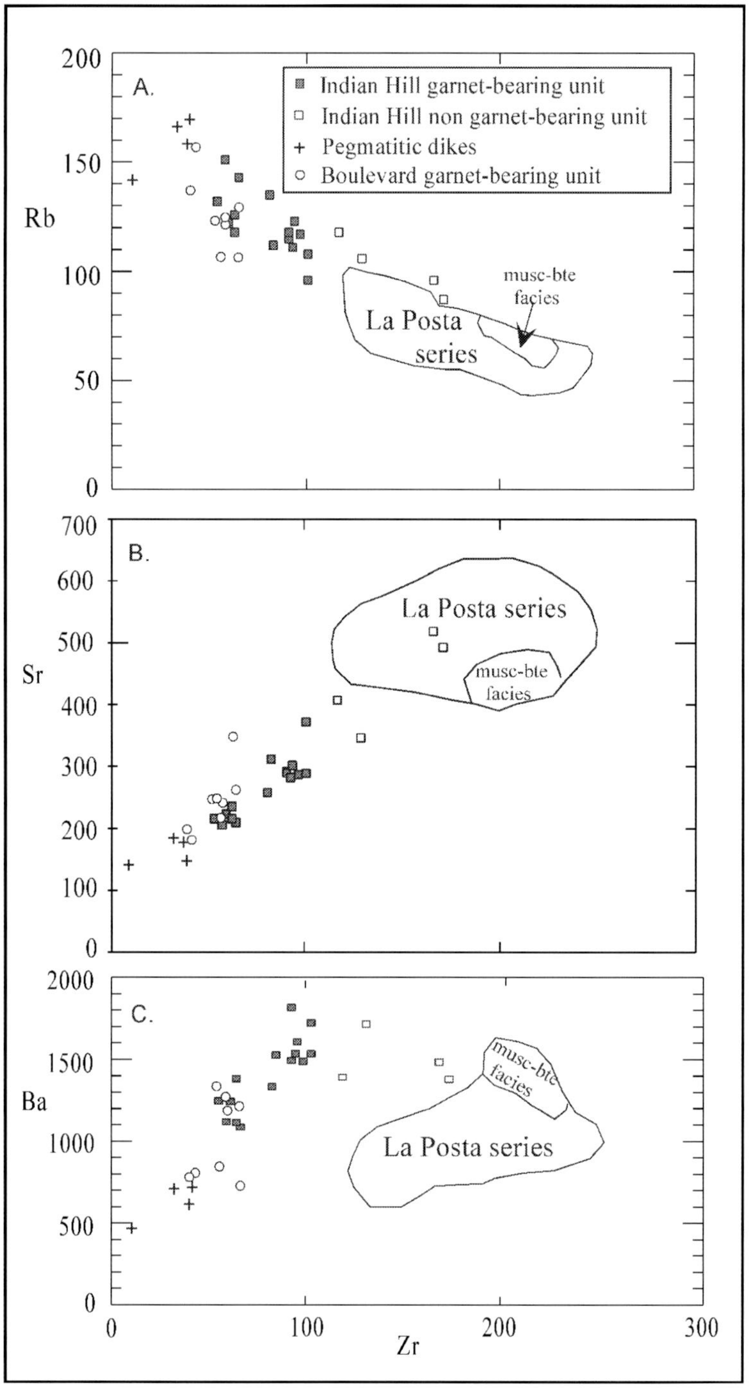

Figure 4. Ba-Zr, Rb-Zr, and Sr-Zr variation diagrams.

the Indian Hill at 0.70628. Although this value is marginal for a crustal-derived melt, it is distinct from the 0.70437 measured in the La Posta pluton and argues for the involvement of a source(s) richer in radiogenic Sr than oceanic crust. Apatite grains recovered from pocket zones of several southern California gem-bearing pegmatites have low Rb concentrations and present-day $^{87/86}$Sr between 0.715 and 0.780 (Table 3). One of these dike systems (Pack Rat dike) lies within the Indian Hill-Tule Mountain roof pendant (Figure 3) about 1 km north of Tule Mountain and has produced economic quantities of gem spodumene, spessartine, and several varieties of beryl. With such low Rb, the present-day ratios must be close to the initial ratios (Gastil and others, 1990) and suggest that the pegmatitic fluids were derived from a source region enriched in radiogenic Sr.

SOURCE REGION CHARACTERISTICS

During anatexis, the development of the REE pattern in the melt is a function of the mineralogy in the source region, the length of time that melt and residue are in contact, and the cooling history of the melt. If equilibrium is attained and the melt does not interact with crustal rocks during ascent, then its REE pattern should reflect the mineralogy of the source region. The consistency of the rock patterns (Figures 5a, b) both in shape and in relative position strongly argues for an equilibrium REE assemblage. If the magma rose through the same broadly homogeneous reservoir, any melt-country rock interaction will not significantly change the original REE abundances. Given that the Sr-Zr, Ba-Zr, and Rb-Zr variation diagrams (Figure 4) are consistently linear and independent of both the La Posta and the host-rock chemical composition, the amount of melt-country rock interaction must have been minimal. The REE patterns, therefore, must directly reflect the source region and can be interpreted in terms of source lithologies.

The REE patterns for the dike system reflect the relative partition coefficient distribution for feldspars, especially in terms of the pronounced negative Eu anomaly and the flat HREEs, and suggest that it was the primary mineral in the source region capable of affecting the relative REE concentrations. The granitoid patterns, however, exhibit a garnet influence characterized by a steep pattern that could also mask the presence of other minerals. The spatial, temporal, and host rock affinities of the two units argue for the presence of a common source region capable of producing high-silica melts and the observed dike-granitoid emplacement sequence. These constraints can be met through melting of a pelitic source initially containing the assemblage biotite-muscovite-plagioclase-quartz.

Patino-Douce and Johnston (1991) found that at 10 kb and 825°C, 18% of the source rock melted to produce peraluminous silica-rich liquids through the reaction:

Musc + Plag + Quartz = ALS + Melt.

Here ALS denotes the appropriate aluminosilicate polymorph. Muscovite would disappear during the melting reaction leaving a residue of feldspar, quartz, and aluminosilicate. The only mineral in this residue capable of leaving an REE signature would be plagioclase. Patino-Douce and Johnston (1991) also recognized that the melt produced by this reaction was richer in H_2O (~4.7%) than would be possible if only muscovite underwent dehydration. They suggested that biotite underwent partial dehydration resulting in the composite reaction shown below (Patino-Douce and Johnston, 1991):

$$Musc + Biotite_1 + Plag + Qtz = Biotite_2 + ALS + Melt.$$

If some of the melt escaped upward to form small dikes, closed-system fractionation could have readily produced the pegmatitic textures and mineral zoning characteristic of the southern California gem-bearing pegmatites.

At temperatures above 850°C, biotite undergoes complete dehydration through the reaction:

$$Biotite + ALS + Plag + Qtz = Garnet + Kfs + Melt.$$

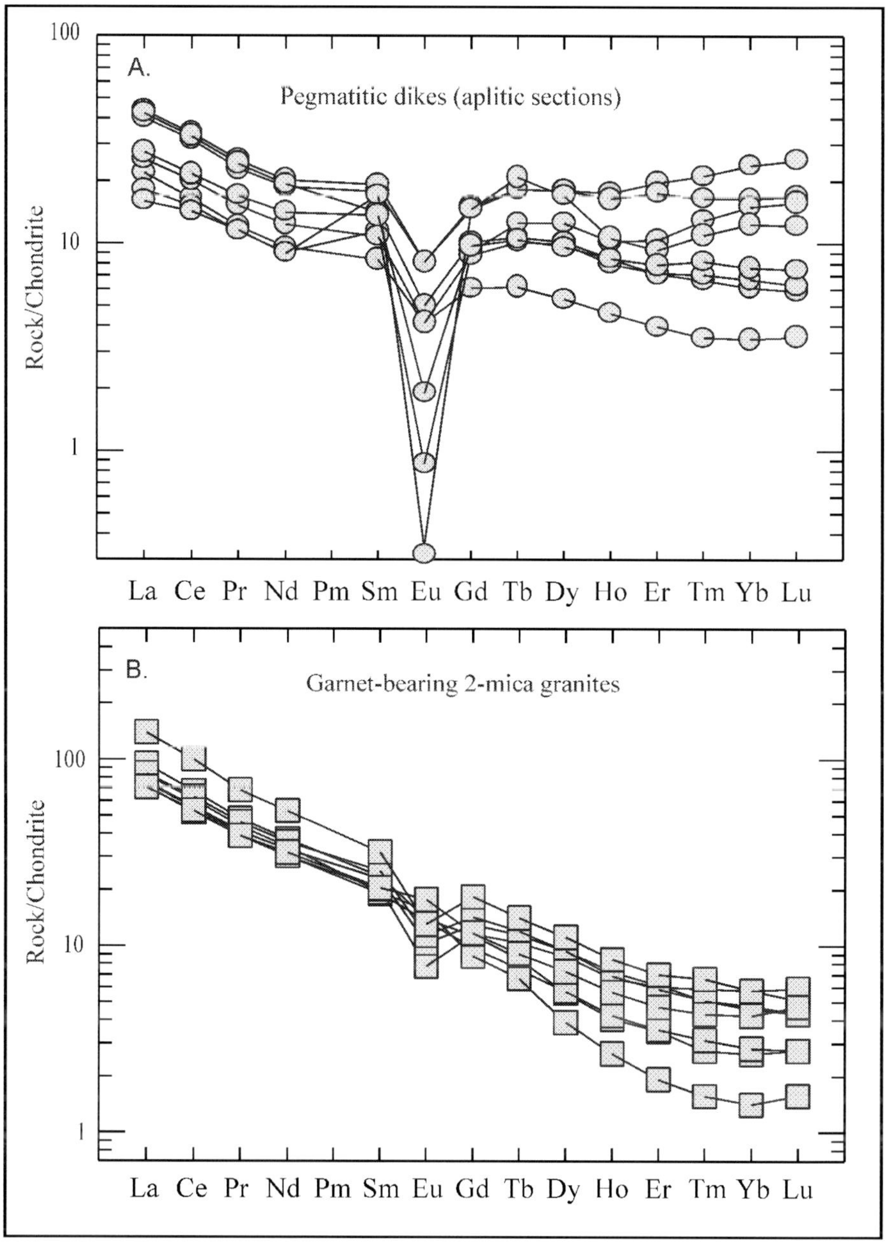

Figure 5. A. REE patterns for the aplitic portions of the pegmatite dikes in the Indian Hill-Tule Mountain roof pendant. B. REE patterns for the garnet-bearing, two-mica granitoids in the Indian Hill-Tule Mountain roof pendant. REE's normalized to Sun and McDonough (1989).

In this reaction the amount of melt produced is between 45 and 50% and is capable of gravitational separation from the source region. Lower water contents (~3.5%) from this reaction would allow the granitic liquid to rise at least short distances from the zone of melting before freezing. Biotite and possibly plagioclase would disappear from the residue so that the melt would inherit a garnet-dominated pattern. Lack of any melt-crystal separation during ascent and solidification in the observed homogeneous granitoid plutons would result in a LREE-enriched pattern reflecting the garnet influence.

CONCLUSIONS

The limited range of radiometric ages within and around the Indian Hill-Tule Mountain roof pendant indicate that all of the associated plutonic rock units were created during a single thermal event between about 90 and 100 Ma. The La Posta pluton, however, is clearly different in chemical composition, isotopic character, and mineralogy from the garnet-bearing granitoid rocks and

Table 3. Rb-Sr SYSTEMATICS FOR "POCKET" APATITE

Pegmatite	District	$^{87/86}Sr$	Sr (ppm)	Rb (ppm)	$^{87}Rb/^{86}Sr$
Little 3	Ramona	0.79039	19	0.167	0.0260
Packrat	Sacatone	0.71547	865	0.680	0.0023
Himilaya	Mesa Grande	0.71095	146	0.047	0.0009
Elizabeth R	Pala	0.76537	56	1.35	0.0071
Maple Lode	Aguanga	0.75434	17	1.64	0.0281

Note: T. Davis, C.S.U. Los Angeles, Analyst

pegmatite dikes. Walawender and others (1990) and Gromet and Silver (1987) argued for its derivation through partial melting of an oceanic crustal source followed by limited interaction with Precambrian continental crust during ascent. The small quantities of the garnet-free granitoid within the Indian Hill pluton most likely formed by mixing of the residual La Posta magma and the garnet-bearing monzogranite melt that solidified to form the bulk of the Indian Hill pluton. This origin by hybridization is consistent with the trace-element patterns and the spatial and temporal association of the two rock units.

The garnet-bearing monzogranites and the pegmatitic dikes appear to be more or less contemporaneous but with the dike system slightly older than the monzogranites. Their chemical composition and spatial association, however, suggest an origin by progressive partial melting of a common source terrane whose mineralogy included muscovite, biotite, quartz, and feldspar. The experimental studies of Patino-Douce and Johnston (1991) on water-absent melting of metagraywacke compositions at 10 kb total pressure show that the first melt generated from this protolith would be in equilibrium with plagioclase, quartz, and an aluminosilicate until muscovite depletion. The higher water content of this derived melt is one of the conditions required to create pegmatitic structures within silicate magmas. A temperature rise of approximately 25°C would initiate the breakdown of biotite whose refractory assemblage would contain garnet. The distinctive REE patterns of the two units reflect, respectively, the plagioclase- and garnet-bearing residues characteristic of sequential melting of a graywacke protolith. The experimental work on which these conclusions are based can be extrapolated to lower total pressures and more geologically attainable temperatures without any change in the relative melting sequence (Patino-Douce and Johnston, 1991).

The question remains, then, as to the heat source responsible for the thermal event that created the dikes and garnet-bearing monzogranites. The La Posta pluton was emplaced at crustal levels at near liquidus temperatures (Walawender and others, 1990). Early crystallization of amphibole followed by biotite and plagioclase suggest water content on the order of 2% and emplacement temperatures near 950°C (Naney, 1983). Given the size of the pluton (>1400 sq km), its ambient heat energy coupled with that released by crystallization would be sufficient to cause anatexis in the muscovite- and biotite-bearing roof rocks. If the La Posta pluton rose as a large diapir into the upper crustal environment, heat loss into the Indian Hill-Tule Mountain roof pendant would first produce small amounts of water-rich silicate melt. These melts, although not water saturated (Patino-Douce and Johnston, 1991), would not rise far from their source terrane before freezing. After a short melting hiatus, temperatures would reach the point of biotite dehydration and produce more water-deficient magmas that could rise further from the source terrane and cut off the earlier dike system.

The occurrence of gem-bearing pegmatite dikes (and associated garnet-bearing monzogranites) thus requires two components:

1. The presence of large La Posta-type plutons to provide the necessary heat energy.
2. Spatially associated metasedimentary assemblages with graywacke protoliths to provide the necessary chemical compositions.

The distribution of gem pegmatite mines and mining district in southern California appears to reflect these conditions in that the bulk of the mining districts occur within either the medial portion or the eastern zone of the batholith (Figure 1). The Pala and Rincon Districts lie west of the boundary but have been subjected to post-solidification translations along the Elsinore fault system. The Ramona District, whose pegmatitic dikes intrude mafic tonalite, does not appear to fit the listed conditions and remains an enigma. All other mining areas meet the above two requirements. This causal relationship then may provide a broader view in any search for gem mining opportunities in southern California.

ACKNOWLEDGMENTS

The author is indebted to Jim Clanin and Chuck Houser for supplying apatite crystals for part of this report. Byron

Weege, Jim Clanin, Roland Reed, Louis Spaulding, Matt Taylor, and Bill Larsen generously allowed access to their mines and shared their knowledge of southern California pegmatites. A review by Peter Weigand greatly improved this manuscript.

REFERENCES CITED

Augustine, E., 1998, Structure and petrology of a portion of the Sacatone Springs roof pendant at Tule Mountain [M.S. Thesis]: San Diego State University, 21 p.

Berggreen, R.G., and Walawender, M.J., 1977, Petrography and metamorphism of the Morena Reservoir Roof Pendant, Southern California: California Division of Mines and Geology Special Report 129, pp. 61-65.

Cerny, P., 1991, Rare-element granitic pegmatites. Part 1: Anatomy and internal evolution of pegmatitic deposits: Geoscience Canada, v. 18, No. 2, pp. 49-67.

Clinkenbeard, J.P., 1987, The mineralogy, geochemistry, and geochronology of the La Posta Pluton, San Diego and Imperial Counties, California [M.S. Thesis]: San Diego State University, 215 p.

Carollo, G.F., 1993, Geochemistry, petrography, and REE analysis of leucogranite plutons from the western zone of the Peninsular Ranges batholith [M.S. Thesis]: San Diego State University, 177 p.

Carollo, G.F., and Walawender, M.J., 1993, Geochemistry and petrography of leucogranite plutons from the western zone of the Peninsular Ranges batholith; in P.L. Abbott, Sangines, E.M., and Redina, M.A., eds., Geologic investigations in Baja California, Mexico, South Coast Geological Society Annual Field Trip Guidebook, pp. 151-162.

DePaolo, D.J., 1980, Sources of continental crust: Neodymium isotope evidence from the Sierra Nevada and Peninsular Ranges: Science, v. 209, pp. 684-687.

DePaolo, D.J., 1981, A neodymium and strontium isotopic study of the Mesozoic calc-alkaline granitic batholiths of the Sierra Nevada and Peninsular Ranges, California: Journal of Geophysical Research, v. 86, pp. 10470-10488.

Dunn, T., and Sen C., 1994, Mineral/matrix partition coefficients for orthopyroxene, plagioclase, and olivine in basaltic to andesitic systems: A combined analytical and experimental study: Geochimica et Cosmochimica Acta, v. 58, pp. 717-733.

Foord, E.E., London, D., Kampf, A.R., Shigley, J.E., and Snee, L.W., 1991, Gem pegmatites of San Diego County, California, in Walawender, M.J. and Hanan, B.B., eds., Geological Excursions in southern California and Mexico: Department of Geological Sciences, San Diego State University, pp. 128-146.

Gastil, R.G., 1975, Plutonic zones in the Peninsular Ranges of southern California and northern Baja California: Geology, v. 3, pp. 361-363.

Gastil, R.G., Phillips, R.P., and Allison, E.C., 1975, Reconnaissance geology of the state of Baja California: Geological Society of America Memoir 140, 170 p.

Gastil, R.G., Diamond, J., Knaack, C., Walawender, M., Marshall, M., Boyles, C., and Chadwick, B., 1990, The problem of the magnetite-ilmenite boundary in southern and Baja California, in Anderson, J.L., ed., The nature and origin of Cordilleran magmatism: Geological Society of America Memoir 174, pp. 19-32.

Germanario, M.P., 1982, The depositional and tectonic environments of the Julian Schist, Julian, California [M.S. Thesis]: San Diego State University. 95 p.

Girty, G.H., Thomson, C.N., Carmichael, D.L., Netto, S.L., and Wagemakers, R., 1998, Normal-sense mylonites of the Scove Canyon segment (SCs), Cuyamaca-Laguna Mountains shear zone (CLMSZ), California, in Snoke, A., Tullis, J.A., and Todd, V.R., eds., Atlas of Mylonitic and Fault-Related Rocks, Princeton Press.

Gromet, L.P., and Silver, L.T., 1987, REE variations across the Peninsular Ranges batholith: Implications for batholithic petrogenesis and crustal growth in magmatic arcs: Journal of Petrology, v. 28, pp. 75-125.

Henninger, C., 1996, Structure of the Mt. Tule dike swarm, Jacumba, California [B.S. Thesis]: San Diego State University, 13 p.

Jahns, R.H., and Burnham, C.W., 1969, Experimental studies of pegmatite genesis. I. A model for the derivation and crystallization of granitic pegmatites. Economic Geology, v. 64, pp. 843-864.

Jahns, R.H., 1979, Gem-bearing pegmatites in San Diego County, California, The Stewart Mine, Pala District, and the Himalaya mine, Mesa Grande District, in Abbott, P.L. and Todd, V.R., eds., Mesozoic Crystalline Rocks. Department of Geological Sciences, San Diego State University, pp. 3-38.

Larsen, E.S, 1948, Batholith and associated rocks of Corona, Elsinore, and San Luis Rey Quadrangles, southern California: Geological Society of America Memoir 29, 182 p.

Lombardi, M.L., 1992, The Alpine tonalite; Image of a gabbroic source? [M.S. Thesis]: San Diego State University, 178 p.

Naney, M.T., 1983, Phase equilibria of rock-forming ferromagnesian silicates in granitic systems: American Journal of Science: v. 283, pp. 993-1033.

Nash, W.P. and Crecraft, H.R., 1985, Partition coefficients for trace elements in silicic magma: Geochimica et Cosmochimica Acta, v. 49, pp. 2309-2322.

Parrish, K.E., Walawender, M.J., Clinkenbeard, J.P., and Wardlaw, M.S., 1986, The Indian Hills granodiorite, a peraluminous, garnet-bearing granitoid, Eastern Peninsular Ranges: Cordilleran Section, Geological Society of America Abstracts with Program, v. 18, no. 2, p.168.

Parrish, K.E., 1990, Geology, petrology, geochemistry, and isotopic character of the Indian Hill pluton, Jacumba Mountains, California [M.S. Thesis]: San Diego State University, 136 p.

Patino-Douce, A.E., and Johnston, A.D., 1991, Phase equilibria and melt productivity in the pelitic system; Implications for the origin of peraluminous granitoids and aluminous granites: Contributions to Mineralogy and Petrology, v. 107, pp. 202-218.

Pimental, M.I., 1998, Blair Valley Tonalite; An east-west magmatic connection? [B.S. Thesis]: San Diego State University, 17 p.

Reid, K., 2000, The Sacatone Springs tonalite; Of eastern or western origin? [B.S. Thesis]: San Diego State University, 14 p.

Schnetzler, C.C., and Philpotts, J.A., 1970, Partition coefficients of rare earth elements between igneous matrix and rock-forming mineral phenocrysts-II; Geochimica et Cosmochimica Acta, v. 34, pp. 331-340.

Siem, M.E., 1988, Comparisons of metamorphic grades across the Peninsular Ranges batholith, San Diego County, California [Senior Thesis]: San Diego State University, 29 p.

Silver, L.T., Taylor, H.P., and Chappell, B.W., 1979, Some petrological, geochemical, and geochronological observations of the Peninsular Ranges batholith near the international border of the U.S.A. and Mexico, in Abbott, P.L. and Todd, V.R., eds., Mesozoic Crystalline Rocks: Department of Geological Sciences, San Diego State University, pp. 83-110.

Silver, L.T., and Chappell, B.W., 1988, The Peninsular Ranges batholith; An insight into the evolution of the Cordilleran batholiths of southwestern North America: Transactions of the Royal Society of Edinburgh, Earth Sciences, v. 79, pp. 105-121.

Smith, T.E., Huang, C.H., Walawender, M.J., Cheung, P., and Wheeler, C., 1983, The gabbroic rocks of the Peninsular Ranges batholith, southern California; Cumulate rocks associated with calc-alkaline basalts and andesites: Journal of Volcanology and Geothermal Research, v. 18, pp. 249-278.

Thomson, C.N., and Girty, G.H., 1994, Early Cretaceous intra-arc ductile strain in Triassic-Jurassic and Cretaceous continental margin arc rocks, Peninsular Ranges, California, Tectonics, v.13, pp.1108-1119.

Todd, V.R., Erskine, B.G., and Morton, D.M., 1988, Metamorphic and tectonic evolution of the northern Peninsular Ranges batholith, southern California, in Ernst, W.G., ed., Metamorphic and crustal evolution of the western United States (Rubey Volume VII): Prentice-Hall Publishing Company, Englewood Cliffs, New Jersey, pp. 894-937.

Walawender, M.J., 1976, Petrology and emplacement of the Los Pinos pluton, southern California: Canadian Journal of Earth Sciences, v. 13, pp. 1288-1300.

Walawender, M.J., 2000, The Peninsular Ranges; A geological guide to San Diego's back country: Kendall/Hunt Publishing company, Dubuque, Iowa, 114 p.

Walawender M.J. and Smith T. E., 1980, Geochemical and petrologic evolution of the basic plutons of the Peninsular Ranges batholith, southern California; Journal of Geology, v. 88, pp. 233-242.

Walawender, M.J., Gastil, R.G., Clinkenbeard, J.P., McCormick, W.V., Eastman, B.G., Wernicke, R.S., Wardlaw, M.S., Gunn, S.H., and Smith, B.M., 1990, Origin and evolution of the zoned La Posta-type plutons, eastern Peninsular Ranges batholith, southern and Baja California, in Anderson, J.L., ed., The nature and origin of Cordilleran magmatism: Geological society of America Memoir 174, pp. 1-18.

Walawender, M.J., Girty, G.H., Lombardi, M.R., Kimbrough, D., Girty, M.S., and Anderson, C., 1991, A synthesis of recent work in the Peninsular Ranges batholith, in Geological Excursions in southern California and Mexico, Walawender, M.J. and Hanan, B.B., eds., Department of Geological Sciences, San Diego State University, pp. 297-318.

The Julian Gold Mining District and Gamma Radioactivity of the Elsinore Fault, San Diego County, California

George C. Copenhaver, Jr., *C.E.G., Project Engineering Geologist*
Geocon Incorporated, 6960 Flanders Drive, San Diego, California 92121-2974

"ELDORADO" DELAYED

The great California gold rush of 1849 established San Diego as an important military port city and supply point for the flood of gold-seekers making the long shipboard and inland migrations to the Sacramento Valley. This was but a rest-stop on the way to "Eldorado"; the fabled gold fields that became reality with the discovery of the large gold-placers in the rivers at the foot of the Sierra Nevada. But San Diego and its hinterland, remained mostly a sleepy collection of ranching and farming settlements until 1869 when a rancher, also a veteran miner of the 1849 rush, discovered placer gold in stream gravels of Coleman Creek near the present town of Julian. The headwaters of this gold-bearing stream and surrounding tributaries are located in the seismically active Elsinore fault zone (Figure 1).

The gold discovery jump-started the growth of the entire area, causing a population of a few hundred to increase to over 6000 by 1873 (Weber, 1963). The town of Julian and the surrounding mining areas actually grew faster than San Diego for a few years, with about 2000 residents just in the Julian area. With a rapidly expanding infrastructure supporting the region and the arrival of the railroad in 1885, the population, countywide, stabilized at about 17,000 by 1890. (Copenhaver, 1991).

Gold mining diminished steadily, along with the depletion of most of the larger profitable deposits. Despite all of the prospecting and mining activity, only a few mines, located in the south or southeast portions of the mineralized zone produced most of the gold. The most productive mines were the Ready Relief group, the Golden Chariot and the Stonewall mines (Weber, 1963). With the exception of a few die-hard prospectors and miners that scratched out a meager living from the remaining narrow veins, the bonanza era of gold mining became history. Nonetheless, the Julian Mining District still ranks as one of the principal gold-producers of the United States with over 219,000 ounces of the yellow metal extracted, most of it before the end of the 20th century (Koschmann and Bergandahl, 1968). After the turn of the 20th century, agriculture in the form of fruit orchards literally blossomed around Julian, with grape vineyards and apple orchards being planted directly over formerly productive gold veins. An example is the Ranchito gold mine that is located on the edge of a small meadow (field trip stop). The mine and mill were shown in an early photograph next to the meadow covered by grape vines. The meadow, incidently, is underlain by a main strand of the Elsinore fault (Figure 2).

At about the same time agriculture was growing, different mineral extraction operations began to thrive with the discovery of gem-bearing pegmatites in similar geologic settings to the gold-bearing veins...again paralleling the Elsinore fault and fringing the gold-producing districts to the north and south. At one point, about 1912, San Diego County became a world-class gem tourmaline and lithium (lepidolite) producer (Weber, 1963). Both these minerals are found nearby in a large pegmatite dike striking parallel to the Elsinore fault approximately one mile southeast of the Ranchito gold mine field trip stop (see "RM" and "GP," Figure 1). To the northwest are the gem-mining districts of Mesa Grande and Pala, where the majority of gem quality tourmaline and pegmatite minerals are produced. Other occurrences also present to the southeast.

Up to the present, San Diego County remains an important source of gem-quality tourmaline, with museum-quality pegmatite mineral specimens still being selectively mined. In addition to the gems and minerals, the region also provides other golden opportunities and attractions for new residents such as a thriving economy based on trade, technology, construction and tourism...all set in a mild, Mediterranean climate. Indeed, for many, San Diego is...Eldorado delayed!

GEOLOGY, PALEOSEISMICITY, AND RADIOACTIVITY

This section summarizes some recent geological, paleoseismic and radioactive studies in the portion of the

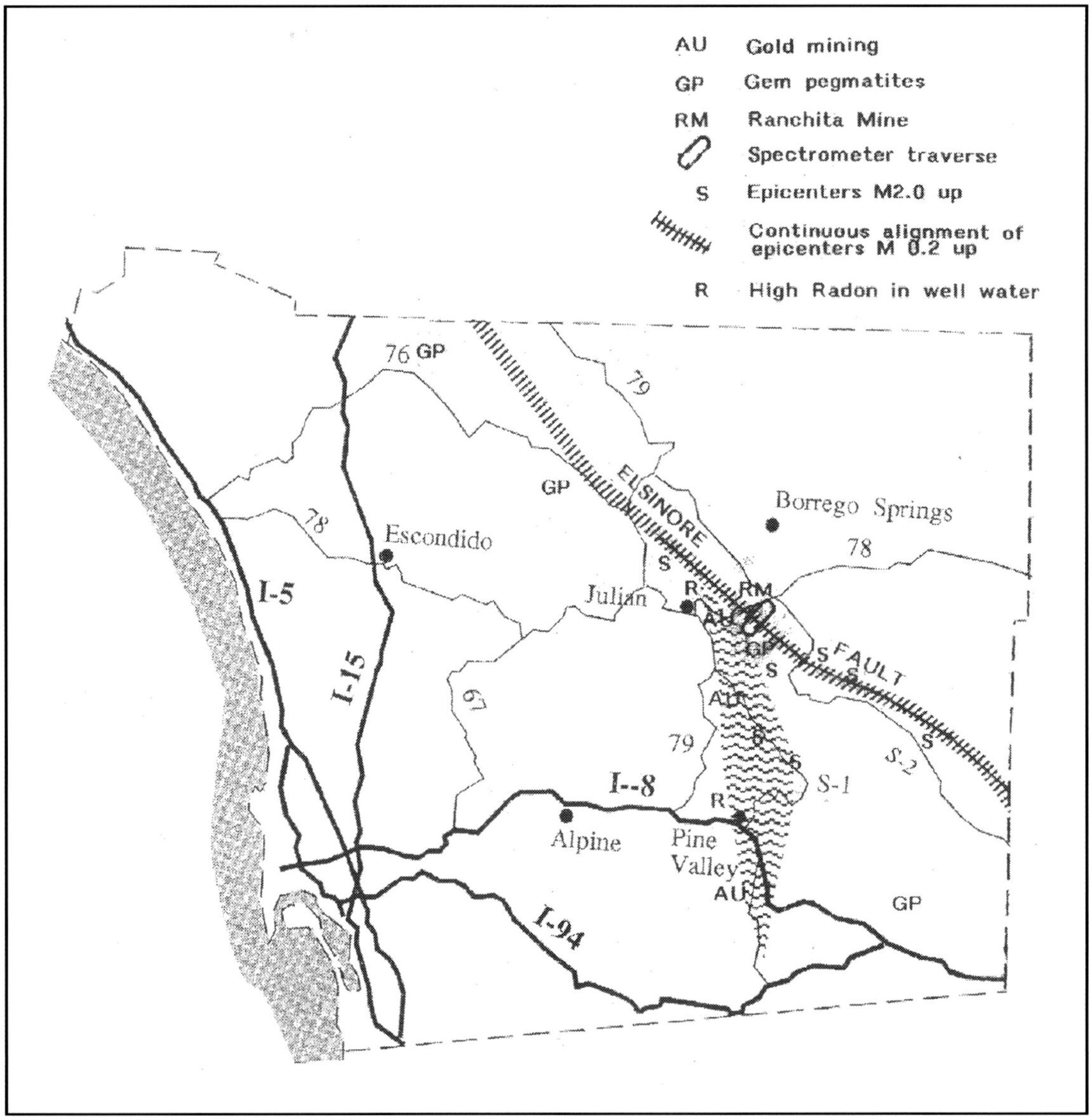

Figure 1. Generalized map of San Diego County showing the intersection of the Elsinore fault zone (hatched) and the Cuyamaca-Laguna Mountain fault zone (wavy lined). The field trip stop at the Ranchito Mine is shown as "RM" to the right of Julian. This area of structural intersection is also a focal point for gold mineralization, gem-pegmatites, and seismicity. Gamma-ray spectrometry studies may be successful in locating previously unknown active fault strands and assist geologic understanding of the Peninsular Range Batholith. (Map base is after Walawender, 2000.)

Elsinore fault zone centering near the Julian area and the field trip stop at the Ranchito gold mine. The diverse geological, mineralogical and active seismic features that exist there include a confluence of several events and structures. Sedimentary and metasedimentary rocks of the Triassic(?) Julian Schist were intruded by mafic and felsic igneous rocks of the Cretaceous Peninsular Range Batholith (primarily the La Posta pluton) with subsequent igneous intrusive heating and melt-differentiation that

Figure 2. View to the northwest from the Ranchito Mine field trip stop (meadow in the foreground) along the main strand of the Elsinore fault. A very linear trace is incised along the fault covered by a "green belt" of vegetation that parallels the zone along Banner Canyon and extending through the saddle between Volcan Mountain and Julian (upper left). The Ranchito mine-shaft tailings are just beyond the right side of the meadow. The intersection between the Cuyamaca-Laguna Mountain fault and the Elsinore fault is located approximately in the low area in the center of the photo, where Chariot Canyon meets Banner Canyon.

created gold-bearing quartz veins followed by gem-bearing pegmatites dikes in surrounding rocks (Walawender, 2000). These events are clustered parallel to the seismically active Elsinore fault, but mostly occurred at great depths and tens of millions of years prior to the fault. Perhaps the most important event was the emplacement of the La Posta pluton of the Peninsular Range Batholith. This intrusion generated enough heat to not only melt preexisting rocks such as the Julian Schist, perhaps liberating and

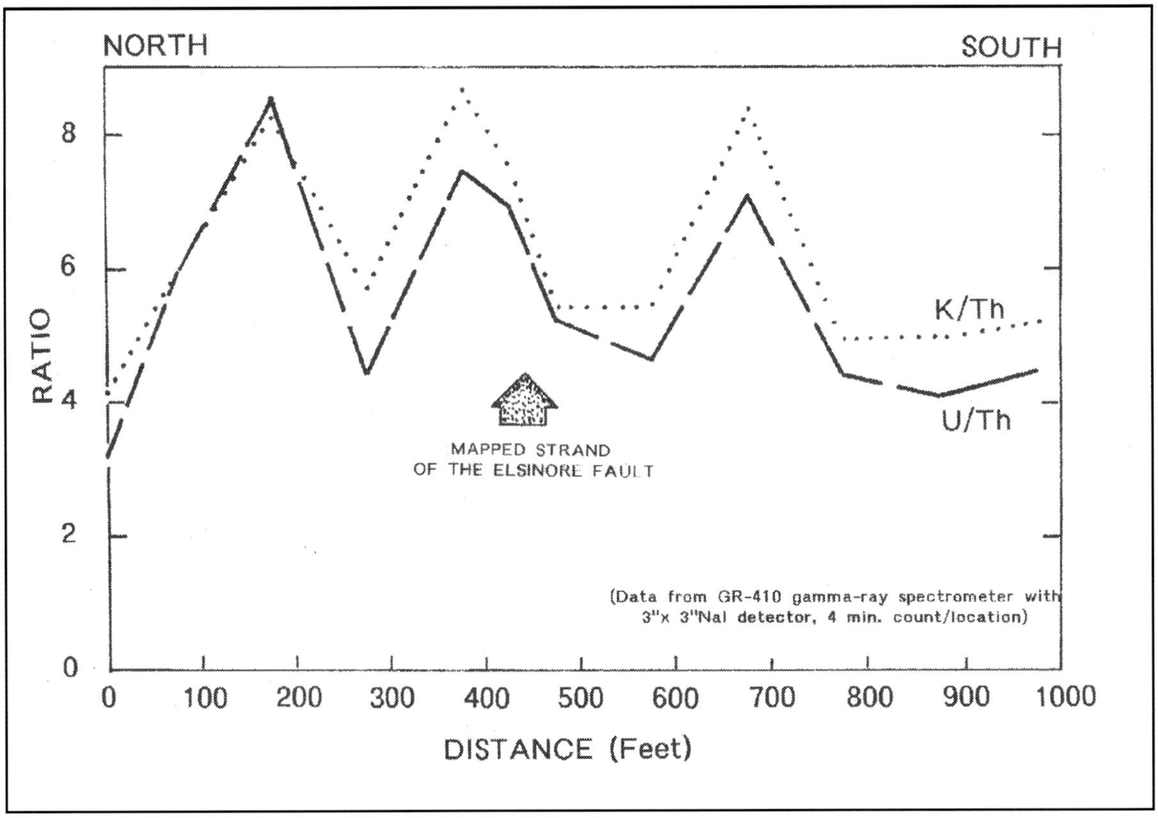

Figure 3. Portable gamma-ray spectrometer measurements along a traverse 1000 feet long across the Elsinore fault zone approximately one-half mile southeast of the Ranchito mine are summarized above. The data is derived from the detection of the three major natural radionuclides, 214Bi (decays from uranium), 40K (decays from potassium), and 208Tl (decays from thorium). Ratios of the raw count totals for four minutes for potassium and uranium and thorium are shown because rations with the relatively uniform thorium levels will serve to normalize results and reduce localized radiation "noise" such as cosmic radiation and instrumental drift. The results indicate three anomalous highs, of which one correlates with a mapped single trace of the Elsinore fault. It is possible the other two anomalies represent additional splays of the fault that have not been mapped. A number of other studies (Steinitz, 2003, Copenhaver, 2001, and Hattori, 2000) suggest that the anomalies could be caused by radon gas seepage up active pull-apart faults.

concentrating ancient placer gold into tabular quartz veins, but also to contribute water and boron-rich residual melt fluids that formed the gem-bearing pegmatite dikes.

As a principal strand striking parallel to the active San Andreas fault system, the Elsinore fault clearly exhibits a linear trace along the south side of Volcan Mountain in the form of Banner Canyon and the linear meadow adjacent to the Ranchito Mine at the field trip stop (Figure 2). The surface trace(s) of this zone are also manifest in the numerous topographic benches, vegetation lineations, springs and northwest-southeast striking shear zones exposed in roadcuts along Highway 78 between Julian and Chariot Truck Trail.

Paleoseismic investigations recently established small, but significant, late Holocene slip rates along the fault zone of 4.9 mm/yr along northern strands of the fault in the Temecula Valley (Rockwell, T.K., 2000). Other studies closer to Julian, and near the Pala pegmatite mines, also suggest activity in late prehistory between A.D. 1655 and A.D. 1810 (Vaughn et al., 1999). The concensus of most recent studies, moreover, seems to conclude that the Elsinore fault is accumulating significant strain that could be released in earthquake events on the order of M6.5-6.8 (Rockwell, T.K., 2000). This prediction is in concert with the magnitudes of historical earthquakes that occurred at distant points along the fault in the late 19th and early 20th centuries. The concept of unpredictable seismic disasters in the "New Eldorado" led the author to apply some new research that uses radioactivity generated by active faults to locate them…and to incite ongoing debates concerning earthquake prediction.

FAULT ACTIVITY = RADIOACTIVITY?

After the disasterous1995 earthquake in Kobe, Japan, research on earthquake prediction in this island nation with too many active faults and too little developable land experienced a rapid increase. Previously installed monitoring stations indicated anomalous radon, a radioactive gaseous decay product of uranium, just prior to the Kobe event (Igarashi, G., 1995). This occurrence, along with others studies suggesting the phenomena of gamma-ray radioactivity fluctuations related to faults, prompted major government funding to implement real-time monitoring networks of gamma-ray and seismic detectors along major active faults (Hattori, Matsuo, et al., 2000).

Not long after the Japanese studies, the author's employer, Geocon, Incorporated, utilized a portable gamma-ray spectrometer to augment its geotechnical and paleoseismic investigations locating buried strands of the active San Jose fault on the California State Polytechnic University, Pomona (Copenhaver, G., Chapin, M., and Brown, J., 2001). It should be emphasized that traditional methods such as trenching and continuous-sample borings were employed to discover most of the buried fault segments; the spectrometer surveys collected data on profiles and grids that assisted in interpreting overall trends and served the planning of future subsurface exploration programs.

Geochemical samples taken from borings during the Cal Poly study confirmed a high trace content of uranium in smectitic clays within the faulted sediments and gouge of the San Jose fault. We concluded that the surface spectrometer anomalies were caused by the radioactivity of the uranium daughter, radon gas, that seeped up through fissures or porous alluvium. A recent extensive study by the Geological Survey of Israel not only reports a similar phenomenon along the active Dead Sea rift fault, but also a strong statistical temporal connection to earthquakes on that fault (Steinitz, G., et al., 2003). After eight years of monitoring radon and hundreds of earthquakes, they found earthquakes occurred preferentially 3 days after the start time of a surge of radon anomalies on the Dead Sea rift fault.

Figure 3 shows the results of a portable gamma-ray spectrometer traverse approximately 1000 feet long and oriented N-S crossing the topographic saddle about one-half mile southeast of the Ranchito Mine field trip stop. Only a single strand of the Elsinore fault is mapped by the California Geological Survey, formerly CDMG, through the saddle into Rodriguez Canyon. However, the results of the spectrometer traverse indicate three distinct radiometric anomalies within a 700-foot-wide interval. This data was collected in a single day, without repeated measurements at the same station locations over extended periods of time (as were conducted by the Israeli studies). However, it would be an interesting project to perform repeated measurements under control conditions on a known active strand of the Elsinore fault, and through the same period, tabulating the time of occurrence of even minor tremors.

Typically, the portable gamma-ray spectrometer is designed to detect the three most common geologic sources of gamma-rays: Potassium (K), Uranium (U), and Thorium (Th). For convenience of interpretation and the reduction of potential "background effects," such as cosmic rays, or artificial sources of gamma-rays, ratios between the major gamma-sources of K and U and the relatively low uniform Th levels, accumulated over 4 minute counting periods for each radionuclide, are graphed. No attempt to convert the data to equivalent chemical concentrations was made. Careful outcrop and geomorphic mapping in the Rodriguez Canyon area may confirm the existence of multiple strands of the Elsinore fault that could help explain the spectrometer data. But other factors such as lithologic variation, meteorological effects, and instrumental drift could also be responsible for the anomalies (Ward, Stanley H., 1981).

Several active faults including the San Andreas rift zone, San Jose fault, Elsinore fault and the Rose Canyon fault have been traversed with the portable gamma-ray spectrometer by the author. All of the traverses yielded anomalous zones with similar characteristics to those measured across the Elsinore fault. Many of the anomalies were also located near faults that were previously mapped, or were physically confirmed by trenching or drilling, lending credence to the use of the portable spectrometer in assisting fault studies, or other geological research.

CONCLUSIONS

The intriguing geology, gold deposits, gem-pegmatites and seismicity of the Julian area continue to invite further study. Peninsular Range Batholith rocks, especially the La Posta pluton, played a unique role in focusing the geologic diversity. Exceptionally high temperatures of the La Posta pluton may have been the sum of radioactive heat generation from the natural radioactive elements (uranium, potassium and thorium) in the crust and heat flow within the mantle. Perhaps gamma-ray spectral analysis of rocks in the various tectonic settings in the Peninsular Range Batholith (and active fault zones) will help evaluate radioactive heat generation and thermal states in the earth's crust.

There may also be hope for resurrecting interest in the United States for continuing research on predicting earthquakes. Would Californians living in active fault zones have an interest? There are many owners whose property bestrides active faults, that do not have the land (or luxury to buy more) to design "setbacks" or "buffer zones" that would avoid surface rupture. Perhaps this is a dilemma in America because we have too much land. We can still continue that tradition, within a better, more secure, and privatized conservation system.

REFERENCES

California Geological Survey, 1958, Geology and Mineral Resources of the Santa Ysabel Quadrangle, San Diego County, California: CDMG Bulletin 177.

Copenhaver, George C., 1991, Mining in the San Diego, California Urban Environment Past, Present and Future: in Environmental Perils San Diego Region, Editors Patrick L. Abbott and William J. Elliott, San Diego Association of Geologists.

__________, Chapin, Michael S., Brown, James L., 2001, Combined Campus Fault Study, California State Polytechnic University Pomona, California, Geocon, Incorporated, unpublished report prepared for the Trustees of the California State University.

Hattori, Mutsuo, Okano, Masaharu, Togawa, Orihiko, 2000, Sea Bottom Gamma Ray Measurement By NaI(Tl) Scintillation Spectrometers Installed On Manned Submersibles, ROV and Sea Bottom Long Term Observatory, Japan Marine Science and Technology Center (JAMSTEC).

Igarashi, G., et al., 1995, Ground-water Radon Anomaly Before the Kobe Earthquake in Japan: Science, v. 269.

Koschmann, A. H. and Bergendahl, M.H., 1968, Principal Gold-Producing Districts of the United States: Geological Suvey Professional Paper 610, p. 76.

Rockwell, T.K., Bergmann, M., and Kinney, M., 2000, Holocene Slip Rate of the Elsinore Fault in Temecula Valley, Riverside County, California: in Geology and Enology of the Temecula Valley, Barbara B. Birnbaum, ed., San Diego Association of Geologists.

Steinitz, G., Begin, Z., and Gazit-Yaari, 2003, Statistically Significant Relation Between Radon Flux and Weak Eathquakes in the Dead Sea Rift Valley: Geology, v. 31; no.6.

Vaughan, P.R., et al., 1999, Paleoseismicity of the Elsinore Fault at Agua Tibia Mountain, Southern California: Bull. Seismological Society of America, v. 89, no. 6.

Walawender, M.J., 2000, The Peninsular Ranges, Kendall/Hunt Publishing Co., Dubuque, Iowa.

Ward, S.H., 1981, Gamma-Ray Spectrometry in Geologic Mapping and Uranium Exploration: Economic Geology, 75th Anniversary Volume, pp. 840-849.

Weber, F.H., Jr., 1963, Geology and Mineral Resources of San Diego County, California: California Geological Survey, formerly California Division of Mines and Geology, County Report 3.

Application of Ground Penetrating Radar (GPR) at the Cryo-Genie Gem Pegmatite Mine, San Diego County, California

***Jeffrey E. Patterson**, University of Calgary*

Discovery of a major gem tourmaline pocket at the Cryo-Genie Pegmatite Mine, San Diego County, in September, 2001, prompted the author to suggest to the mine owner that Ground Penetrating Radar (GPR) might be a useful tool to assist in the exploration efforts at the mine (Kampf, et al., 2003, Figure 1). Previous investigations at other San Diego County gem pegmatite mines demonstrated the efficacy of GPR for delineating pegmatite orientation and finding gem pockets (Patterson, 1996, Patterson and Cook, 1999, 2002). However, not all gem pegmatites are amenable to the use of GPR (Lees, pc, 2003). High clay contents in the overburden and wet clay seams within a dyke limits the depth of penetration and reduces the GPR signal quality. Poor understanding of the mineralization sequence and internal structure of a pegmatite hampers interpretation. Therefore, this study was undertaken to establish whether GPR would be as effective a geophysical exploration tool at the Cryo-Genie Pegmatite Mine as it has been at other San Diego County mines.

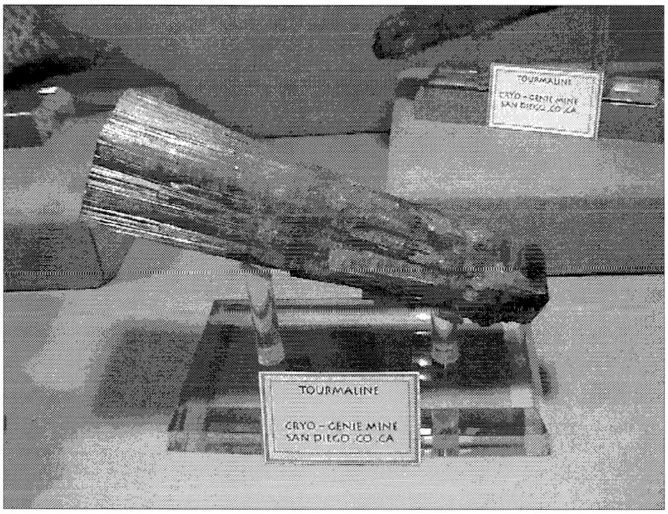

Figure 1. Gemmy Elbaite Tourmaline from the Big Monday Pocket, Cryo-Genie Mine. (Gochenour's Minerals showcase, Mineral Society of Southern California Show, December, 2001.)

The Cryo-Genie Mine is located in northern San Diego County, about 2 miles north-northwest of the village of Warner Springs (Agua Caliente) (Figure 2). The locale was mentioned in the first geologic investigation of San Diego County by Blake, as part of the railroad surveys near the 32nd parallel (Parke, 1854-1855). Blake reports lithia mica in the sediments of the canyon near Agua Caliente (Figure 4). This particular site was first prospected as the Lost Valley Truck Trail Prospect in 1904 for gem tourmaline and beryl by Bert Simmons with little success (Gochenour, 2002). The claim was located as the Lindy B Mine by the San Diego Mineral and Gem Society in 1962 and as the Cryo-Genie in 1974 by Bart Cannon. Since 1994, Dana Gochenour has held the claim to the mine. A complete history of the current mining operations and mineralogy is given in Kampf, et al. (2003), and at the Cryo-Genie Website at: http://home.earthlink.net/~goke/Cryo-Genie.htm. A mining diary of activities in 2002 can also be found on the MSA Pegmatite Interest Group Website at: http://www.minsocam.org/MSA/Special/Pig/PIG_articles/PIG_articles.html.

Basic information on the geology and mineralogy of the pegmatite is available in Weber (1963, p. 106). He describes the location, geology, and mineralization as follows:

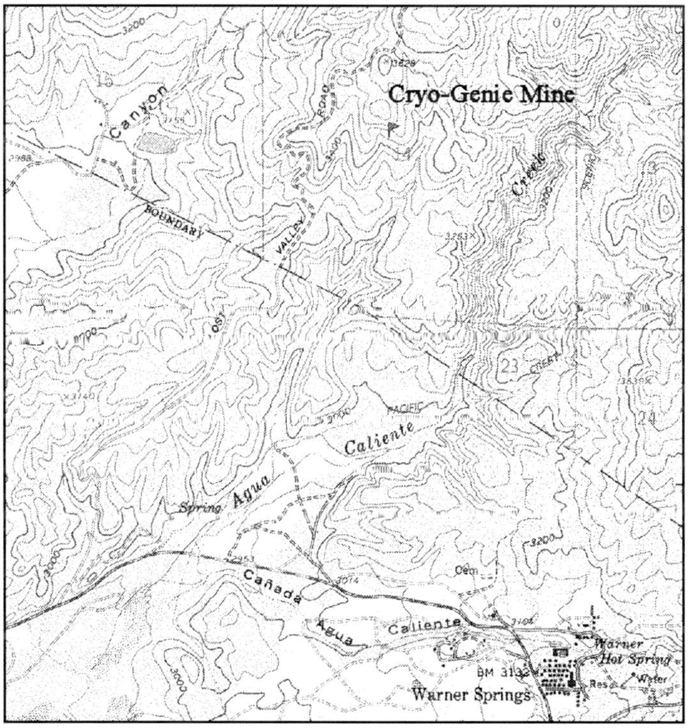

Figure 2. Cryo-Genie Pegmatite Mine, Site Location Map.

Deposit consists of a granite pegmatite dike, enclosed in hybrid rocks, which is exposed just southwest of the crest of a small hill. The dike strikes northward, dips about 30 degrees to the east, is about 10 ft. thick, and can be traced laterally for about 200 ft. The dike can be divided into 3 zones: (1) a lower zone, about 4 to 5 feet thick, which is chiefly graphic granite; (2) a core, about 1 ft. thick, which is composed of quartz, albite (including variety Clevelandite), muscovite, lepidolite, tourmaline, garnet, and allanite; and (3) an upper zone, 4 to 5 ft. thick, which is medium to coarse grained graphic granite with muscovite and black tourmaline crystals as long as 9 in. The tourmaline crystals of the core commonly range in color from solid black to crystals with black cores and grass green exteriors. Also noted were pale blue, pale pink, and colorless crystals. Pale crystals are thin and probably average less than 1/2 to 1 in. in length. Some of the smallest crystals are gemmy. (Figure 3.)

During October, 2001, the author visited the site to conduct a geological reconnaissance of the underground workings with a lighted display of the First Pocket, a major Quartz and Lepidolite zone. We also wanted to review

Figure 3. Outcrop southwest of crest of small hill. The dyke strikes north and dips both west and east. Green and Pink Elbaite Tourmaline and Morganite Beryl were recovered from this section of Dyke 2.

the general geology as presented during a Smithsonian Museum Pegmatite Camp, June, 1997, because in 1997, the site was overgrown with brush and only small typical Orthoclase, Quartz, Muscovite, and Schorl blocks were visible on the dump of Dyke 1. Dyke 1, containing the primary mine workings, generally strikes north, dipping 26 to 36 degrees to the west. Underground, in the recently excavated stopes, the dip is west 36°. In the region of the current working face, File 104, the dip is west 26°. A typical surface exposure of Dyke 1 is shown in Figure 5. For the most part, the dyke resembles the other classic LCT (Cerny, 1982), pegmatites of San Diego County. But, there are significant differences that make the Cryo-Genie pegmatite unique. These differences will be discussed below. Two other dykes outcrop on the claim. Dyke 2 has been mined, in the southwest outcrop, for gem tourmaline and morganite. Dyke 3 has only been exposed in a small glory hole and appears to be barren.

The preliminary investigation of Dyke 1 was conducted on 6 January, 2002. Data were collected with a GSSI SIR-4400R GPR unit and recorded on a model SR-8100H Graphic Recorder (Figure 7). The antenna used was a GSSI Model 3105, operating at a central frequency of 300 MHz (Figure 6). This equipment was manufactured in 1984 and uses Intel-286 processor technology. Hard-copy, analogue, output is available on heat sensitive carbon paper. The Model SIR-4400R has limited data acquisition capability, being used originally for the identification of very near-surface (10 to 50 cm) concrete reinforcing rods and conduit. The maximum scan range is 400 ns. No digital signal enhancement is available on this early model. The mode of recording was 'continuous' for all scans.

For the purposes of the preliminary investigation, the 300 MHz, central frequency, broadband antenna was selected to provide a reasonable image of the near surface region (less than 5 m). No underground scans were attempted at that time due to the noxious fumes generated by the heat sensitive carbon paper output. A/C power, 110 volt, for the graphic recorder was provided by the on-site electric generator. This limited the area to be scanned by the length of available power cords. The SIR-4400R GPR is powered by a 12 volt gel-cell motorcycle battery and has an operating time of about 4 hours. The combined weight of both units is over 40 kg, so the equipment was transported in the bed of a 4x4 truck (Figures 6 and 7).

During the surface survey of Dyke 1, marking flags were positioned every 1 m along the route of scan. These line of scan marks are shown on Figure 8 as vertical dashed lines. Correlation of the flags/marks with the known position of the underground workings is very close, giving strong support to the interpretation. The results of GPR scan line 1 are shown in Figure 9.

Interpretation of the GPR output shows that there are several discernible features on the scan. A previously excavated surface cavity, or 'coyote hole', is clearly shown

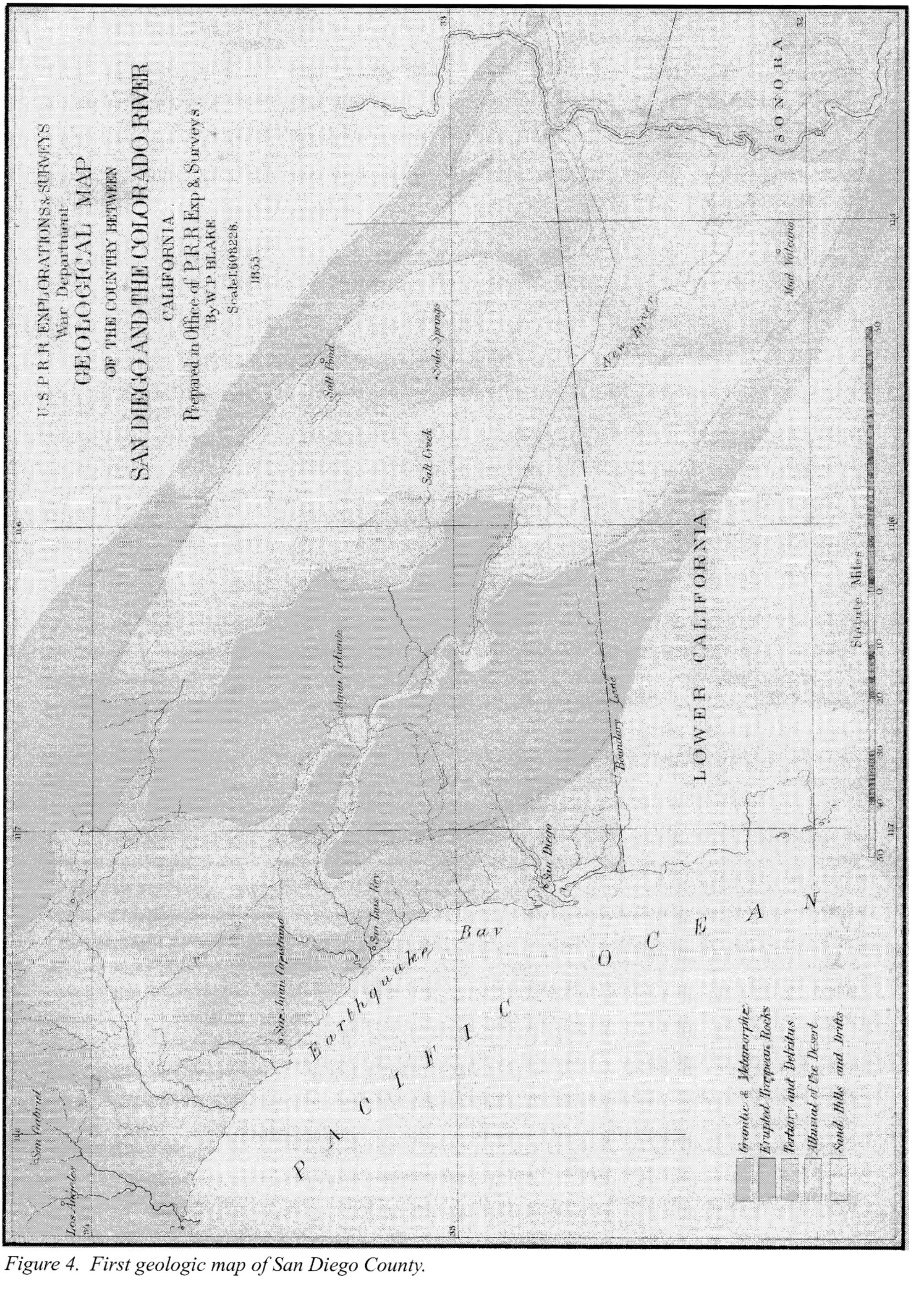

Figure 4. First geologic map of San Diego County.

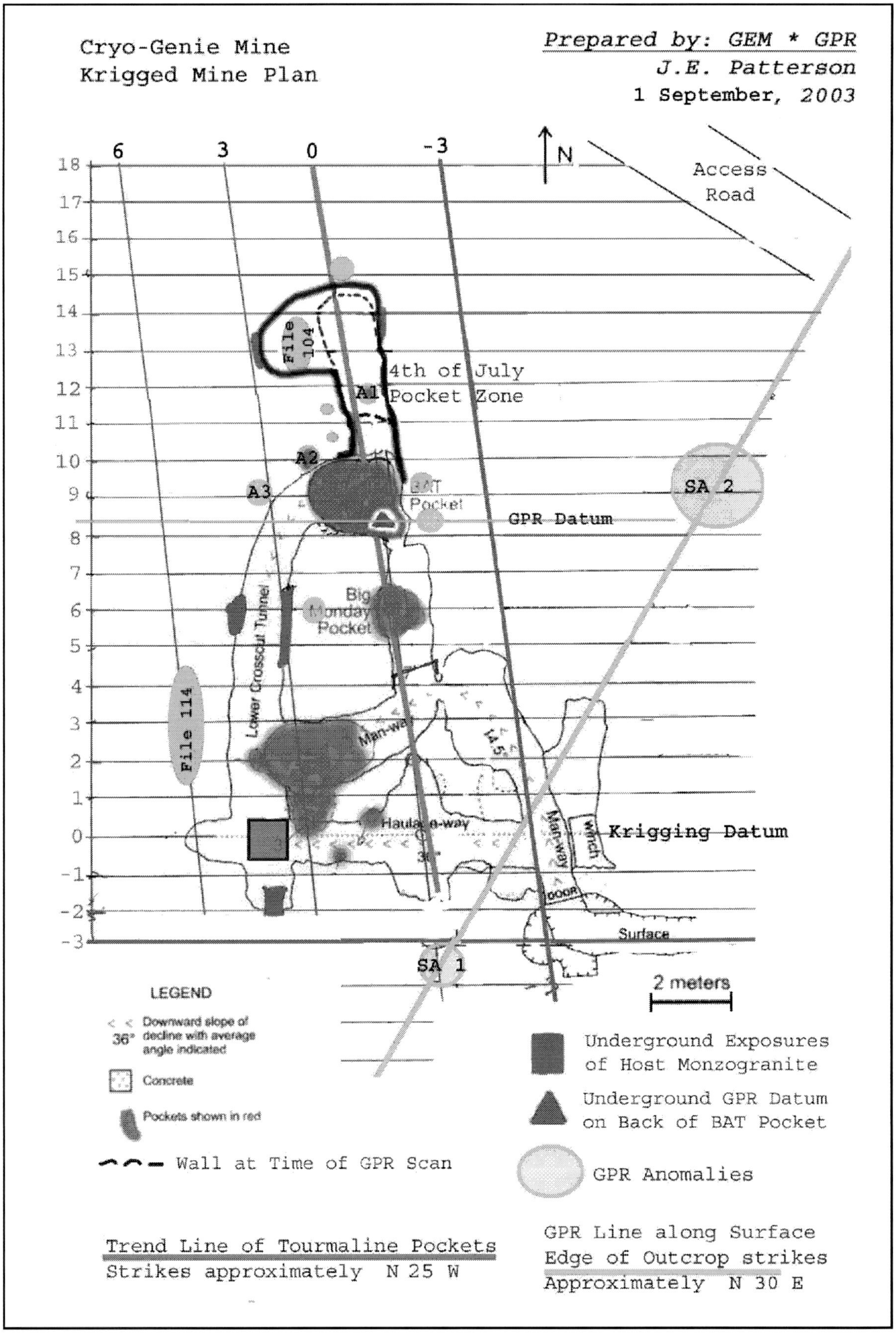

Figure 5. Sketch plan of the Cryo-Genie Pegmatite Mine showing areas of GPR Scans on Dyke 1.

Figure 6. Outcrop of the pegmatite, northeast of the current underground workings. Dyke 1 is west-dipping and approximately 1 meter in thickness at this point. The 300 MHz antenna is to the left front of the truck.

Figure 7. GSSI Model 4400R GPR unit and Model SR-8100H Graphic Recorder. The GPR unit is powered by a 12 volt D/C gel-cell battery. The Graphic Recorder is powered by 110 volts A/C.

Figure 8. Outcrop of the preliminary investigation scan line area of Dyke 1, looking southwest.

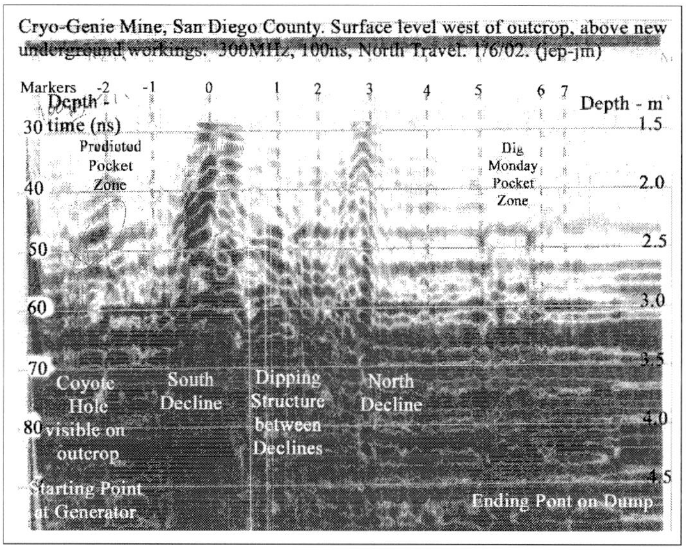

Figure 9. Interpreted Scan 1 on Dyke 1. Compare with Figures 5 and 8.

at mark -2. Note that the electrical generator, near this position, has no obvious effect on the recorded signal. A significant tunnel-type anomaly occurs at mark 0, which matches the measured location of the South Decline. A second tunnel-type anomaly occurs at mark 3, which matches the measured location of an excavated zone now used for storage. The sloping anomaly between the declines appears to be the return from the fractured pocket zone.

On 16 June 2003 we scanned again, this time underground using only the SIR-4400R GPR unit. The transient, single line "oscilloscope," output was surprisingly easy to interpret in the darkness of the drift, even without automatic gain control (Figure 10). We were able to discern three small anomalies as follows: A1) on the face of the working drift, was excavated and found to be the 4th of July Pocket, containing a nice bi-colored gem tourmaline, an etched blocky pale purple aquamarine, and a large, 25 cm quartz crystal; A2) on the wall between the main drift and the BAT Pocket area, was excavated and found to contain a small gemmy thumbnail sized tourmaline; A3) along the wall of the main drift has yet to be excavated. This gave us another successful trial, so decided to upgrade our equipment to a proper, military hardened, GSSI SIR-II GPR unit, with digital storage and data processing capability (Figure 11).

To test this new unit, we chose the cleared and leveled staging zone overlying the southwest surface workings (Figures 3 and 12). The scan of Dyke 2, File 53, was conducted 13 July, 2003. This scan was aligned along an apparent dip of Dyke 2, from south to north. The time-depth

Figure 10. The Author reads the visual output from the GSSI SIR-4400R, excavated BAT Pocket area, view S 40 E.

Figure 11. GSSI SIR-II GPR and 900 MHz antenna, northwest corner of excavated BAT Pocket area, view N 40 W.

recording of this scan was set to 100 ns, which provided an interpretable resolution at the maximum target depth of 5 m. The interpreted line of scan 2, File 53, is shown on Figure 13.

Figure 12. Southwest outcrop zone showing the line of scan for Dyke 2, File 53, looking southwest.

In order to correlate and verify our Cryo-Genie interpretation, we returned to our previous work at the Little 3 Mine, Ramona (Patterson, 1996). This is a similar tabular LCT pegmatite with a dip of 32°. Figures 14, 15, and 16 from the Little 3, show the line of scan, the excavated line of scan, and the interpreted GPR scan. The similarities between the Cryo-Genie and Little 3 dykes are very clear. Based on these similarities in the scans, we proceeded to conduct the underground survey at the Cryo-Genie as described below.

Our next challenge was to scan the underground workings with the SIR-II GPR, to determine if we could identify any pocket anomalies. Since correct geophysical interpretation requires a superb understanding of the structural geology as well as mineralogy of a pegmatite system, we started with a

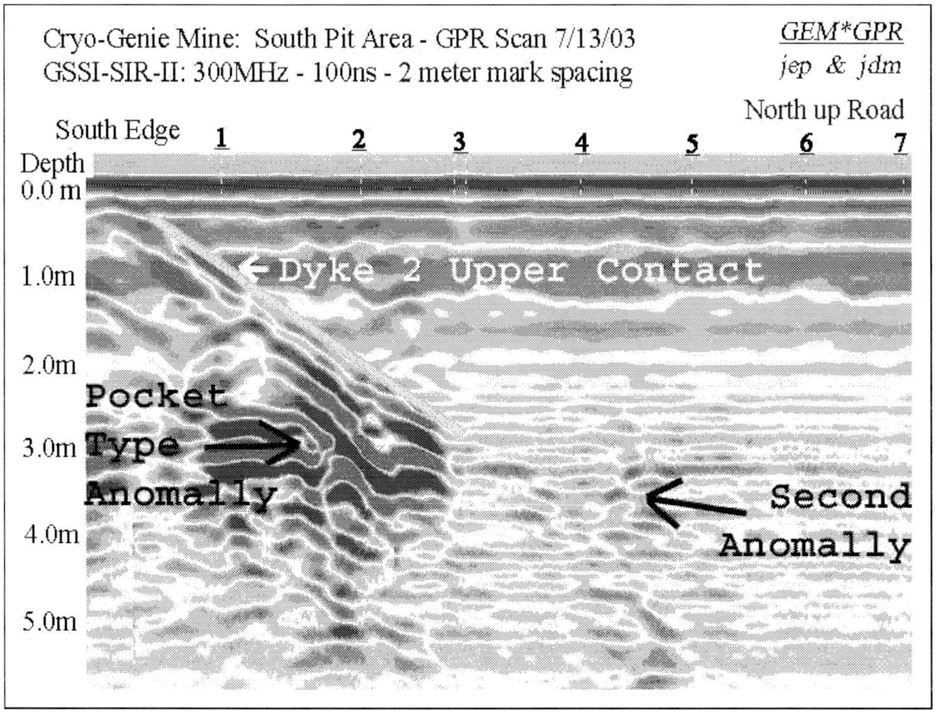

Figure 13. Interpreted Scan Line 2, File 53. The dipping dyke structure is clearly visible. A potential pocket zone anomaly appears near marker 2 at about 3.0m depth. A second anomaly zone appears between markers 4 and 5 at about 4.5m depth. This second anomaly could be a pocket zone or the intersection with another dyke. Compare with figure 12.

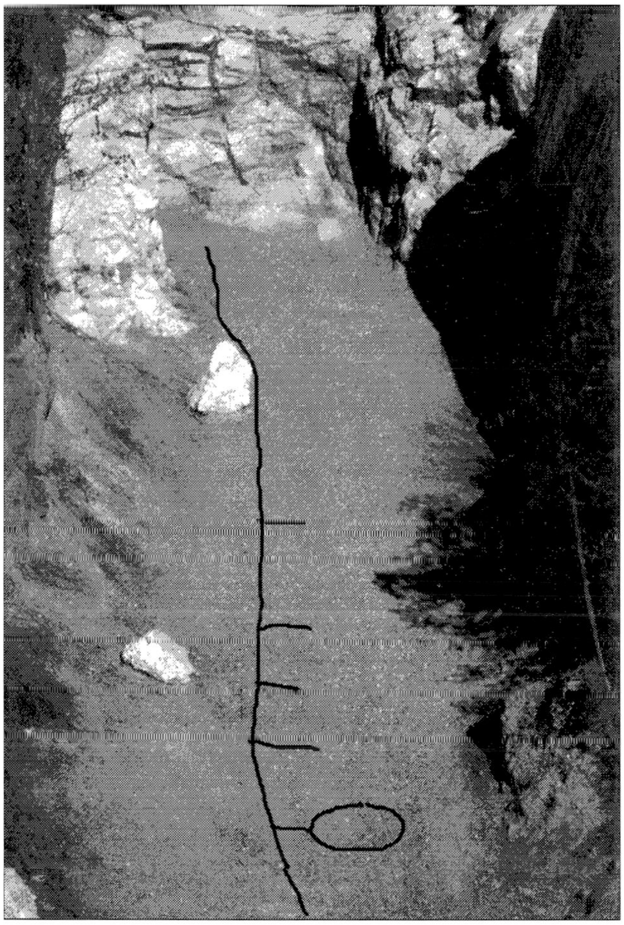

Figure 14. The GPR Line of Scan from our original work at the Little 3 in 1996.

Figure 15. The excavated area from Figure 14, with GPR line of scan painted in.

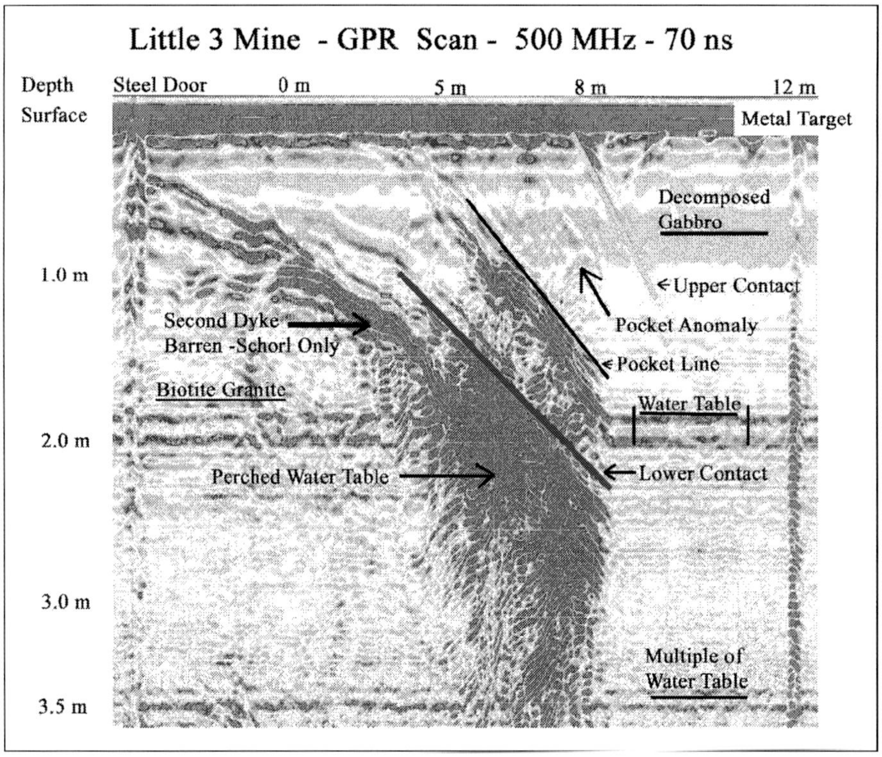

Figure 16. Interpreted GPR scan from the Little 3 Mine. Compare with Figures 13, 14, and 15.

detailed review of the structure of the dyke as observed in the underground excavations. The dyke dilates from about 1 meter at the surface outcrop near the mine entrance and as shown on Figure 6, to about 2.5 meters at the zone of pockets shown on Figures 5, 19 to 23, 29 to 32, and 36 to 38.

The following geologic description is preliminary, due to the limited exposures available in the current surface and underground workings. Since each LCT pegmatite has its own unique fingerprint, based on the slight differences in initial melt composition, melt volume, pressure at the time of intrusion, and subsequent cooling and crystallization, continued mining will allow for additional further investigation and clarification of interpretation. The Cryo-Genie offers a rare opportunity to study this type of gem-bearing pegmatite as mining progresses. Already unique rare minerals such as Löllingite, Bismuthinite, and Pääkkönenite have been identified and described (Kampf, et al, 2003).

A cross-section for typical barren zones is shown in Figures 19 and 20. The upper contact of the pegmatite, with the overlying Quartz-Biotite gneiss and monzogranite, consists of a Biotite-rich Quartz-Oligoclase border for about 5 cm. The a-axis of the Biotite is normal to the contact. This grades rapidly into a massive Orthoclase, Quartz, Muscovite, and downwardly flaring Schorl unit for about 1 m (Figures 19, 20, 29, and 30). In places there are horses (xenolithic inliers), of the overlying gneiss and monzogranite, pods of quartz with the a-axis of muscovite parallel to the contact, 1 to 3 mm almandine garnets, sparse nodules of hydrothermally altered Mn,Fe Phosphates, and pods of granitic textured pegmatite (Figures 29 to 32). Near the middle of the dyke, the Schorl crystals may start flaring upwards and Lepidolite may appear in interstitial feldspar zones (Figures 20, 24, and 30). Beneath this central zone is a single, prominent, undulating band of aplitic grained Garnet, Schorl, Quartz, Albite, and Muscovite. Additional coarse bands of rough textured pegmatite constitute the next zone for about 75 cm. The lower contact zone, exemplified by the outcrop near the bottom of the haulageway, is characterized by dark brown Mn,Fe phosphate stained, typical pegmatite minerals having a granular texture, with a sharp contact with the host gneiss — granite (Figure 19).

In the pocket regions (First Quartz, Big Monday, BAT, and 4th of July), things change significantly, with the perthitic Orthoclase changing to perthitic Microcline in which the plagioclase has been leached out and re-crystallized elsewhere as Clevelandite roses (Figure 1). The Potassium Feldspar also changes color from off-white/grey to buff as a pocket zone is approached (Figures 31 and 36). Within a pocket, the majority of the massive K-Feldspar is fractured along cleavage planes and suspended in an ocher brown sticky mud (Figure 37).

The pink Elbaite tourmaline, with late stage blue-green caps, does not grow in a fractionated form along the C-axis, as at the Himalaya Mine, rather it grows as free flaring crystals in the open spaces of a pocket, on flat-lying interstitial faces of Microcline crystals, or as epitaxial continuations from previously crystallized Schorl terminations (Figures 1, 17, and 24). The finest pink Elbaite tourmaline specimen found to date is on display in the Hall of Gems and Minerals at the Natural History Museum of Los Angeles County (http://www.nhm.org/calendar/home.html), and is featured on the cover of the May/June issue of Rocks and Minerals (Kampf, et al., 2003).

In the pocket regions, the central portion of Dyke 1 may dilate an additional 1 meter. However, until complete contact to contact excavations are made, or diamond core

Figure 17. Microcline with leached and re-crystallized Albite (Clevelandite), Quartz, and flat-lying, gemmy, polychrome Elbaite tourmaline. (BAT Pocket.) (Gochenour's Minerals showcase, Mineral Society of Southern California Show, December, 2001.)

drilling is conducted, the exact thickness of the pegmatite will remain unknown. In the region between the BAT Pocket and the current working face, File 104, there is a second pocket zone underneath the aplitic band. This lower pocket region is enriched with lithium and contains small vugs with gemmy blue-green Elbaite tourmaline. Beryl maybe found in both pocket regions of the pegmatite.

There is no central pocket line in Dyke 1, as typified by the historical drawings in Figure 26 or the photograph of the cross-section of the gem-bearing Topaz Ledge at the Little 3 Mine, Ramona (Figure 28). Nor does the pegmatite fractionate continuously inward from contact to contact, creating oblate spheroid pockets as at the Himalaya Mine, Figure 27. Rather, two primary discrete pulses make up the majority of Dyke 1. An apparent third pulse appears to have brought in the majority of the Lithium, Beryllium, and Boron enriched pocket mineralization. This created an interesting internal texture to many of the Microcline matrix specimens (Figure 18).

Figure 18. Perthitic Microcline, completely leached of all plagioclase. This mineral re-crystallizes elsewhere as Clevelandite. An apparent 3rd pulse of microcrystalline pink and green Elbaite tourmaline and laths of Albite crosscut the leached microcline crystal and then crystallize as euhedral micro-crystals with quartz crystals in the open interstitial zones between microcline blocks.

*Figure 19. Barren zone at the south end of the main drift. A horse of the country rock is exposed just above the 117 marking. A downward flaring schorl is exposed just left of the *. The band of Garnet, Schorl, Quartz, Albite, and white Mica is exposed along the lower section.*

Figure 20. Northwest wall of the main drift leading to the BAT Pocket area. Note upwardly flaring Schorl above the banded zone with a small patch of lepidolite above it.

Figure 22. Mn,Fe Phosphate nodule and epitaxial Lepidolite on Muscovite.

The majority of the pegmatite consists of a first pulse, which appears to be a high temperature Orthoclase, Quartz, Oligoclase, Schorl, and Muscovite phase. Plagioclase (Oligoclase) is contained within the perthitic microcline and graphic granite of the upper 1 m and portions of the lower sections. Very little, if any, Albite was observed in this upper central zone of the pegmatite. The emplacement of the second pulse brought in a lower temperature Albitic phase and the initial lithium mineralization, as evidenced by graphic Albite-Quartz intergrowths in the bottom of several pockets and the epitaxial rim of Lepidolite on Muscovite (Figure 22). It also lifted the lower contact into the middle portion of the pegmatite and re-crystallized this biotite border zone to a schorl and garnet rich aplite band. A third pulse appears to have brought in an extremely hydrous salic, lithium, boron, beryllium, phosphate rich melt, that crystallized as the giant Quartz crystals, large polychrome Elbaite tourmalines (Figures 1, 17, and 18), the massive compact lepidolite masses, and the montebrasite pods (Figure 25). A post-emplacement, probably Elsinore Shear Zone time, clay filled fracture can generally be followed from pocket to pocket along the current main drift (Figures 23 and 38).

Figure 21. Downwardly flaring Schorl with a second growth of graphic Quartz-Schorl.

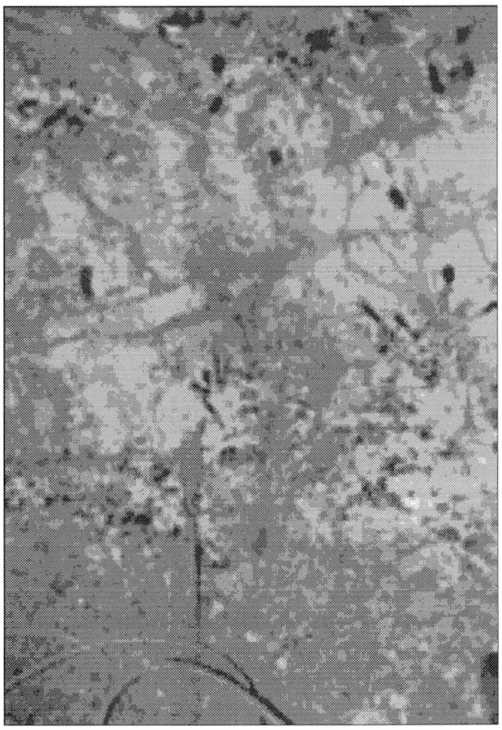

Figure 23. Central region of Dyke 1, underground, with upwardly flaring Schorl and a median fracture network. The miners follow this clay filled fracture network between some pockets.

Figure 24. Epitaxial pink Elbaite tourmaline overgrowth on the flat pedion termination of a pulse 1 Schorl crystal.

Figure 25. Montebrasite pod in zone of compact Lepidolite. Main Haulage-way stope, view N 20 W.

The underground GPR survey was conducted on 19 July, 2003. The areas previously scanned were re-visited to verify the presence or absence of any noted anomalies. A Datum, marking dots, and start-stop-turnaround lines were painted on the walls and back with bright pink fluorescent spray paint. After each scan, the File number was also sprayed on the wall. For the most part, the GPR surveys are accomplished following industry standard data collection and verification procedures. Each scan line is established, measured, photographed and noted so that the processed data can be positively correlated back to the specific section of pegmatite.

Scanning in the dip direction of a pegmatite presents the most complex condition, as the distance to the upper or lower contact must be considered, and the depth of scan time adjusted accordingly. A nominal 30° dip was assumed

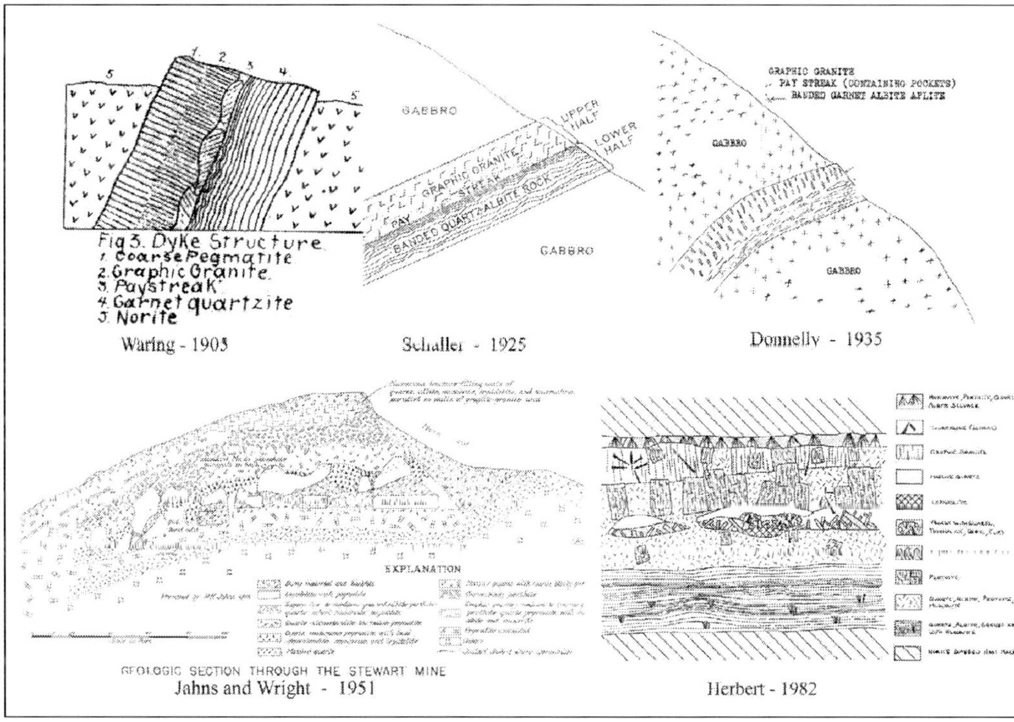

Figure 26. Pegmatite diagrams from various investigators at Pala. Note the simplicity of the early diagrams versus the detail in the later work.

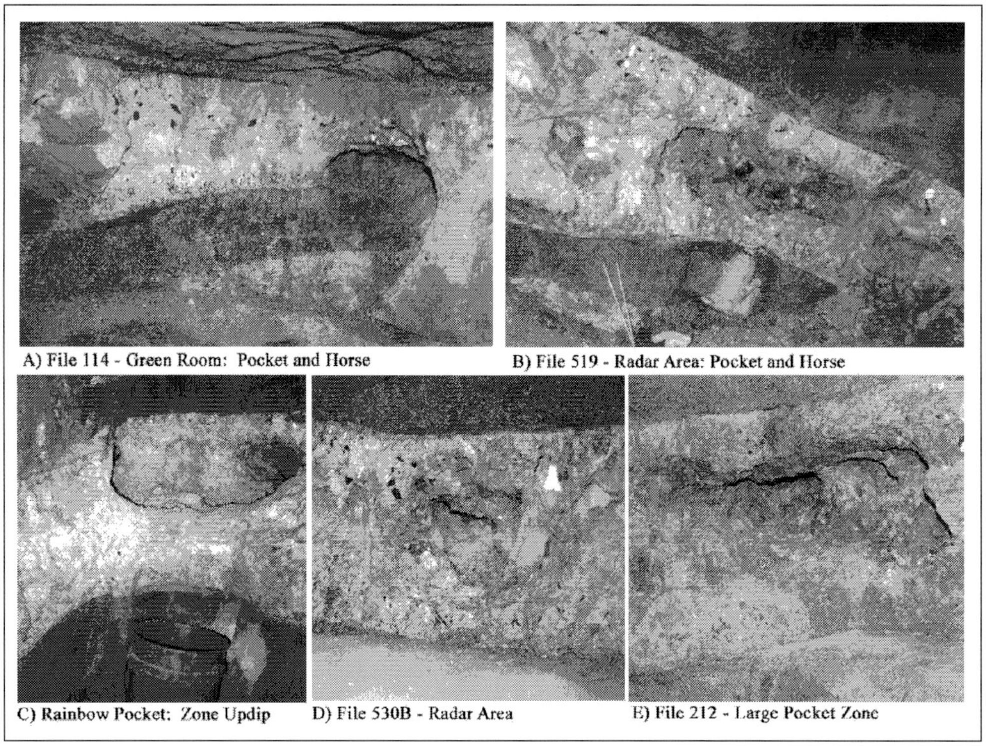

Figure 27. Photographs of pockets investigated during the GPR studies at the Himalaya Mine. The Rainbow Pocket in (C) was discovered on the 4th of July, 1996. The other pockets were discovered using GPR in June and July, 1998.

Figure 28. Main Topaz Ledge investigated during the GPR studies at the Little 3 Mine, Ramona. This composite dyke consists of an upper gem-bearing dyke with Elbaite, Lepidolite, Topaz, and Hambergite, followed by "Line Rock" with Biotite Granite inliers, an aplitic lower contact, and then another Schorl/Garnet-rich lower dyke. It is also the type locality for the rare mineral Boro-Muscovite.

Figure 29. Mine owner, Dana Gochenour, points to zone of potential pocket mineralization, working face 6/6/03, looking N 20 W.

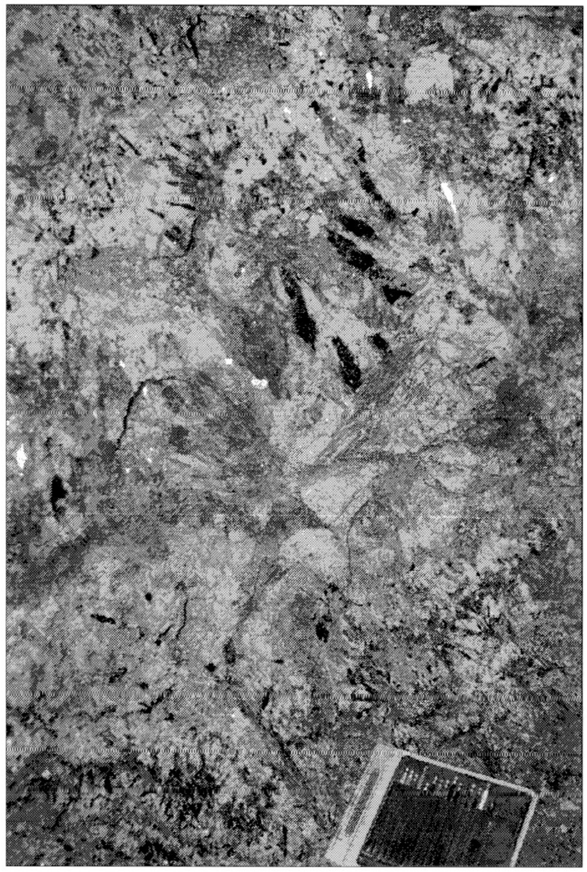

Figure 30. Same face as Figure 28, washed and cleaned. Note the downward flaring Schorl crystals, the small central pod of Lepidolite in an interstice with Microcline and Albite (Clevelandite), and the aplitic Schorl-Garnet band at the level of the notebook. The Lepidolite mass is considered a good geologic indicator for potential pocket mineralization. Also note the ***ocher*** *staining coating the minerals in the central to lower portion of the photograph. This coating is a distinctive signature of the Cryo-Genie pegmatite.*

Figure 31. Working Face, 6/26/03, view N 20 W. The 4th of July Pocket is exposed in the drill hole at the lower left.

for all scans. This gave an approximate distance to the upper contact of 2 meters. The time of the scan, for a dielectric of 8, was set to 30 ns. From our experience, this time provides for a 1.5 meter scan depth. Based on our observations and interpretations, no upper or lower contacts were intercepted in any of the scans. Since scanning along the strike of the dyke presents the simplest case. A nominal 30 ns was selected for all underground scans using a 900 MHz, central frequency, broadband antenna.

Previously, on 26 June, 2003, we were present in the underground during excavation of the working face of the drift containing anomaly A1 (Figure 31). Two lifts were blasted between the faces seen in Figures 29 and 30, and the face shown in Figure 31. Several rare Mn,Fe Phosphate alteration

Figure 32. Working face as scanned on 7/26/03, GPR File 104. Dots indicate marker points. Xs indicated anomalies.

Figure 33. Quartz crystal removed from the 4th of July Pocket, with typical pocket wall matrix of Cleve-landite, Lepidolite, and blue-green Elbaite tourmaline on upper left. The Quartz crystal is sitting on a typical ocher stained matrix from the floor of the BAT Pocket. This floor matrix consists of small crystalline nodules of Lepidolite cemented in Quartz with a covering of gemmy pink and polychrome Elbaite tourmalines.

minerals were found in the wall, just above and to the left of the hammer handle. The color of the Potassium Feldspar had changed from grey to buff. Also, the relative location of the Lepidolite mass had dropped downward on the wall face. On the 4th of July, anomaly A1 was opened and found to be a complete pocket, containing a gemmy, doubly terminated, polychrome Elbaite tourmaline, a pale purple mass of Morganite Beryl, and a 30 cm terminated Quartz crystal. The high energy response from this pocket zone may have been caused by the piezeo-electric response from either the Tourmaline, Quartz, or both.

The working face was advanced to the position seen in Figures 5 and 32. The first region to be scanned was the current working face, Files 102 to 109. Since anomaly A1 had been mined through (Figure 31), we proceeded to scan the extended face of the drift (Figure 32). For the scan of File 104, at the suggestion of Dave Kalamas, mine security, I directed the antenna downward, towards the downdip continuation of the pocket zone, on the west wall. The antenna direction was adjusted to the mid-level on the north wall and up-dip on the east wall. The raw data results are shown in Figure 33. For this group of scans, the brightest return was observed in File 104 (Figures 32, 34, and 35). A significant angled line was observed in the raw data (Figure 34), and this was interpreted in real-time as a potential pocket.

The raw data for Files 53, 104, and 114 were printed and presented to Dana Gochenour and Jim Clanin on 29 July, 2003. The nature and location of all the anomalies was discussed with respect to future mining operations. The gemmy tourmaline contents of anomaly A2 was also verified. The physical condition of the west wall at File 114 was inspected, to rule out the possibility that blast fractures were causing the angled anomalies in the raw data

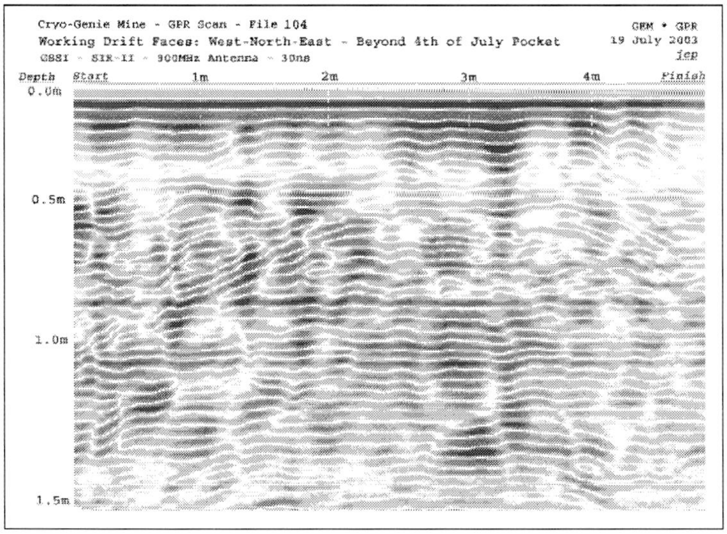

Figure 34. Raw Data GPR Scan of the Working Drift, File 104. The pocket region is interpreted to lie beneath the sloping line between markers 1m and 2m, and between depths 0.5m and 1.0m. Compare with Figures 29 to 32.

(Figure 39). The remnant of a shallow single drill hole was found on the surface between markers 2 m and 3 m (Figure 38). Further excavation, of File 104, at that time was delayed by the construction of a pneumatically operated hoist system so that the gemmy specimens, pocket materials, and mine muck can be efficiently removed.

Subsequently, the working zone for File 104 was excavated between 4 and 8 August, 2003 (Figure 36). A significant pocket was exposed, as shown in Figure 37. After processing the GPR data (Figure 35), the nature and extent of the near region of the pocket was visualized. In Figure 35, the bright energy regions are interpreted to be Quartz and Tourmaline crystals, sheared free from the walls of the pocket and currently "floating" in a morass of ocher brown sticky clay, small nodules of Lepidolite, and sheared and broken fragments of the pegmatite, primarily Microcline and Quartz. A major matrix specimen of Aquamarine Beryl was recovered from this pocket during the week of 25 August, 2003.

Once the pocket at File 104 is excavated, perhaps attention will be turned to the potential pocket at File 114. According to Dana Gochenour, there is no clear geologic evidence that a potential pocket exists as shown in Figure 38 and interpreted from Figures 39 and 40, especially since the First Quartz Pocket is only 2 meters updip of the interpreted pocket zone (Figure 5). When both the raw data and interpreted scans of File 104 (Figures 34 and 35) and File 114 (Figures 39 and 40) are compared, the similarity between them is distinct. Our experience at the Himalaya Mine (Patterson and Cook, 2002) indicates that bright energy returns can be interpreted as potential pockets with reasonable certainty. Barren zones, even if fractured, do not return bright energy. Clay filled fractures generally return linear images as those observed in Figures 13 and 16.

The discovery of pocket mineralization has historically been proven simply by excavating significant portions of a given pegmatite. Certain geological indicators, such as structural changes as seen in rolls and terraces, clay filled fractures between pockets, and solid-frozen pegmatite grading into highly cavernous-vuggy zones, and mineralogical changes such as zoning differences in tourmaline going from Iron-rich Schorl to Lithium-rich Elbaite, or Muscovite grading to Lepidolite, or color changes in

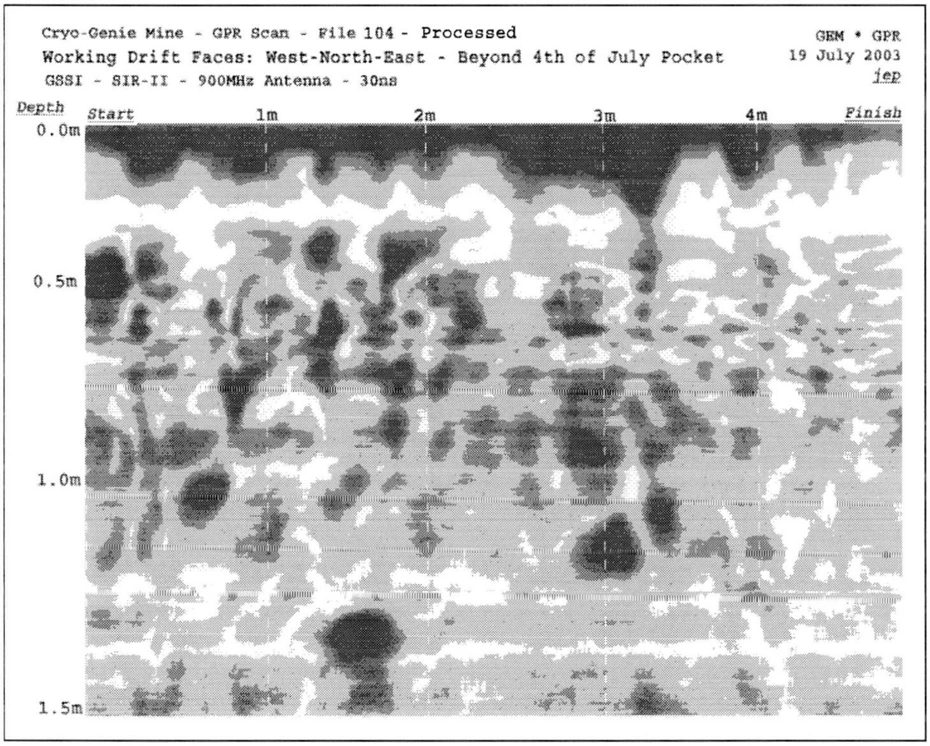

Figure 35. Processed GPR Scan of the Working Drift, File 104. The bright energy regions are interpreted to be crystals of Quartz and Tourmaline. Compare with Figures 29 to 32 and 34.

Figure 36. Continuation of the Working face, N 20 W. The face of the wall for File 104 has been removed for a distance of about 30 cm.

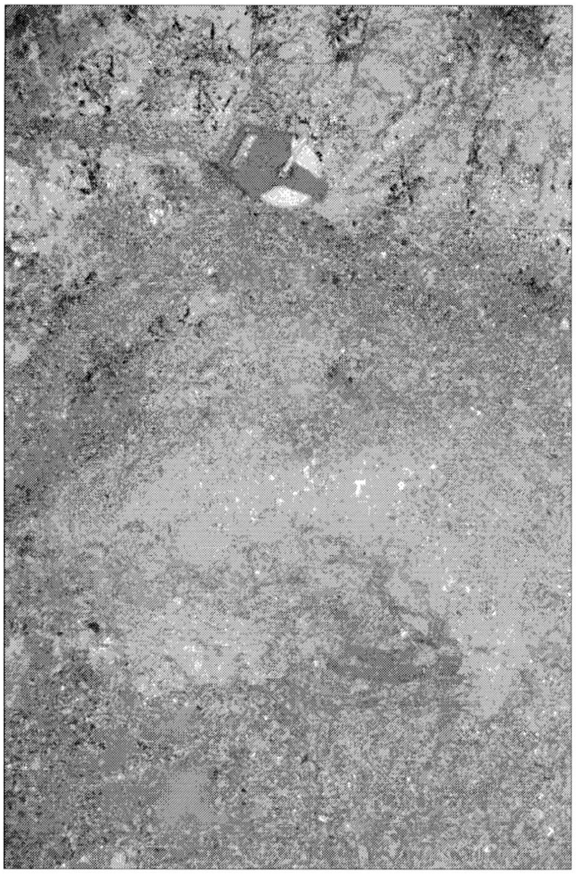

Figure 37. Top of Pocket, File 104, exposed on 8/8/03. This zone is shown as the anomalous regions in File 104 in Figure 23, as the sloping line just before marker 1m and ending just after marker 2m, and as the angled line of bright energy zones between markers 1m and 2m in Figure 24. Additional bright energy zones are now clearly shown between markers 1m and 2m at 1.35m depth, and at marker 3m at 1.15m depth.

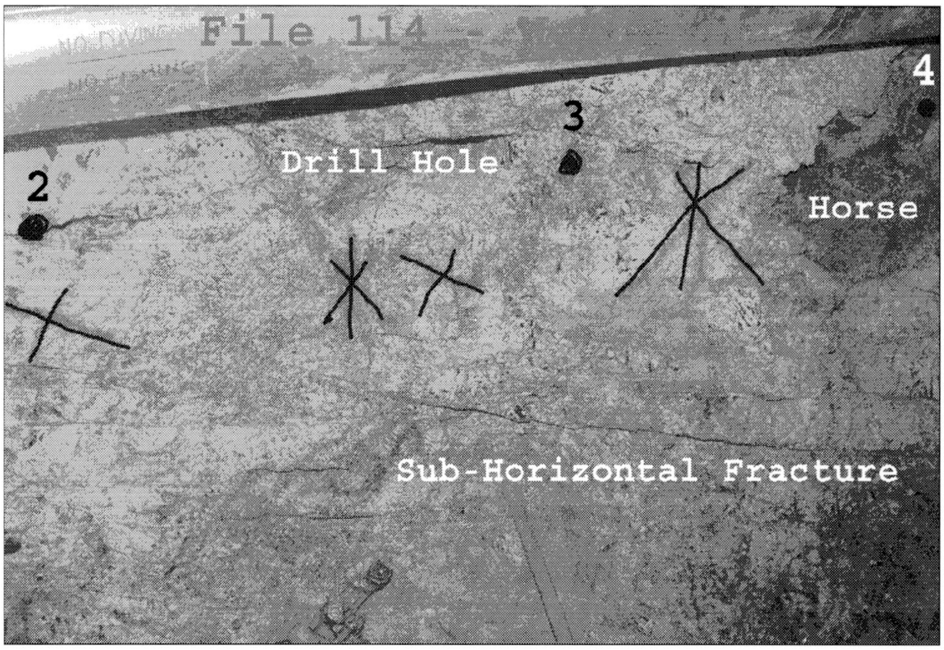

Figure 38. The west wall of the main drift, File 114. Anomaly markers are based on a real-time interpretation of the raw data at the time of the scan. Note that there are no clear geologic indicators for potential pocket mineralization along the line of scan. At the 4th of July Pocket zone and at File 104, Lepidolite was evident in the central portion of the pegmatite. This is considered a good geologic indicator at this mine for pockets.

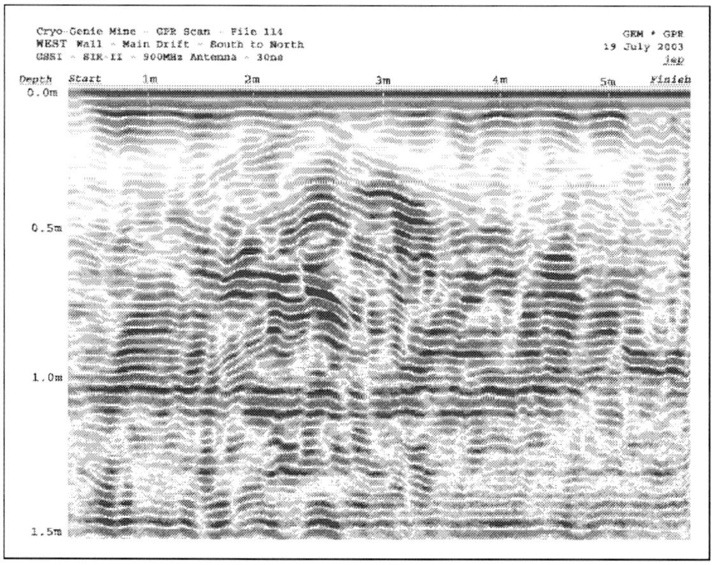

Figure 39. Raw GPR data scan for the west wall of the main drift, File 114. The bright area between markers 2 m and 4 m appears to be a major pocket type anomaly. Since the geology of this portion of the pegmatite is not well understood, only a thorough excavation will provide verification of this preliminary interpretation.

Potassium Feldspar, are all seen as positive indicators of potential pocket-bearing sections. Since the trends in the indicators at the Cryo-Genie have yet to be statistically documented, Dana Gochenour and his mine operator, Jim Clannin, have found that their 30+ years of pegmatite mining experience is

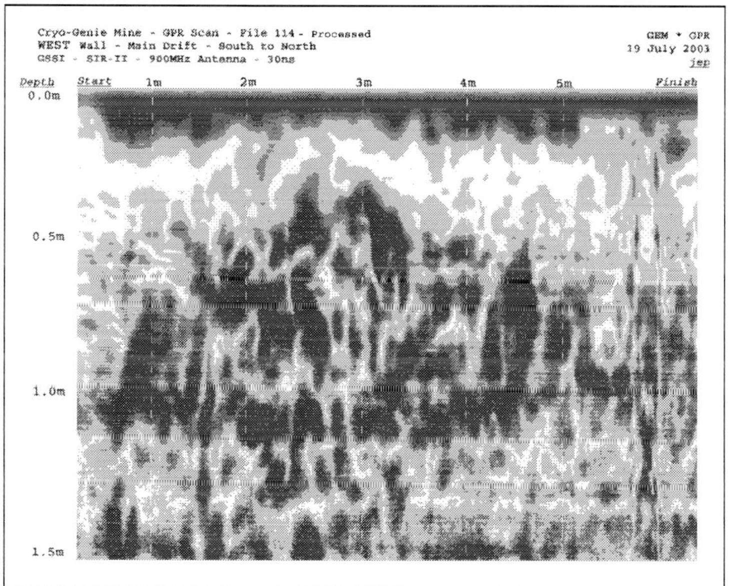

Figure 40. Processed data scan for the west wall of the main drift, File 114. The excavation of this anomaly will verify the accuracy of the GPR method at the Cryo-Genie Pegmatite Mine.

presently adequate to observe the signs of pocket formation and to follow them to their ultimate productive conclusion. This may be the case with the pocket at File 104. However, until the application of GPR for the identification of anomalies associated with pockets, excavation has been the only method of verification.

In this study we have demonstrated that GPR can easily identify pocket and structural anomalies within pegmatites. This will aid the mining effort by defining the blast zone before the pocket is excavated, thus reducing costs and minimizing possible damage to the specimens. In sections of the pegmatite where the normal indicators are not present, as at File 114, GPR has proven to be invaluable in outlining the face of the pocket region (Patterson and Cook, 2002). We have also demonstrated that near surface anomalies can be imaged, should the operators choose to conduct open-pit excavations. The depth of the interpreted region is currently limited only by the frequency of the antenna. A high-energy, low-frequency antenna (35-70 MHz) may be able to image the pegmatite contacts at depth. Alternative geophysical methods, such as shallow seismic or VLF, could also be employed to determine the subsurface structure.

The interpreted surface scan for Dyke 1 matches very well the underlying mine workings. Both open drifts or declines and geologic features are clearly visible on the recorded scan. (Figure 9). The strength and clarity of the scan of Dyke 1 is very impressive, given the limitations of the recording equipment. Subsequent re-scanning, from the surface, of Dyke 1 with the SIR-II GPR, indicated several surface anomalies which have either been excavated or could be excavated easily. The scans of the underground workings are very impressive, given that we have had a 100% positive correlation with our interpretations so far. As mining progresses, and we are able to closely measure the exact distances of the working face to the anomalies, we should be able to improve the accuracy of our interpretation and complete a detailed geologic mapping of the complex features of the Cryo-Genie Pegmatite. The basic statistical approach, plotted on Figure 5, should also aid in planning for continued operations.

In our opinion, GPR is proving to be a useful geophysical prospecting tool at the Cryo-Genie Pegmatite Mine. A 3-D survey, using our current technology recording equipment, would provide a complete image of the subsurface. The preliminary complete underground survey appears to have revealed several pocket anomalies. The GPR method was successful in finding pockets at the Himalaya Mine, and it appears that it is also successful at the Cryo-Genie, as suggested in our first trial investigation conducted in January, 2002 (Black, 2002).

ACKNOWLEDGMENTS

The author would like to thank the mine owner, Dana Gochenour for the opportunity to continue our GPR research on his claim and for providing geological and mineralogical samples for analyses. Jim Clanin is especially thanked for his driving assistance during the initial survey scan of Dyke 1. Dave Kalamas is thanked for his assistance during the scanning of File 104. Jim Means and Ed Hansen provided field assistance for the GPR investigations. Financial support was provided by the Natural Sciences and Engineering Research Council of Canada in the form of a Research Grant to my advisor at the University of Calgary, Frederick A. Cook, Ph.D., Department Geology and Geophysics, and by a grant from the E.E. Patterson Family Trust.

REFERENCES

Black, C., 2002, Cryo-Genie Mine Website: http://home.earthlink.net/~goke/GPR.htm.

Cerny, P., 1982, *Anatomy and Classification of granitic pegmatites*, in Granitic pegmatites in science and industry, Cerny, P., ed., Mineralogical Association of Canada Short Course Handbook, 8, 1-40.

Donnelly, M.G., 1935, *The lithia pegmatites of Pala and Mesa Grande, San Diego County, California*, unpublished Ph.D. Thesis, California Institute of Technology, Pasadena, California, 235 p.

Gochenour, K., 2002, *Tourmaline Discovery at the Cryo-Genie Mine*, Friends of Mineralogy Symposium, Pasadena, California.

Herbert, E.L., 1982, *Clay Mineralogy of Granitic Pegmatites in the Pala District, San Diego County, California*, unpublished M.Sc. Thesis, San Diego State University, 73 p.

Jahns, R.H. and Wright, L.A., 1951, *Gem and Lithium Bearing Pegmatites of the Pala District, San Diego County, California*, California Division of Mines Special Report 7A, 72 p.

Kampf, A.R., Gochenour, K., and Clanin, J., 2003, *Tourmaline Discovery at the Cryo-Genie Mine, San Diego County, California*, Rocks and Minerals, 78, 3, 156-168.

Lees, B., 2003, *Failure of GPR to find economic gem bearing pockets at the Morro Redondo Pegmatite, Minas Gerais, Brazil*, Collector's Edge Minerals, Colorado, Personal Communication.

Parke, J.G., Lt., and Campbell, A.H., 1854-1855, *Report of Explorations for Railroad Routes from San Francisco Bay to Los Angeles, California, West of the Coast Range, and from The Pimas Villages on the Gila to the Rio Grande, near the 32d Parallel of North Latitude*, in Reports of Explorations and Surveys to Ascertain the Most Practicable and Economical Route for a Railroad from the Mississippi Rover to the Pacific Ocean, made under the direction of the Secretary of War, in 1853-6, According to Acts of Congress of Mar 3, 1853, May 31, 1854, and August 4, 1854, Volume VII., Washington, Chapter XVIII — Geology of the District from San Diego to Fort Yuma, 119-129, Chapter XXVIII — List of Minerals Collected, 187-188, Plate VIII.

Patterson, J.E., 1996, *Modeling of Layered Aplitic Pegmatite Dikes Using Ground Penetrating Radar, Little Three Mine, Ramona District, San Diego County, California*, The Spirit of Inquiry, University of Arizona Undergraduate Research Program, 17.

Patterson, J.E., and Cook, F.A., 1999, *Successful Application of Ground Penetrating Radar to Gem Tourmaline Exploration*, poster, Friends of Mineralogy, NYF Pegmatite Symposium, Denver, Colorado.

Patterson, J.E., and Cook, F.A., 2002, *Successful application of ground-penetrating radar in the exploration of gem tourmaline pegmatites of southern California*, Geophysical Prospecting, 50, 2, 107-117 (http://www.pegmatite.com/Gem-Star-GPR.pdf).

Schaller, W.T., 1925, *The genesis of lithium pegmatites*, American Journal of Science, 5th series, 210, 269-279.

Waring, G.A., 1905, *The pegmatite veins of Pala, San Diego County, California*, American Geologist, 35, 356-369.

Weber, F.H., Jr., 1963, *Geology and Mineral Resources of San Diego County, California*; California Division of Mines and Geology County Report, no. 3, 309 p.

The Pack Rat Mine

Charles E. Houser, *Petra Geotechnical, Inc.*
12225 World Trade Drive, Suite P, San Diego, California 92128

Philip T. Farquharson, *CG-Squared Productions*
3547 Lowell Street, San Diego, California 92106

INTRODUCTION

The Pack Rat Mine is located in one of the many gem-bearing, granitic pegmatites in San Diego County (Figure 1). The mine is in eastern McCain Valley overlooking Carrizo Gorge and the Imperial Valley of southern California (Figure 2). A portion of the railroad tracks associated with the San Diego and Arizona Eastern Railroad is visible to the northeast of the mine (Figure 3). Several derailed freight cars can even be seen downhill from the tracks. The Salton Sea and Chocolate Mountains are visible to the northeast. Topography in the vicinity of the mine ranges from moderate to steep, as the mine is situated on the edge of the eastern slope of the Peninsular Ranges.

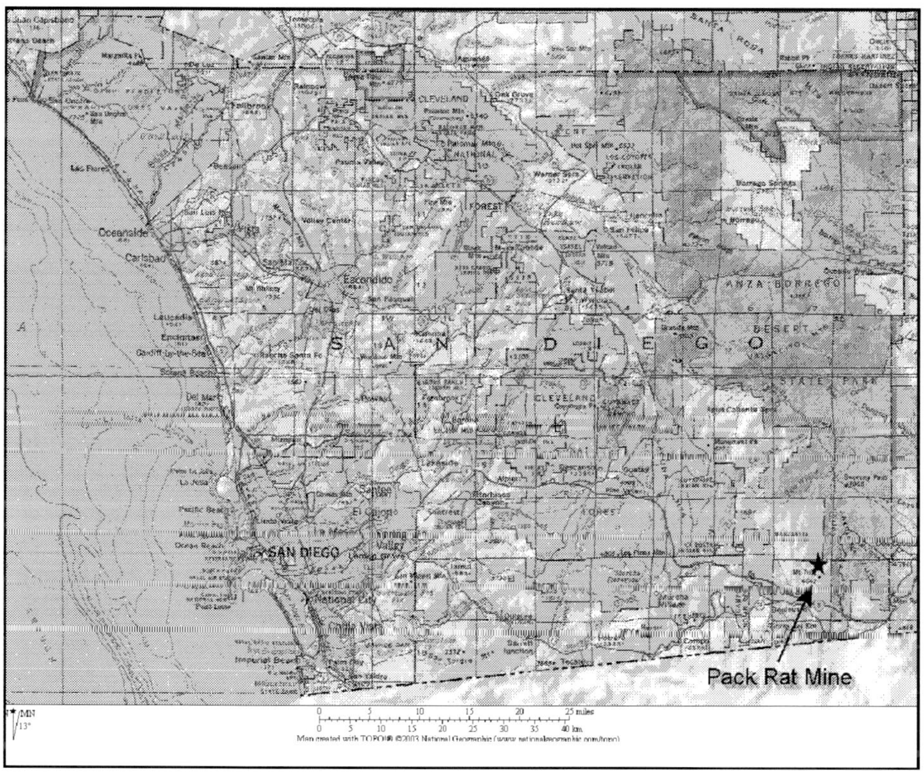

Figure 1. San Diego County, showing location of the Pack Rat Mine.

GEOLOGIC SETTING

The geology of the region around the mine consists of crystalline granitic rocks of the southern California Peninsular Ranges batholith, and pre-batholitic metamorphosed granitic and sedimentary rocks. The mine is located near the eastern edge of the La Posta Granodiorite — a biotite, hornblende granodiorite comprising one of the largest plutons in the batholith. From the northern end of the mine claims, the east-dipping contact of the La Posta Granodiorite is clearly visible in the steep canyons (Figure 4).

The area contains numerous pegmatite and aplite granitic dikes, many of which are visible throughout the terrain in the vicinity of the mine (Figure 5). The mine claim appears to contain principally two significant dikes, an upper or structurally higher dike exhibiting a thickness up to approximately 40 feet, and a lower dike with an average thickness of 8 to 10 feet (Figure 6). Both pegmatite dikes strike roughly north-south and dip to the west. Average dip is in the range of about 25 to 35 degrees although localized portions of the dikes may dip much steeper or shallower. The host rocks surrounding the pegmatites include biotite schist and metamorphosed granitic rock, possibly diorite. The pegmatites are granitic in composition, with quartz, potassium feldspar (microcline, orthoclase), albite, muscovite, and schorl the primary minerals present.

MINE HISTORY AND PRODUCTION

The mine complex known as the Pack Rat Mine formerly

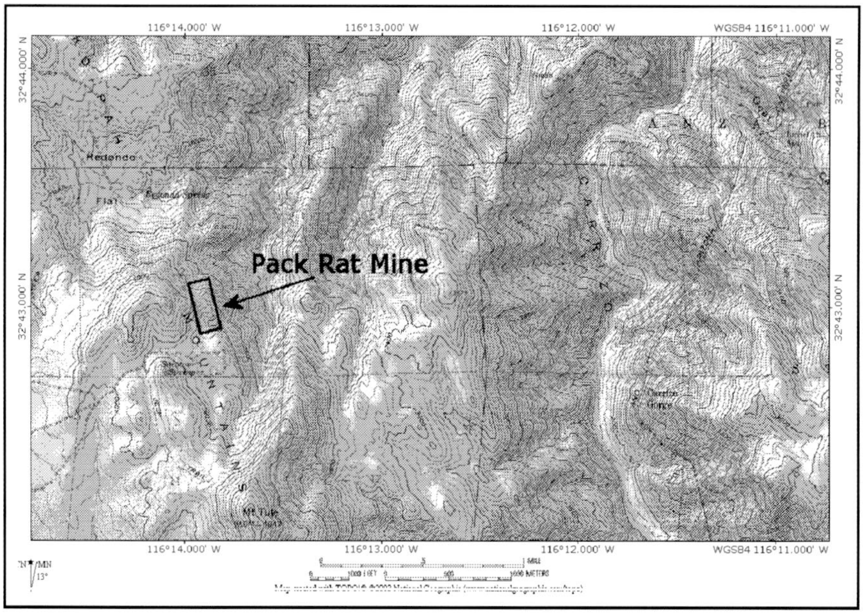

Figure 2. Location of the Pack Rat Mine.

consisted of two mining claims, the Pack Rat Mine in the southern portion of the complex and the Beebe Hole in the northern portion. The Beebe Hole was discovered in the mid-1970s when a prospector named Loren Beebe encountered a large pocket containing spodumene of purple/lilac (kunzite) and yellow (triphane) coloration. Sinkankas (1976) reports that approximately 24 kilograms of spodumene was recovered, and that the colors included lilac, blue, yellow, and colorless. In addition to spodumene, the Beebe Hole produced numerous microcline crystals of relatively good quality. Many of the recovered microcline specimens are twinned in the baveno habit

Figure 3. View to the east-northeast from the Pack Rat Mine vicinity. Note the tracks and trestles of the Carrizo Gorge section of the San Diego and Arizona Eastern Railroad, the Coyote and Fish Creek Mountains, and the Salton Sea and Chocolate Mountains beyond.

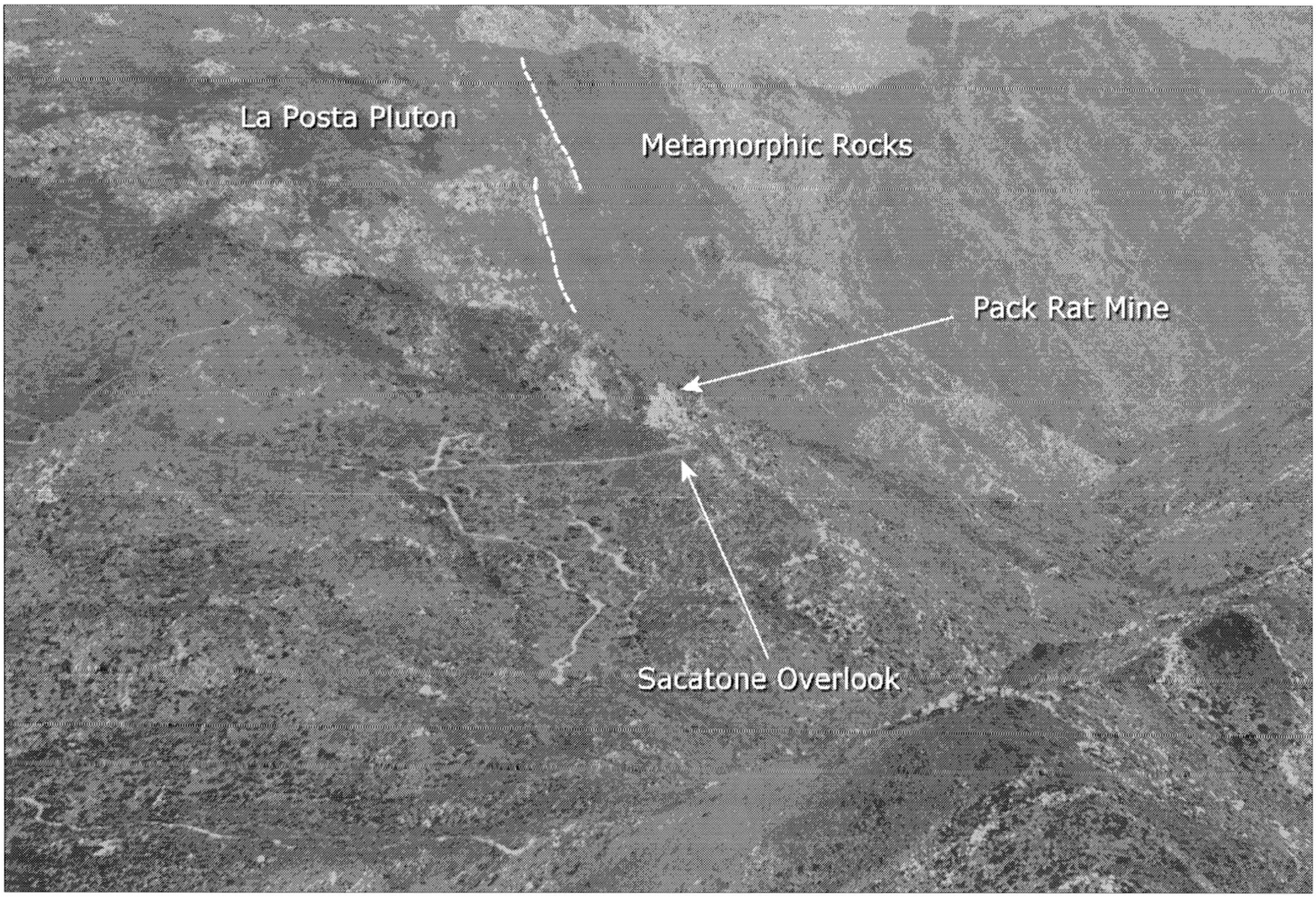

Figure 4. Aerial view to the north from above Tule Canyon. The contact between the La Posta Pluton and the overlying metamorphic roof pendant is indicated by a dashed line. (Pilot: Frank Morris, North Island Navy Flying Club. Photographer: Philip Farquharson, CG-Squared Productions.)

of orthoclase twinning. The Beebe Hole has also produced minor beryl of pale blue (aquamarine), pale pink (morganite), and colorless varieties. Rarely, columbite and cassiterite have been recovered from the Beebe Hole.

Around 1981, Fred Stevens began mining the Pack Rat Mine south of the Beebe Hole. He named the mine for a family of pack rats he was forced to "evict" from the mine cut where he initiated his work. Some small-scale prospecting had been done historically at the mine with previous references to beryl and spessartine garnet (Kunz, 1905). These early references pertained to a prospect called the Crystal Gem Mine. Whether the Crystal Gem Mine was in the location of the Pack Rat Mine or some other location in the Jacumba area is not known. Mr. Stevens subsequently obtained the Beebe Hole claim and merged the two claims into one claim named Pack Rat. He eventually sold the claim to Mr. Gary Wallace, who worked the mine on a limited basis for about 10 years and sold it to Mr. Jim Means and Mr. Blue Sheppard.

As Mr. Stevens began work on the upper dike, he encountered a large pocket containing abundant quartz crystals and some beryl of aquamarine variety. Subsequent work on the upper dike uncovered another large pocket (approximately 8 feet long, 5 feet wide, and 4 feet high) with very dark smoky quartz crystals of essentially black coloration. These quartz crystals ranged in size up to approximately 2 feet long by 10 inches in diameter! A number of smaller pockets have been recovered from the upper dike. One such pocket contained apatite crystals with pale purple coloration.

Mr. Stevens also conducted work on the lower pegmatite. A number of smaller (up to approximately 3 feet in diameter) pockets have been recovered from the lower dike. These pockets typically yield spessartine garnet, well-crystallized albite, quartz crystals, and very sharply crystallized muscovite. Some of the quartz crystals have been of relatively high quality, exhibiting pale smoky coloration and exceptional clarity. One quartz crystal, now in the author's collection, measures approximately 20

Figure 5. Aerial view to the north from above Tule Canyon. Note the numerous pegmatitic and aplitic granitic dikes on Mount Tule. (Pilot: Frank Morris. Photographer: Philip Farquharson, CG-Squared Productions.)

inches in length and 6 inches in diameter, and weighs approximately 40 pounds. The crystal was found and self-collected by the author, and was shoulder-carried roughly one-eighth mile back to the vehicle, thus providing the appropriate amount of pain for the gain.

Minerals rarely recovered at the Pack Rat Mine include helvite, columbite, cassiterite, and various phosphates including herderite and purpurite. Of some interest is the fact that, while schorl tourmaline is abundant in nonpocketed portions of the pegmatite dikes at the mine, very little crystallized schorl or other tourmaline has been recovered from the pockets.

THE MINERAL COLLECTION

Many of the San Diego County pegmatite mines produce a relatively wide variety of mineral species; however, few mines produce a suite of minerals in sufficient abundance that one might expect to develop a mineral collection based solely on that mine. The Pack Rat Mine does produce such a suite of minerals, and several private and public collections contain excellent representations of specimens from the mine. Most notably, the mineral collection of the Los Angeles County Museum of Natural History (LACMNH) includes such a representative collection of Pack Rat Mine material. The large black smoky quartz crystal discussed earlier resides in that collection, along with a roughly 6-inch-diameter opaque aquamarine crystal, an exceptional quality gemmy aquamarine that is about 2 inches long by 0.5-inch diameter, and the second best spessartine garnet specimen recovered at the mine. (The best spessartine from the mine resides in the author's collection, in the author's humble opinion.)

Regarding collecting at the Pack Rat Mine, the mine is located on Bureau of Land Management (BLM) land, but it is a valid and current mining claim thereby giving all mineral rights to the claim owners. Access to the area may not be restricted, but collecting can only be done with the permission of the claim owner. Any mineral finds must be reported to the owner for documentation purposes and so that, if minerals of significant economic value are recovered, an appropriate price can be established for purchase of the specimens. The author has

Figure 6. View of the Pack Rat Mine from Sacatone Springs Overlook.

approached several mines in San Diego County under such pretenses. As a result, the author's collection includes several specimens self-collected and purchased for a reasonable and fair (usually, anyway) price.

CONCLUSION

The pegmatites of San Diego County have been an important source of minerals such as tourmaline, beryl, spodumene, topaz, and garnet for over 100 years. Many important mineral collections throughout the world include San Diego County pegmatite specimens. For collections of San Diego County minerals, the Pack Rat Mine provides an important source of typical and rare mineral species found in local pegmatites. Hopefully, future mining efforts at the Pack Rat Mine will continue to provide a source of such minerals.

AUTHOR'S END NOTE

Mining a pegmatite such as the Pack Rat at a small scale is a labor-intensive prospect. Mining operations are not conducted at a level that makes acquisition and installation of automated or heavy equipment (backhoes, slushers, ore carts, etc.) feasible. As a result, a great deal of energy and sweat is expended during a typical day of mining. Under such circumstances, proper nutrition is essential. The author had the privilege of spending countless hours assisting Mr. Stevens with mining during his tenure at the mine. Mr. Stevens was a cook in the U.S. Coast Guard for many years. As a result, some of the finest meals enjoyed by the author were at the Pack Rat Mine. You haven't had carne asada until you've enjoyed it with Mr. Stevens at the Pack Rat Mine. Unfortunately, Mr. Stevens now resides in Auburn, California. So if you are going up that way…

REFERENCES

Fisher, J., 2002, *Gem and Rare Element Pegmatites of Southern California*, Mineralogical Record, Vol. 33, pp. 363-407.

Kunz, G.F., 1905, *Gems, Jeweler's Material, and Ornamental Stones of California*, California State Mining Bureau Bulletin 37, 171 p.

Sinkankas, J., 1976, *Gemstones of North America*, Vol. 2, Van Nostrand Reinhold, New York, 494 p.

Rehabilitation of an Old Hydraulic Fill Dam for Stability Against Earthquakes

***John A. Bischoff**, Associate*
Woodward-Clyde Consultants, San Francisco, California

***Thomas C. MacDonald**, Principal Engineer*
Leedshill-Herkenhoff, San Francisco, California

***Thomas M. Wilson**, General Manager — Chief Engineer*
Vista Irrigation District, Vista, California

Condensed from an article originally published in:
Commission Internationale des Grands Barrages
Quinzième Congrès des Grands Barrages
Oausanne, 1985

INTRODUCTION

Near-catastrophic failure of the Lower San Fernando Dam, a hydraulic fill dam, during the 1971 San Fernando Earthquake dramatically emphasized the need to reevaluate the stability of other hydraulic fill dams under seismic loadings and to develop rehabilitation measures to ensure their safety. As a result, in December 1971, the State of California Division of Safety of Dams (DSOD) requested that a comprehensive program be initiated for investigation the seismic safety of Henshaw Dam, a hydraulic fill structure located on the San Luis Rey River in San Diego County, California (Figure 1). Henshaw Dam, which was built in 1922, is owned by Vista Irrigation District (VID), which uses the impounded water for irrigation and domestic water supply. The originally constructed facilities consist of a main and two secondary dams, a spillway through the right abutment of the main dam, and an outlet tunnel through the left abutment of the main dam. A general plan of the original Henshaw Dam facilities and a cross-section view through the main dam at the maximum section are shown on Figures 2 and 3, respectively.

Because the dam is located within the seismically active Elsinore fault zone, the initial investigations undertaken as a result of DSOD's request were directed toward the collection and analysis of sufficient soils, geologic, seismic, and other information to determine whether it would be necessary to carry out an extensive dynamic stability analysis to assess the dam's stability under seismic loadings. As a result of these investigations, it was concluded that the hydraulic fill materials in the dame would probably liquefy under earthquake loadings produced by the Elsinore fault zone and, that with a full reservoir, the dam would not be safe. Thus, a full dynamic analysis of Henshaw Dam was not warranted.

Consequently, in June 1973, it was proposed to proceed with investigations directed toward construction of modifications to ensure that embankment failure would not cause catastrophic flooding of the city of Oceanside and the other areas downstream of the dam. Modifications would include construction of a berm or other work on the downstream toe of the main dam to limit reservoir releases and stabilize the downstream slope and modifications to the spillway to reduce the normal operating level in the reservoir. Investigations of various alternative concepts proceeded, and in 1974 a proposed program for modifications to Henshaw Dam was presented to DSOD. This program included:

1. Modifications of the existing spillway by lowering its crest and modifying its configuration, thus reducing the storage capacity of the reservoir to 6.2×10^7 m^3 (50,000 acre-feet).

2. Construction of a flow attenuating reinforced rockfill berm structure — or Flow Retardation Structure (FRS) — against the downstream face of the existing dam.

Investigations, design, and construction of the modifications outlined above were completed in September 1981. the modified spillway reduces the maximum storage to a level that in turn substantially reduces the net cross-sectional area of the main dam subject to liquefaction. The

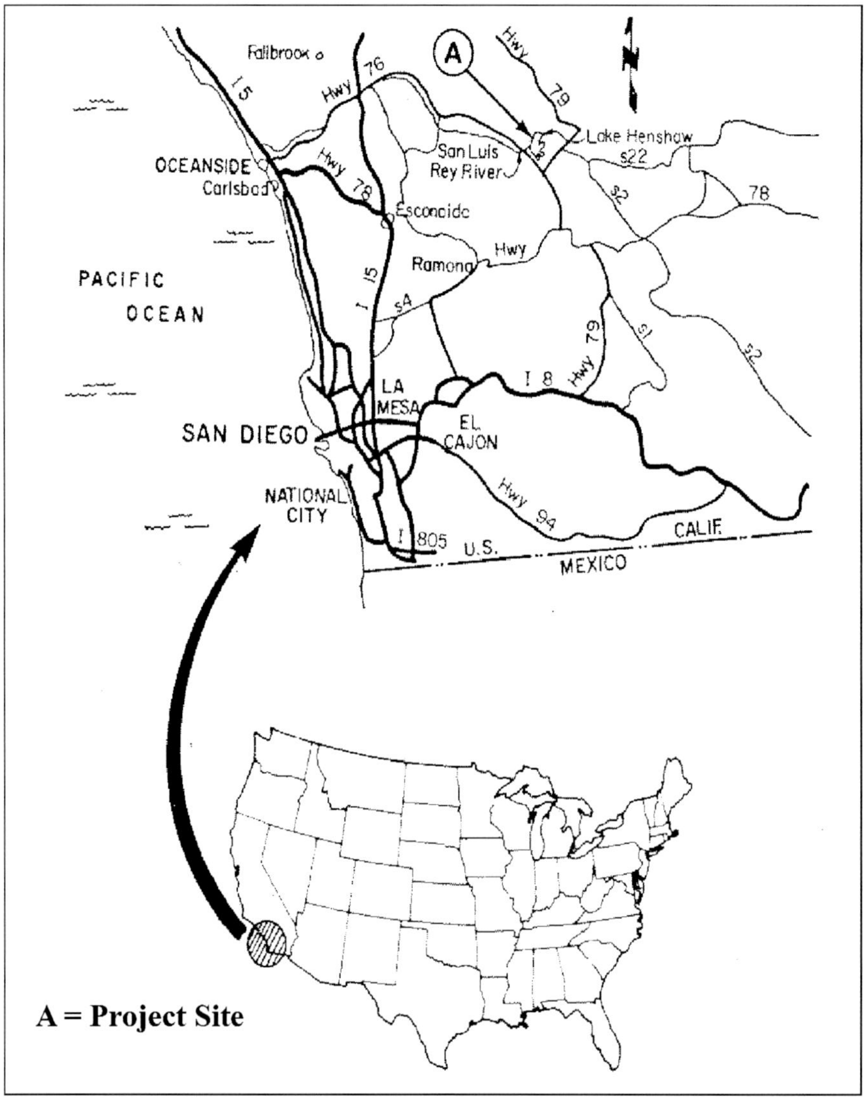

Figure 1.
Project Location Map.

reinforced FRS serves to buttress the existing dam and improves its stability under static and earthquake loadings. In addition, in the event of liquefaction and failure of the dam during an extreme seismic event, the FRS is designed to attenuate, even if overtopped, any flood that may occur during or subsequent to failure of the existing dam.

A history of Henshaw Dam and details of the development of the unique features incorporated into the design of the modifications made to Henshaw Dam, including design considerations and details of the construction work, are presented.

CONSTRUCTION HISTORY

Construction of the original Henshaw Dam began in 1922 and was completed in early 1923. The dam was constructed by the hydraulic fill process, a method of dam construction sometimes used in the early part of the 20th century because it was cost-effective and allowed rapid construction.

The original spillway consisted of a 76 m (250 ft.) long arched overflow weir and a 175 m (575 ft.) long chute and stilling basin. The original spillway weir crest was at elevation 829.3 m (2720 ft.). As part of the raising of the dam in 1928, the spillway crest was also raised approximately 2 m (7 ft.) to elevation 831.4 m (2727 ft.). After these modifications, the reservoir storage capacity was approximately 2.4 x 10^8 m^3 (194,000 acre-feet). It is significant that at no time during the 60 years of operation prior to the recent modifications of Henshaw Reservoir has any water flowed over the spillway.

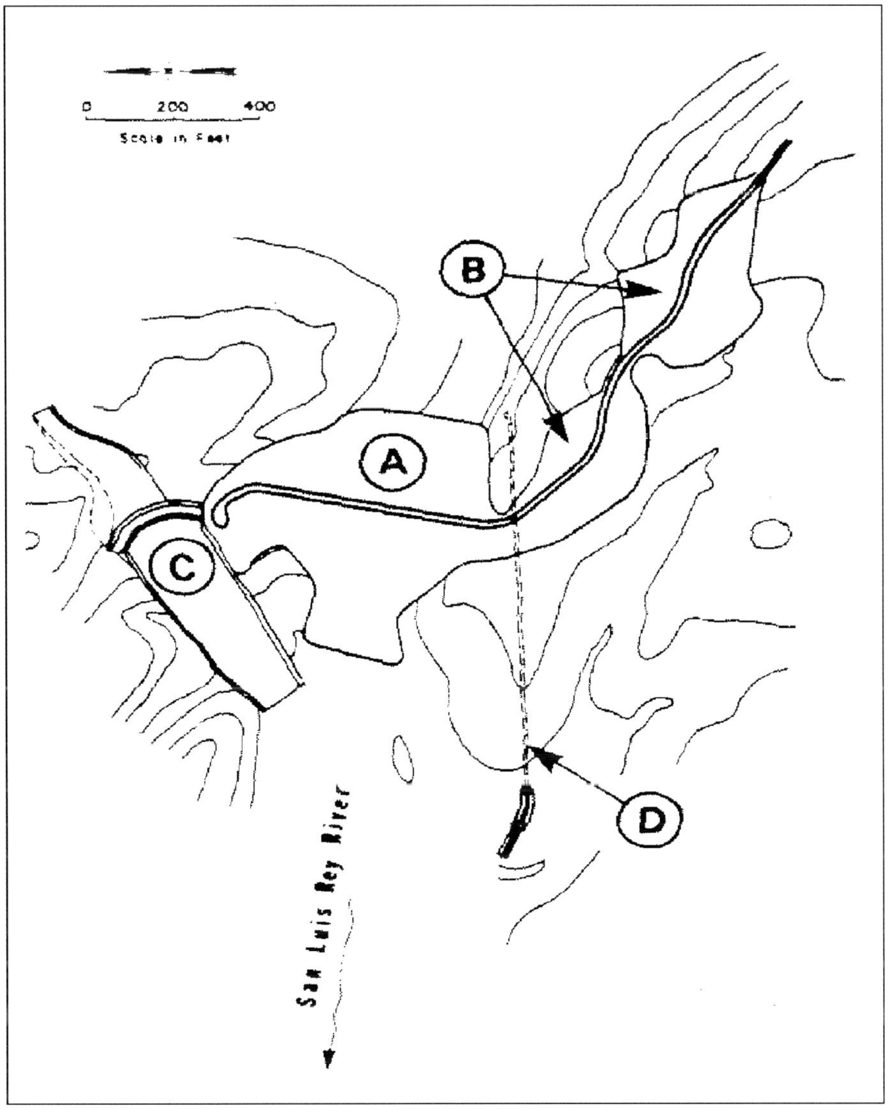

Figure 2. Plan of Henshaw Dam.
A = Main Dam *C = Spillway*
B = Secondary Dams *D = Outlet Tunnel*

CROSS SECTION OF MAIN EMBANKMENT

As indicated on Figure 3, the original Henshaw Dam is fundamentally a "zoned" embankment comprising:

a. Outer shells or rock and disintegrated granite obtained from the spillway excavation.
b. An inner puddle core of silt and clay.
c. An intermediate zone, mostly sand, between the central core and each outer shell, transported and placed hydraulically along the inner face of each outer shell.

Wagonfill Zones

All, or practically all, of the material in the Wagonfill zones (outer shells) was obtained from excavation for the original spillway. The excavation was carried out by drilling and blasting, and the loosened material was hauled to the embankment by horse-drawn wagons and dumped in thin lifts to create a series of berms six to eight feet in thickness. The materials consisted of sound granite up to 12 cm (6 in) in diameter and sandy material, mostly weathered and decomposed granite.

The serrated inner face of each Wagonfill zone (Figure 3) stems from the construction procedures used. The outer edge of each longitudinal berm had to coincide with the outer designed slope of the embankment. To maintain adequate width for construction, sufficient berm materials had to be extended inward over the previously deposited coarse materials in the hydraulic beach zone to support wagons traveling longitudinally over the berms.

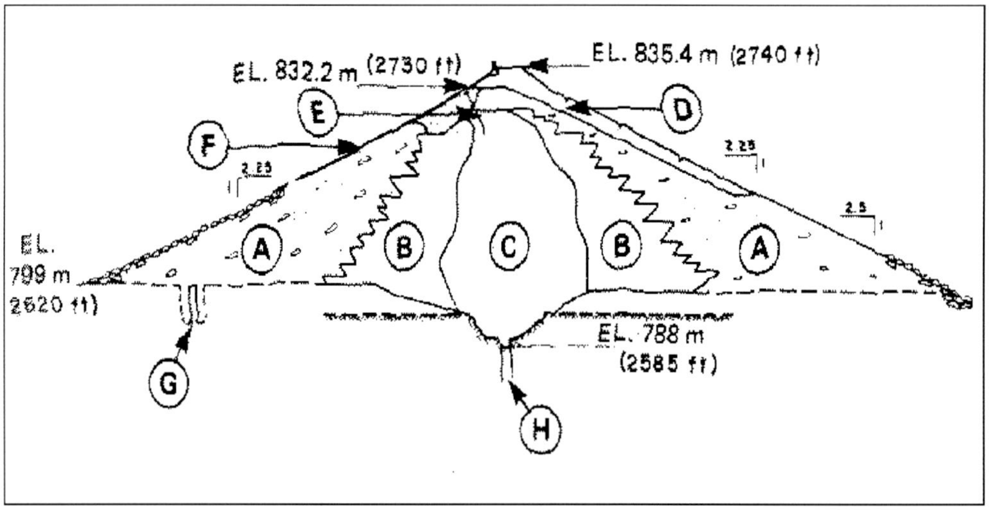

Figure 3. Maximum Section through Henshaw Dam.

A = Wagonfill	*E = Sheet Pile Cut Off*
B = Hydraulic Beach	*F = Concrete Face*
C = Hydraulic Core	*G = Cut Off Wall*
D = Rolled Fill	*H = Grout Holes*

Thereafter, until the full height of each berm was reached, the inner face of the berm was sloped back at the normal angle of repose of the berm material.

Hydraulic Beach Zones

Materials for the beach zones and central core were obtained from a borrow pit located about 610 m (2000 ft.) distant from the dam. The borrow pit material was excavated with hydraulic monitors, collected in a sump, and then pumped through a pipeline to points of discharge. The coarser fractions deposited quickly near each point of discharge, the finer fractions were carried toward the pool between the two wagonfill zones, and the fine silts and clays then settled in the pool.

Although the record is not clear, it is highly probably that deposition of materials along the face of one outer shell was carried on simultaneously with the construction of the next major lift of the opposite wagonfill zone. Hence, it is probable that the contact between the beach zones and the central core is also serrated without a sharp line of demarcation at any point.

Central Hydraulic Core

The material deposited in the central core does not include the coarse fractions deposited in the beach zones, or the very fine materials that were carried off as slimes in the disposal of surplus water, which tended to pool in the central core area.

GEOLOGY AND SEISMICITY

Geology

Henshaw Dam is situated within the Peninsular Range province, a geologic and physiographic unit that occupies the southwester corner of California and the Baja California peninsula. The higher parts of the province are characterized by northwesterly trending, elongated mountain ranges and valleys underlain by pre-Cenzoic age igneous and metamorphic rocks of the southern California batholith. The subparallel traces of several major fault zones also trend northwesterly in the higher parts of the provinces.

The rocks exposed in the mountains surrounding and underlying the sediments of the basin are primarily tonalite (quartz diorite), a granite-like rock of Cretaceous age. Jurassic age metamorphic rocks are exposed in the mountains on the southwest flank of the basin. The tonalite rocks have been deeply weathered in much of the area to uncemented silt and fine to coarse sand. Overlying the intrusive rocks are the Pleistocene age Temecula arkose, which is the aquifer supplying the VID's well field adjacent to Henshaw Dam; the Upper Pleistocene Pauba formation, a thin, terrace-forming material; and the recent alluvium occupying the lower portions of the basin.

Bedrock in the dam site area is also quartz-diorite, termed hereinafter granite. Throughout most of the area, including the dam site, the rock is very weathered, friable, and weak.

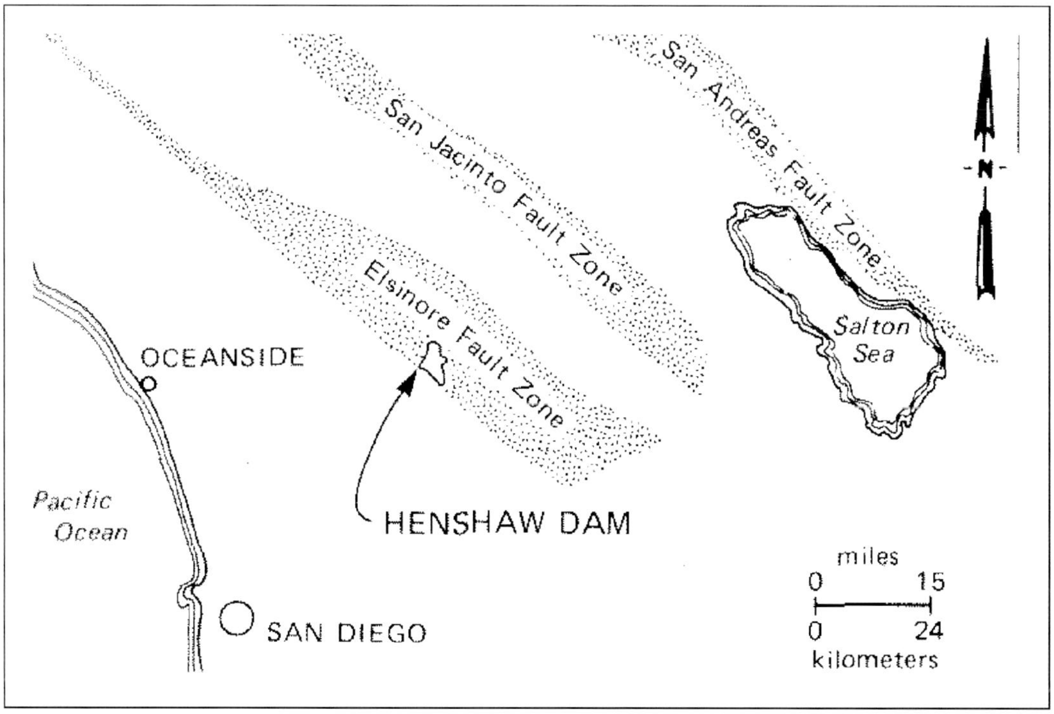

Figure 4. Major Fault Zones.

Seismicity

There are several major faults in the vicinity of Henshaw Dam, including the Elsinore fault on the southwest edge of Henshaw Reservoir, the Aqua Tibia and Aguanga faults near the northeastern edge of the basin, and Agua Caliente, Lancaster, and Hot Springs faults farther northeast. All these faults, plus numerous other subparallel faults, are collectively designated by some geologists and seismologists as the Elsinore fault zone. The zone is 12.9 to 19.3 km (8 to 12 mi.) wide in the vicinity of Lake Henshaw (Figure 4). Subsurface exploration in the foundation of the main dam, together with recent investigations by the California Division of Mines and Geology, indicate the presence of traces of the fault trending diagonally beneath the main dam. On the basis of geomorphic evidence, the recency of movement of most of the Elsinore fault traces is dated as Holocene age (last 11,000 years) and thus is considered "active." From seismicity and geologic investigations. the Elsinore fault is capable of producing a M 6 to M 7.5 earthquake with potential fault rupture displacements at the site having a net offset of up to 1.5 m (5 ft.) in either a vertical or horizontal direction or up to 2.1 m (7 ft.) in an oblique direction.

The more seismically active, yet less critical, San Jacinto and San Andreas fault zones parallel the Elsinore zone at distances of 32 to 77 km (20 to 48 mi., respectively, to the northeast [Figure 4]).

HAZARDS OF EXISTING DAM

The preliminary studies of Henshaw Dam following the San Fernando earthquake revealed several major problems associated with the dam. First, on the basis of data from exploratory investigations and an assessment of the method of placement of the shell materials, the relative density of the materials forming the shells of the dam was estimated to be on the order of 60 to 65 percent. As such, under conditions of high reservoir storage and moderate earthquake shaking, it appeared probable that liquefaction of at least a portion of the main embankment would occur and failure of the dam could be expected. Second, the dam would also be subject to piping if movements were to occur along the traces of the faults underlying the dam. Such fault displacements could lead to failure by piping, quite apart from movements resulting from shaking.

The general solutions selected for mitigating the earthquake hazards consisted of permanently reducing the storage capacity of the reservoir by lowering the spillway crest elevation, and strengthening the main dam by constructing a flow attenuating berm on its downstream face to make it resistant to earthquake forces and fault movements. Based on operation studies, it was determined that a 6.2×10^7 m^3 (50,000 acre-feet) reservoir would be sufficient to meet the District's needs and would not significantly reduce the usable yields of the watershed. Remedial designs,

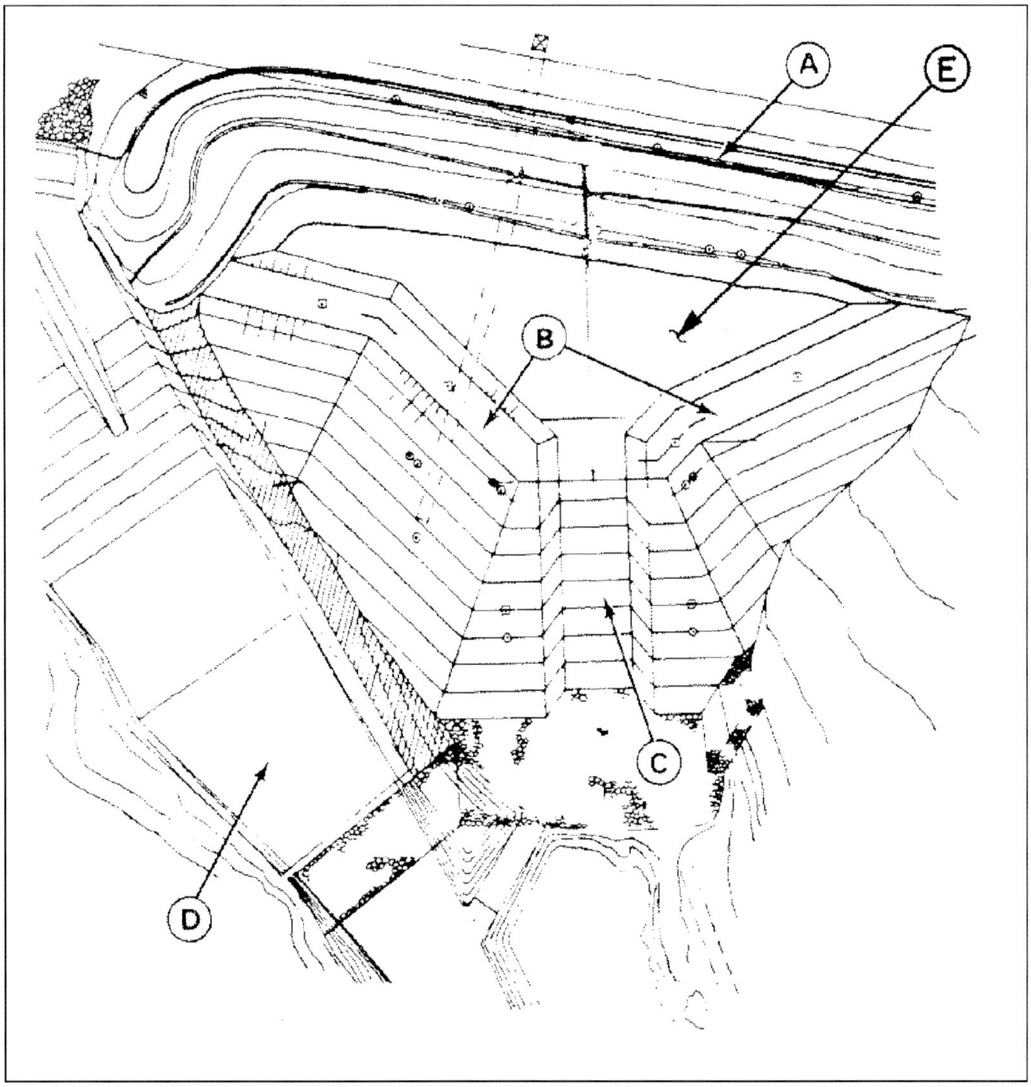

Figure 5. Plan of Flow Retardation Structure.

A = Crest of Existing Dam
B = Banquettes
C = FRS Flow Channel
D = Spillway
E = Crest of FRS

developed and implemented for Henshaw Dam, were aimed at retaining this level of capacity.

MODIFICATIONS TO SPILLWAY

Modifications were made to the spillway to reduce the normal operating level of the reservoir. By reducing the operating level, the cross-sectional area of the main dam that is potentially subject to liquefaction is also reduced, thus improving the overall stability of the dam. From stage-storage relationships, a reduction in the spillway crest level to El. 820 m^3 (2690 ft.) will reduce the maximum reservoir storage from 2.4 x 10^8 m^3 (194,000 acre-feet) to the desired 6.2 x 10^7 m^3 (50,000 acre-feet) needed to meet water supply demands. In developing the modifications, DSOD required that the Probably Maximum Flood (PMF) be used as the design flood.

Probable Maximum Flood

The estimated PMF hydrograph for the 535 km^2 (206 sq. mi.) drainage area has a peak discharge of 5667 m^3/s (200,000 cfs) and a 72-hour runoff volume of 3.58 x 10^8 m^3 (290,000 acre-feet).

FLOW RETARDATION STRUCTURE

As part of the restoration of Henshaw Dam to provide it with a sufficient margin of safety against failure under

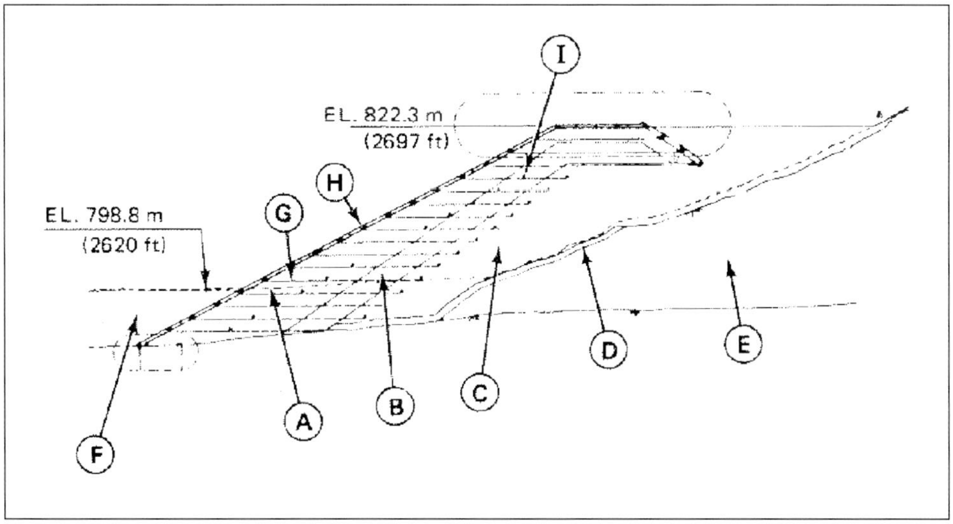

Figure 6. Section Along Centerline of Flow Retardation Structure.

A = Rockfill	*F = Random Rockfill*
B = Transition	*G = Horizontal Rockfill*
C = Random Fill	*H = Surface Reinforcement*
D = Blanket Drain	*I = Concrete Anchor Blocks*
E = Existing Dam	

earthquake loadings, the concept of a reinforced rockfill flow attenuating berm, or FRS, constructed on the downstream toe of the main embankment was developed. In the event of a failure of the main embankment, the FRS would attenuate the release of water from the reservoir by allowing it to flow through or over the reinforced rockfill.

The basic objectives at Henshaw were (a) to provide protection against potential catastrophic liquefaction and piping failures of the main embankment due to earthquake loadings; (b) to provide a structure capable of passing possible reservoir discharges without instability of the structure and danger to downstream improvements; and (c) to make optimum use of available construction materials in the vicinity of the site.

With these functions adequately served, the main dam would not have to be taken out of service, and the FRS would only go into use in the event of failure of the main dam following a major earthquake.

Configuration of FRS

The final configuration of the FRS that was constructed is shown in plan and section on Figures 5 and 6, respectively.

The top of the main portion of the FRS is at elevation 822.3 m (2697 ft.), which provides a freeboard of 2.1 cm (7 ft.) above the new reduced full reservoir operating level at elevation 820 m (2690 ft.). This crest elevation and corresponding freeboard was selected on the basis of the potential adverse vertical offset that could occur on the Elsinore fault, passing diagonally beneath the main dam and the FRS.

The surface and embedded reinforcement for the FRS was designed as shown to cope with sloughing or raveling of the surface of the rockfill under overtopping or through-flow conditions and to prevent deep-seated shear failure within the FRS.

Stability Analyses for Final Layout

Analyses of the final layout of the FRS were made both for static loadings corresponding to a range of reservoir levels and earthquake loadings corresponding to a M 6.5 event on the Elsinore fault system. The seismic analyses were made using pseudo-static analysis procedures considering both initial major earthquake loadings and "aftershock" earthquake loadings. The design of the FRS, including the reinforcement of the rockfill materials, provided for overall safety factors of 1.5 for static loadings and 1.2 for earthquake loadings or overtopping.

Rockfill Reinforcement

A continuous pattern of 25 mm (1 in) diameter surface reinforcement (supplemented with heavy-duty chain-link fabric) oriented horizontally and vertically on the exposed surfaces of the FRS flow channel was provided to cope with sloughing or raveling of the surfaces of the FRS. Similarly, a continuous pattern of 25 mm (1 in) anchor bars embedded deep into the FRS embankment and anchored with concrete anchor blocks was provided to cope with deep-seated shear failure within the embankment under the prescribed loading.

Transition Zone Materials

Provisions were made to cope with the potential for offset movements along the Elsinore fault, which passes directly under Henshaw Dam. For design purposes, it was assumed that displacement would occur along a single and discrete fault plane (even though it is much more likely that the total displacement would be spread relatively uniformly over the 90 m [300 ft.] to 120 m [400 ft.] width of the fault zone). It was considered essential that the design maximum fault displacement be conservative. The design displacements are 2.1 m (7 ft.) in an oblique direction and 1.5 m (5 ft.) in both the vertical and horizontal directions.

The transition zone of the FRS (which is of critical importance in preventing piping, either through the wagonfill of the main dam embankment or through the FRS) has minimum dimensions of 6 m (20 ft.) horizontally and 3 m (10 ft.) vertically. These dimensions were selected to ensure that, in the event of displacements of the magnitude mentioned above, there will be an adequate overlap of transition material to prevent piping.

The gradation specified for the transition materials was developed to ensure that the materials would be self-healing; that is, would seal any fissures developing in the FRS as a result of displacements and that the materials would serve as a filter between the rockfill and the random materials placed against the main dam.

Foundation

A 1 m (3 ft.) thick blanket drain was provided under the random fill materials between the wagonfill of the main embankment and the transition zone materials to ensure continued drainage of the wagonfill.

CONSTRUCTION AND OPERATION

Construction of the spillway modifications and FRS begin in mid-September 1980 and was completed in September 1981. The cost of the construction work, which was bid on a unit-price basis, amounted to approximately $6.5 million. Since completion of the construction, the reservoir has been operated in the full condition at the new spillway crest elevation of 820 m (2690 ft.). Discharges through the modified spillway occurred during the winters of 1982-83 and 1983-84 without any apparent damage to the new work or flooding of downstream areas.

SUMMARY

Major modifications were made to Henshaw Dam, a hydraulic fill dam, between 1980 and 1981, to increase its stability under earthquake loadings. The modified spillway, which is sized to safely pass the PMF, lowers the normal operating level of the reservoir and thereby reduces the cross-sectional area of the dam subject to liquefaction. the new reinforced FRS, constructed on the downstream toe of the dam, serves to buttress the existing dam and improves its stability under earthquake loadings. In addition, in the event of liquefaction and failure of the dam, the FRS will attenuate, even if overtopped, any reservoir flows.

Figure 7. Dam Construction, 1922.

Sturzstrom Deposit Caprocks

Patrick L. Abbott*, Department of Geological Sciences*
San Diego State University, San Diego, California 92182-1020

Steven E. Borron*, American Geotechnical, Inc.*
5764 Pacific Center Boulevard, Suite 112, San Diego, California 92121

Jacobe L. Washburn*, Petra Geotechnical, Inc.*
41640 Corning Place, Suite 107, Murrieta, California 92562

ABSTRACT

Split Mountain has world-famous subaerial and subaqueous sturzstrom deposits that have been carefully studied to learn their distinguishing characteristics (e.g. in Abbott and Seymour, 1996; Abbott et al., 2002). We consider here what caps the sturzstrom deposits. The upper Miocene subaerial-sturzstrom deposit is capped by aerial fallout found as thinly bedded, coarse sandstone. The lower Pliocene subaqueous-sturzstrom deposit is capped by tsunami erosion and deposition in discontinuous graded beds.

INTRODUCTION

The Elsinore fault system is a prominent fact of life in southern California, soon to be realized by millions more people. One of the fastest growing housing markets in the United States is bringing hundreds of thousands of people to live in new houses built along the Elsinore fault zone ready to experience its next magnitude 6 to 7 earthquake. Over geologic time, the *en echelon* faults of the Elsinore system have created some interesting pieces of topography such as the pull-apart basin filled by Lake Elsinore and the push-up topography of Split Mountain.

Split Mountain lies south of the town of Ocotillo Wells on Highway 78 (see Thomas Bros. Map, p. 410, F10). Between the Vallecito and Fish Creek Mountains lies Split Mountain where Neogene basin-fill strata have been upwarped to expose some intriguing geologic history. The awe-inspiring Split Mountain outcrops of the catastrophic rock fall and flow masses known as sturzstroms were a prominent part of a recent South Coast Geological Society field trip (Abbott and Seymour, 1996).

In Miocene time, extension in the Salton Trough region was accommodated on linked normal faults. The extension created high-standing mountain blocks with pervasively fractured mountaintops such as the Vallecito and Fish Creek Mountains (Frost et al., 1996). The isolated mountains had intervening basins such as the Fish Creek-Vallecito basin, which contains a sedimentary fill that is 5 km thick (Kerr, 1982; Winker, 1987). The Fish Creek-Vallecito basin has undergone an inversion of topography as its sedimentary strata infill has been squeezed up to make Split Mountain. The exposed strata include both subaerial and subaqueous sturzstrom deposits.

OVERVIEW OF STURZSTROM DEPOSITS

In latest Miocene time, the heavily fractured Vallecito Mountain top dropped a ~300 x $10^6 m^3$ volume of tonalite that fell more than 300 meters, shattered, and flowed eastward for 12 km down an alluvial fan and then up a braided stream valley as the Split Mountain sturzstrom (Abbott et al., 2002). In earliest Pliocene time, a heavily fractured mountaintop in the northern Fish Creek Mountains dropped a ~300 x $10^6 m^3$ volume of plutonic rocks, pegmatite, gneiss, and schist that fell, shattered, and flowed southward for several km along the seafloor as the Fish Creek sturzstrom (Abbott et al., 2002).

The sturzstrom deposits have very distinct and different features from strata deposited in other depositional environments. Yet, careful inspection of the subaerial and subaqueous sturzstrom deposits reveals no obvious textural or fabric differences between them despite the fact that one ran under air and the other ran underwater. In regard to both sturzstrom deposits:

1. They preserve the lithologic domains and rock distributions that existed in the original bedrock sources despite having fallen, shattered, and flowed for up to 12 km.
2. They are mostly unmixed.
3. They have jigsaw-puzzle fabric wherein the shattered pieces of bedrock are not scattered and can be visually put back together.
4. They have shattered pegmatites that step up in the direction of transport.

5. They have megaclasts concentrated at the top of the deposit.

The differences found when comparing subaerial and subaqueous sturzstrom deposits occur at the base of each deposit. Real differences occur when sturzstroms travel over a firm and dry surface versus a soft and wet surface. The subaerial sturzstrom had the following effects:

1. Boulders sitting on the ground were sliced through, thus decapitated stones were produced.
2. Basal surfaces were grooved and striated.
3. The substrate was disturbed for less than 1 m depth.

The subaqueous sturzstrom had profound effects on the substrate and within the basal sturzstrom deposit:

1. Muddy substrate was folded to depths >35 m.
2. Diapirs of substrate strata rose into the sturzstrom mass.
3. Lobes of sturzstrom material injected into the substrate.

In sum, the main bodies of subaerial and subaqueous deposits are identical, but marked differences are seen at their bases. What about the tops of sturzstrom deposits? Are there differences in the immediately overlying sediments? Are sturzstrom flow masses capped by genetically related sediments?

SPLIT MOUNTAIN STURZSTROM CAPROCK

The Split Mountain sturzstrom (SMS) deposit begins on the eastern side of Split Mountain at approximately the midpoint of Fish Creek Wash and extends eastward to, and partly around, the Fish Creek Mountains. A left-lateral fault runs subparallel to the wash and has provided the path for the erosional development of the gorge. The faults have offset portions of the SMS to the south as seen on the west wall of the gorge. Typically, the tributary canyon exposures on the west side of the gorge tend to be steeper, narrower and more difficult to explore. In the case of Fallout Canyon (Figure 1) and most of the other west-side tributary canyons, this can be attributed to the fact that steeper narrower canyons are younger, in the sense that they have suffered less from erosion than the wider, longer, less steep and more exposed canyons that are typical on the eastern side of the gorge. Fallout Canyon on the west wall of the gorge initially opens easily at the bottom but quickly becomes steep and narrow with increasing loose rock debris as elevation increases. This condition has fortunately protected a sensitive deposit of sandstone lying on the upper surface of the SMS. In other words, the overburden of material that has covered this deposit for millions of years has protected it while the recent erosion has, for the moment, exposed much of the deposit intact.

The SMS originated by fall of a huge mass of tonalite and granodiorite from high up the Vallecito Mountains west of the gorge. The fall shattered the plutonic rocks, producing broken rock and an accompanying aerial cloud of finer debris (Borron, 1999). Aerial fallout from the massive dust cloud associated with the SMS is clearly preserved in an east-flowing tributary canyon we call Fallout Canyon. Granular sand from the aerial fallout occurs in a 20 cm thick interval comprised of thinly bedded, coarse sandstone separated by thin biotite-rich layers (Figure 2) reflecting the complexities within the roiling cloud of shattered debris. The fallout beds are indurated and composed of individual grains of quartz, plagioclase, and plutonic rock detritus (Figure 3). The grain size varies from small gravel to silt and is a normally graded deposit. Grain-size analyses were completed on two samples retrieved from Fallout Canyon. The samples were soaked in an HCl solution to free the grains from cement. The grains were then sieved through a stack of 12 sieves to determine the grain-size distributions. A similar, but smaller, plutonic rock-fall event occurred at the Happy Isles Campground in Yosemite in 1996. The campground area was covered with shattered plutonic rock debris that was carefully sieved (Wieczorek et al., 2000). The grain-size distributions of the 1996 Yosemite samples are compared to the latest Miocene Fallout Canyon samples (Figure 4). Results of the sieve analyses are consistent, suggesting that the Split Mountain sturzstrom event was accompanied by roiling clouds of aerial debris. The curves show granular, silty, coarse sandstones that are very poorly sorted. The sediments are angular, reflecting their impact origin and short time of transport.

FISH CREEK STURZSTROM CAPROCK

What sits on top of the Fish Creek sturzstrom? Winker (1987) defined a new stratigraphic unit called the Wind Caves member. He defined the member as sitting on top of the Upper Boulder Bed (our Fish Creek sturzstrom) and being discontinuous with thicknesses varying from 0 to 200 m of thin sandstone beds interbedded with gray claystone beds. Winker describes the lower Wind Caves as being dominated by "L" suite (Local) sediments derived from local sources and the upper Wind Caves as "C" suite

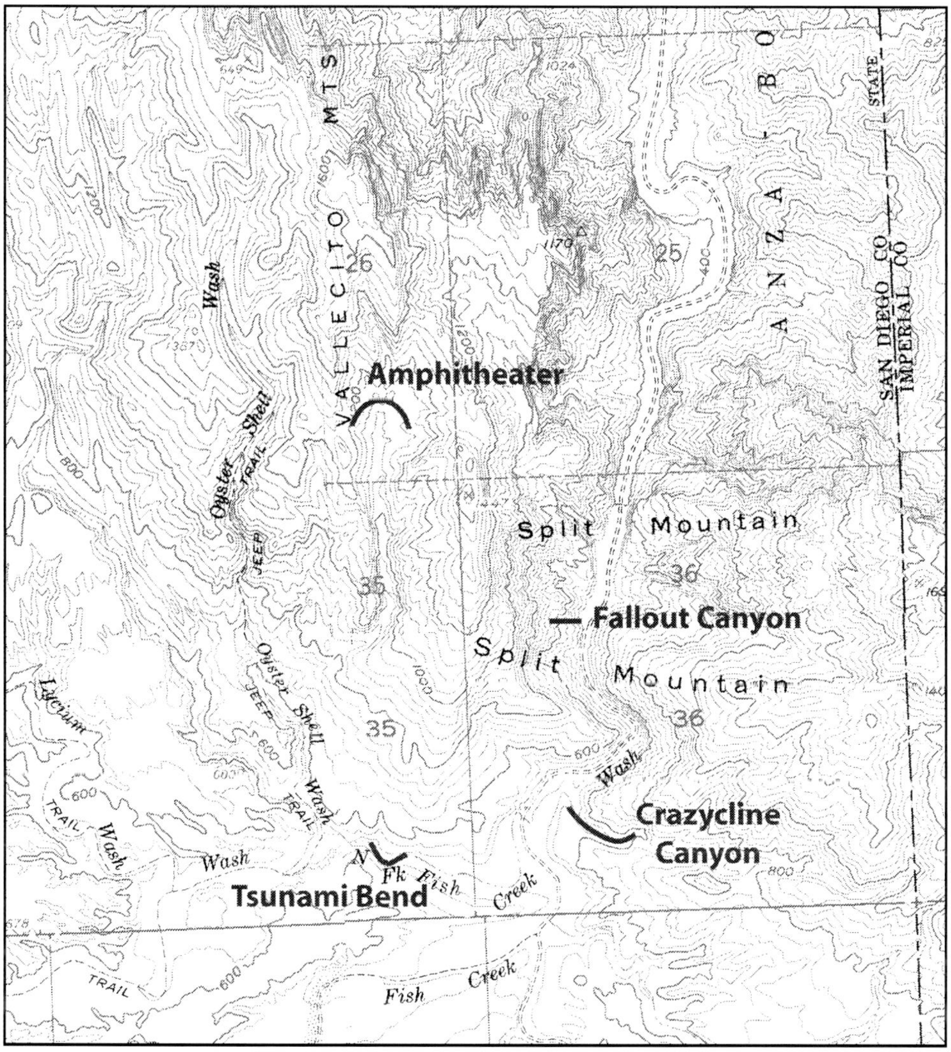

Figure 1. Location map. Area studied is on four U.S.G.S. 7.5' topographic quadrangle maps: Harper Canyon (upper left), Borrego Mountain SE (upper right), Carrizo Mountain NE (lower right), and Arroyo Tapiado (lower left).

(Colorado) sediments brought from a wide region by the Colorado River. Winker states (1987, p. 153): "No lithologic equivalents of the Wind Caves member are known to crop out elsewhere in the Salton Trough region." Winker tells of southerly directed paleocurrent indicators, and postulates a submarine-fan depositional system. Another cogent observation by Winker needs to be mentioned (1987, p. 149): "One boulder, exposed in lower Wind Caves Wash, consists of a large reworked clast of the underlying Upper Boulder Bed."

The Wind Caves member also can be studied via magnetic susceptibility. This is based on the original discovery by Gastil (1975) that the Peninsular Ranges have magnetite-bearing plutonic rocks on their western side and non-magnetite-bearing plutonic rocks on their eastern side. Thus, Salton Trough sediments derived locally from the eastern side of the Peninsular Ranges are nonmagnetic. This is in contrast to the magnetite-rich sediments delivered to the Salton Trough by the Colorado River. This difference in magnetite content allows recognition of the arrival of Colorado River-delivered sediment to the Split Mountain region (Gastil et al., 1996). Washburn (2001) used a magnetic-susceptibility meter to map the Wind Caves member (Figure 5). The Wind Caves member has a lower part (Pwl) made of locally derived nonmagnetic sediments and an upper part (Pwc) comprised mostly of sediments delivered by the Colorado River. The magnetic-susceptibility mapping clearly shows that all the sedimentary features immediately on top of the Fish Creek sturzstrom deposit were caused by local processes. The Colorado River could not have been involved; it arrived later in time.

Imbricated conglomerates in the southwesternmost exposure of the Fish Creek sturzstrom deposit yielded southerly flow directions to Washburn (2001). At two localities, imbricated clasts yielded average flow directions of 176° and 167° [Bed 1: n = 30, range = 150-196°; Bed 2: n = 15, range = 149-194°]. Apparently the Fish Creek sturzstrom began as a rock fall in the northern Fish Creek Mountains and continued as a sturzstrom flowing southward over the seafloor. The overlying lower beds of the Wind Caves member also contain southerly flow indicators. Grooved boulders in the FCS also provide some general idea of direction of transport (Figure 6).

Sedimentary Features in the Basal Wind Caves Member

At the mouth of Crazycline Canyon (Figure 1), the top of the Fish Creek sturzstrom deposit has significant relief due to hummocky topography formed during slowing of the sturzstrom and/or erosion immediately after deposition of the sturzstrom (Figure 7). Against and on top of

Figure 2. Layers of fallout sandstone deposited on top of the Split Mountain sturzstrom from the accompanying cloud of aerial debris.

this erosion surface is laminated coarse sandstone deposited quickly from a high-energy current.

In the Amphitheater (Figure 1), the top of the Fish Creek sturzstrom has a 7 m deep, concave-up channel either formed as a swale between hummocks and/or erosion by a high-energy current (Figure 8). The channel is filled by a normally graded sequence of coarse sediments deposited by a strong but rapidly ending current.

As described by Winker (1987), a large rip-up clast of Fish Creek sturzstrom sits within the lower Wind Caves member. A high-energy, short-lived current was necessary to erode and move this penecontemporary block of sediment.

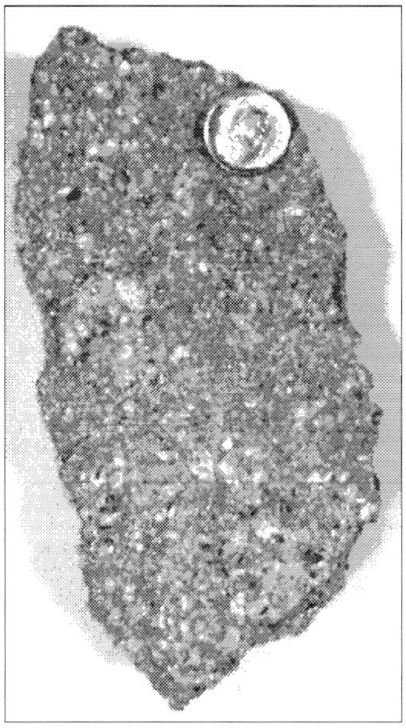

Figure 3. Fallout sandstone. An angular, very poorly sorted, granular, silty, coarse plutonic litharenite.

At Tsunami Bend (Figure 1), the modern stream has carved a meander that provides an excellent 3D exposure of the basal Wind Caves member. A basal layer of cobble conglomerate of 1.5 m thickness is overlain by a 1.84 m thick graded bed of granular conglomerate — coarse sandstone, laminated medium sandstone, rippled sandstone, and mudstone. The 3.34 m thick section is a Bouma sequence of coarse sediments deposited from a short-lived, powerful current. The clasts in the conglomerate are polylithologic, indicating erosion from the Fish Creek sturzstrom deposit. If the conglomerate had been washed down the Vallecito Mountains alluvial fan/fan delta after a heavy rain, it would have produced a nearly monolithologic conglomerate.

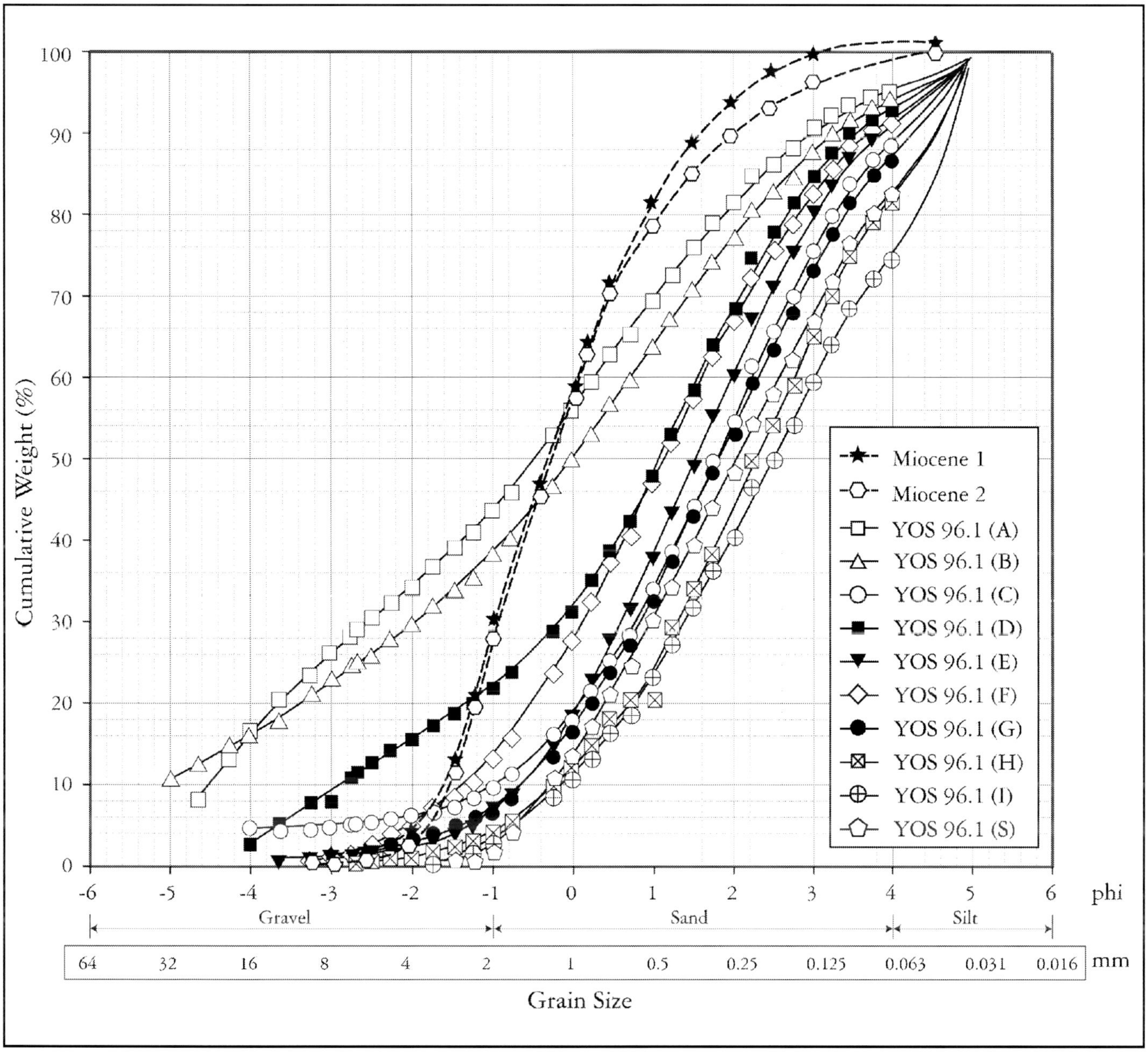

Figure 4. Grain-size distributions for two uppermost Miocene sandstones from Fallout Canyon compared to sands from the 1996 Happy Isles rock fall in Yosemite analyzed in Wieczorek et al. (2000).

What created the unique deposits that discontinuously cap the Fish Creek sturzstrom? The evidence points to tsunami caused by the passage of the Fish Creek sturzstrom into the sea.

Was it necessary for the rock fall that became the Fish Creek sturzstrom to drop directly into the sea in order to create tsunami? No, there are case histories of rock avalanches flowing over land and then creating tsunami after they reached the shoreline. For example, on 21 May 1792, a lava dome collapse from Mt. Unzen in Japan created a rock avalanche of 0.3 km^3 volume that flowed 6.4 km to the sea where it hit the water with enough impact to create tsunami that killed 15,000 people.

CONCLUSION

The caprocks on top of sturzstrom deposits are markedly different. The subaerial Split Mountain sturzstrom deposit was sprinkled with poorly sorted, angular, plutonic rock sediment that moved as an aerial cloud above the flowing sturzstrom and then settled out on top of it in thinly bedded layers. The subaqueous Fish Creek sturzstrom deposit is cut by channels and had material eroded from its top; material was transported by high-energy tsunami that crowned the sturzstrom mass with rip-up clasts and filled concave-up channels with discontinuous masses of graded sediments.

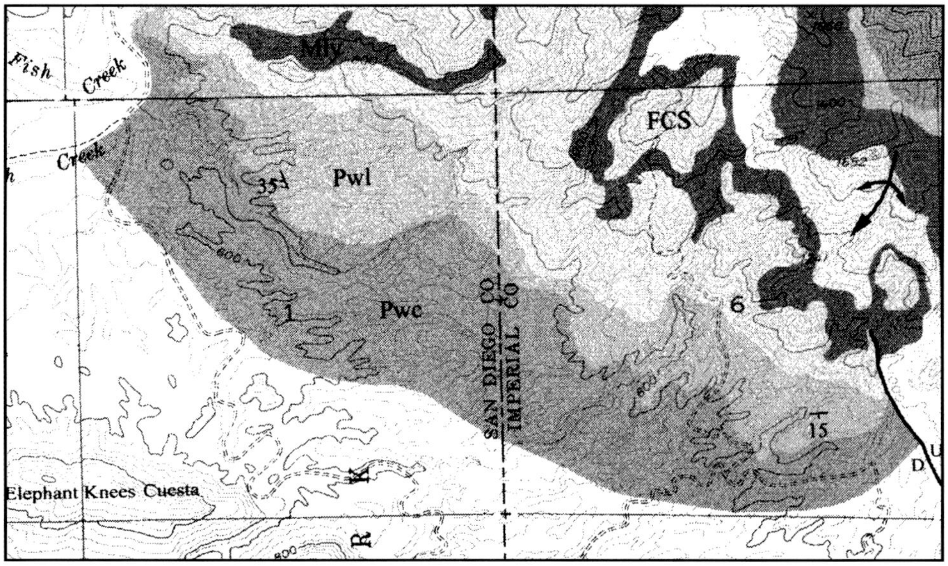

Figure 5. Geologic map from head of Split Mountain Gorge (Washburn, 2001). The Fish Creek sturzstrom (FCS) is overlain by nonmagnetic, locally derived, lower Wind Caves member (Pwl). Upper Wind Caves member (Pwc) is magnetic and was eroded and delivered from a broad region by the Colorado River.

Figure 6. Large boulder in the Fish Creek sturzstrom, showing grooves scratched during transport. Flow direction was southwest.

Figure 7. Outcrop near mouth of Crazycline Canyon. Quadrant in lower left is Fish Creek sturzstrom. Note near-vertical erosional wall eroded by tsunami then abutted and buried by tsunami sandstone. White dashed line indicates channel base.

ACKNOWLEDGMENTS

We thank Kathie Marsaglia of CSU Northridge for the stimulating discussion in the field where the idea of tsunami deposition first arose.

We appreciate the computer skills and friendly help always provided by Tony Carrasco of SDSU.

REFERENCES CITED

Abbott, P.L., and Seymour, D.C., 1996, Sturzstroms and Detachment Faults, Anza-Borrego Desert State Park, California: South Coast Geological Society Guide Book No. 24, 256 p.

Abbott, P.L., Kerr, D.R., Borron, S.E., Washburn, J.L., and Rightmer, D.A., 2002, Neogene sturzstrom deposits, Split Mountain area, Anza-Borrego Desert State Park, California, in Evans, S.G., and DeGraff, J.V., eds., Catastrophic Landslides: Effects, Occurrence, and Mechanisms: Boulder, Colorado, Geological Society of America, Reviews in Engineering Geology, v. XV, pp. 379-400.

Borron, S.E., 1999, Geology of the Split Mountain sturzstrom, Anza-Borrego Desert State Park, California [M.S. thesis]: San Diego State University, 96 p.

Frost, E.G., Suitt, S.C., and Fattahipour, M., 1996, Emerging perspectives of the Salton Trough region with an emphasis on extensional faulting and its implications for later San Andreas deformation, in Abbott, P.L., and Seymour, D.C., eds., Sturzstroms and Detachment Faults: South Coast Geological Society Guide Book No. 24, pp. 81-122.

Figure 8. In the Amphitheater. Note Fish Creek sturzstrom all the way up the right-hand side of the photo. See tsunami-carved channel in FCS filled with tsunami-deposited gravelly sandstone. White dashed line indicates channel base.

Gastil, R.G., 1975, Plutonic zones in the Peninsular Ranges of southern California and Baja California: Geology, v. 3, pp. 361-363.

Gastil, R.G., Kimbrough, D.L., and Abbott, P.L., 1996, Arrival of the Colorado River in the Salton Trough defined by magnetic susceptibility, in Abbott, P.L., and Seymour, D.C., eds., Sturzstroms and Detachment Faults: South Coast Geological Society Guide Book No. 24, pp. 203-207.

Kerr, D.R., 1982, Early Neogene continental sedimentation, western Salton Trough, California [M.S. thesis]: San Diego State University, 138 p.

Washburn, J.L., 2001, The geometry and geologic history of Neogene rocks between Fish Creek Wash and the Fish Creek Mountains [M.S. thesis]: San Diego State University, 109 p.

Wieczorek, G.F., et al., 2000, Unusual July 10, 1996, rock fall at Happy Isles, Yosemite National Park, California: Geological Society of America Bulletin, v. 112, pp. 75-85.

Winker, C.D., 1987, Neogene stratigraphy of the Fish Creek-Vallecito section, southern California: Implications for early history of the northern Gulf of California and Colorado delta [Ph.D. dissertation]: Tucson, University of Arizona, 494 p.

Geology of a Portion of the Southern Coyote Mountains

***Kenneth C. Schulte**, 1245 Elizabeth Street, Barstow, California 92311*

INTRODUCTION

This paper was extracted from a Bureau of Land Management (BLM) mineral report I prepared in 1986 for a validity examination of six placer mining claims totaling 960 acres in the southern part of the Coyote Mountains in Imperial County, California. Although the mining claims were originally located in 1952 for clay, they also contained valuable deposits of sand and gravel. Since passage of the Common Variety Act on July 8, 1955, mining claims can no longer be located for common variety minerals such as sand and gravel and volcanic cinders.

The area of this investigation (T.16 S., R.9 E., Sec. 8 & N 1/2 Sec. 9), shown in Figure 1, may be reached from Interstate 8 by taking the Ocotillo off-ramp (north side) and heading west 3.7 miles on Imperial Highway (S-2), then north 1.25 miles on the paved road leading to the Massey sand and gravel pit, then 3000 feet west, northwest up a tributary of Palm Canyon Wash, then nearly a mile north, northwest along a dirt trail. A 4-wheel-drive vehicle is preferable for travel along the dirt trails.

There are about 18,000 acres of Wilderness in the Coyote Mountains (Figure 2). Vehicular access is restricted to the "cherry-stem" route in Painted Gorge leading to patented mining claims. For more detailed information or to obtain Desert Access Guide (DAG) maps showing land status and routes, the reader is referred to the El Centro Field Office of the BLM (760) 337-4422.

REGIONAL GEOLOGY

The area under investigation lies at the transition between the Peninsular Ranges Province and the Salton Trough Province which is part of the Colorado Desert. The Peninsular Ranges are part of a large block uplifted abruptly along the eastern edge and tilted westward (Sharp, 1967, p. 17). The Salton Trough is basically a landward continuation of the Gulf of California extending northwestward all the way to San Gorgonio Pass in Riverside County (Sharp, 1976, p. 34). The major drainage is into the Salton Sea. The northwest-trending Elsinore fault zone forms the boundary between the two provinces. Figure 1 is a geologic map of the area at a scale of 1:250,000 (Strand, 1962).

The oldest rocks in the Peninsular Ranges, including the Coyote Mountains, are the pre-Tertiary metamorphic complex, mostly schist and gneiss overlain or flanked by Tertiary-age volcanic and sedimentary rocks.

Tertiary stratigraphy in the southwestern Imperial Valley area includes the following formations, from oldest to youngest: Split Mountain Formation, Alverson Andesite, and Fish Creek Gypsum of Miocene age, and the Imperial Formation, Palm Spring Formation and Canebrake Conglomerate of Pliocene age. A columnar section showing the relationships of these Formations may be found in Bulletin 170, Geology of Southern California (Dibblee, 1954, Ch. II, Part 2, p. 22).

The principal structural feature in the area is the Elsinore fault zone which forms the southwest boundary of the Coyote Mountains. To the northwest the Elsinore fault zone passes through the Anza-Borrego Desert State Park as shown on the San Diego-El Centro Sheet of the Geologic Map of the California (Strand, 1962). In the Yuha Desert, several miles southeast of the Coyote Mountains, the Laguna Salada fault appears as an extension of the Elsinore fault and continues southeast into Mexico.

The Coyote Mountains range was elevated along the Elsinore fault and tilted northward (Dibblee, 1954, p. 27). Dibblee states that vertical displacements along the Elsinore fault are locally great and that a great amount of right lateral movement is indicated by tight, east-trending drag folds in the Cenozoic sediments on the south side of the Coyote Mountains.

LOCAL GEOLOGY

A geologic map of the subject 960 acres may be found as Figure 3. The higher elevations are predominantly pre-Tertiary metamorphic rocks comprised mostly of biotite schist and gneiss with lesser marble and fine-grained metasedimentary rocks including quartzite. Most of the pre-Tertiary metamorphic complex lies northeast of the Elsinore fault, whereas the Tertiary and Quaternary sedimentary units lie to the southwest.

Volcanic rocks in the Coyote Mountains are represented by the Alverson Andesite Formation of probable upper Miocene age (Dibblee, 1954, p. 22). The Alverson

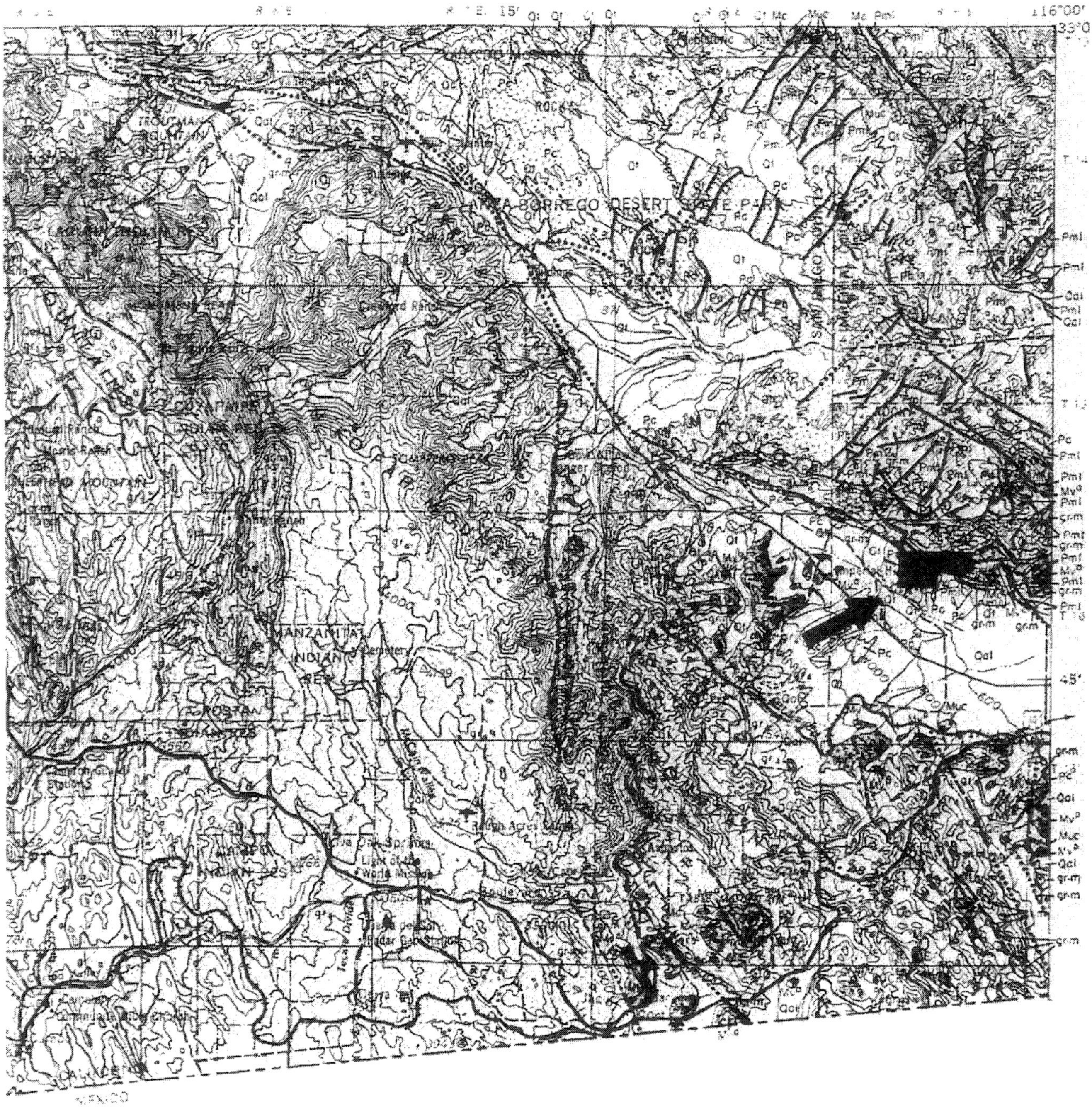

Figure I. Location of study area in relation to the Elsinore Fault Zone.

Scale 1 :250,000
Compiled by R.B. Strand, 1962
Calif. Div. of Mines & Geology

0 1 2 3 4 5 6
Miles

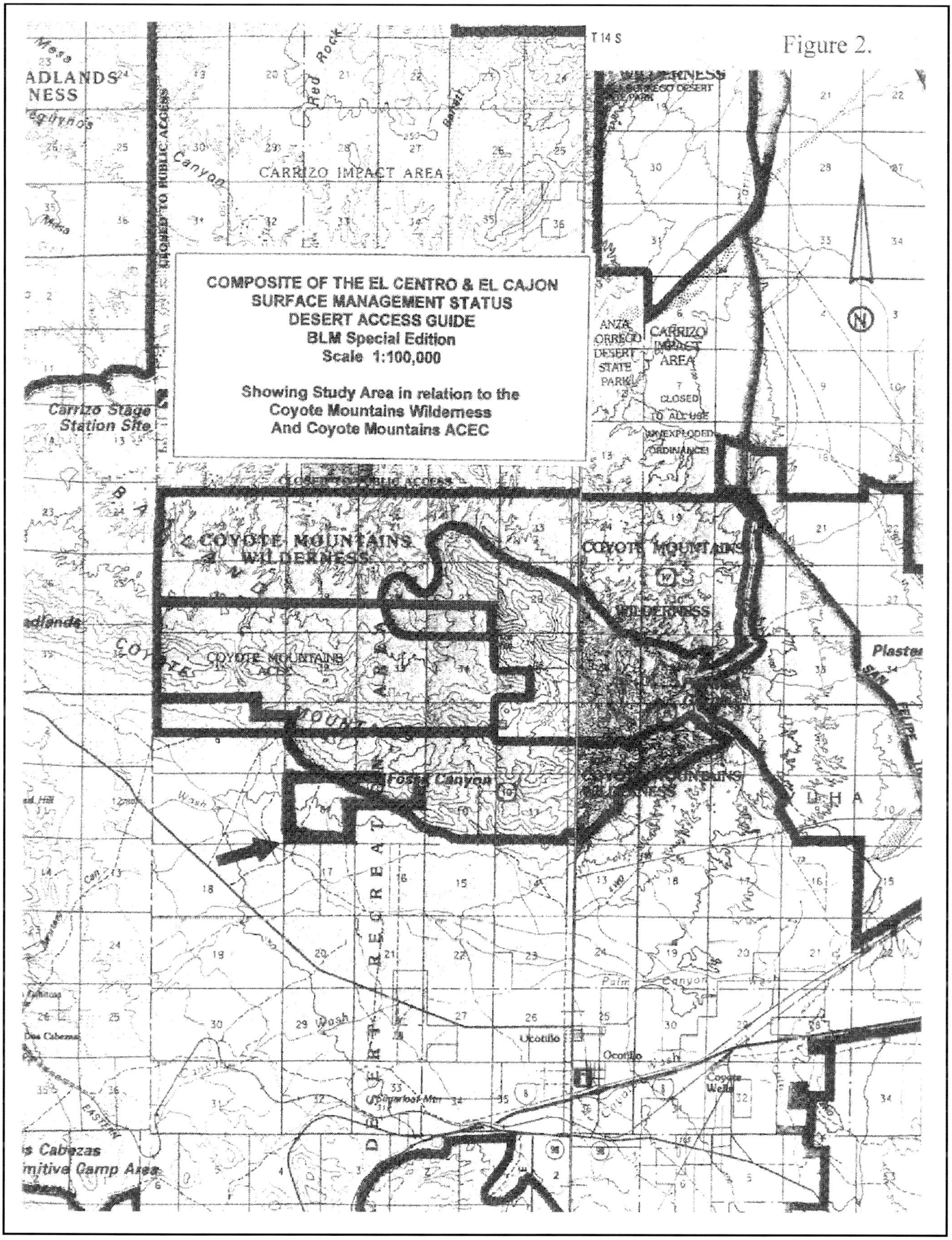
COMPOSITE OF THE EL CENTRO & EL CAJON
SURFACE MANAGEMENT STATUS
DESERT ACCESS GUIDE
BLM Special Edition
Scale 1:100,000
Showing Study Area in relation to the
Coyote Mountains Wilderness
And Coyote Mountains ACEC
CARRIZO IMPACT AREA
CLOSED TO PUBLIC ACCESS
Carrizo Stage
Station Site
COYOTE MOUNTAINS
WILDERNESS
COYOTE MOUNTAINS
ACEC
Fossil Canyon
ANZA
BORREGO
DESERT
STATE
PARK
CARRIZO
IMPACT
AREA
CLOSED
TO ALL USE
UNEXPLODED
ORDINANCE
Ocotillo
Coyote
Wells
Palm Canyon Wash
T 14 S

Figure 2.

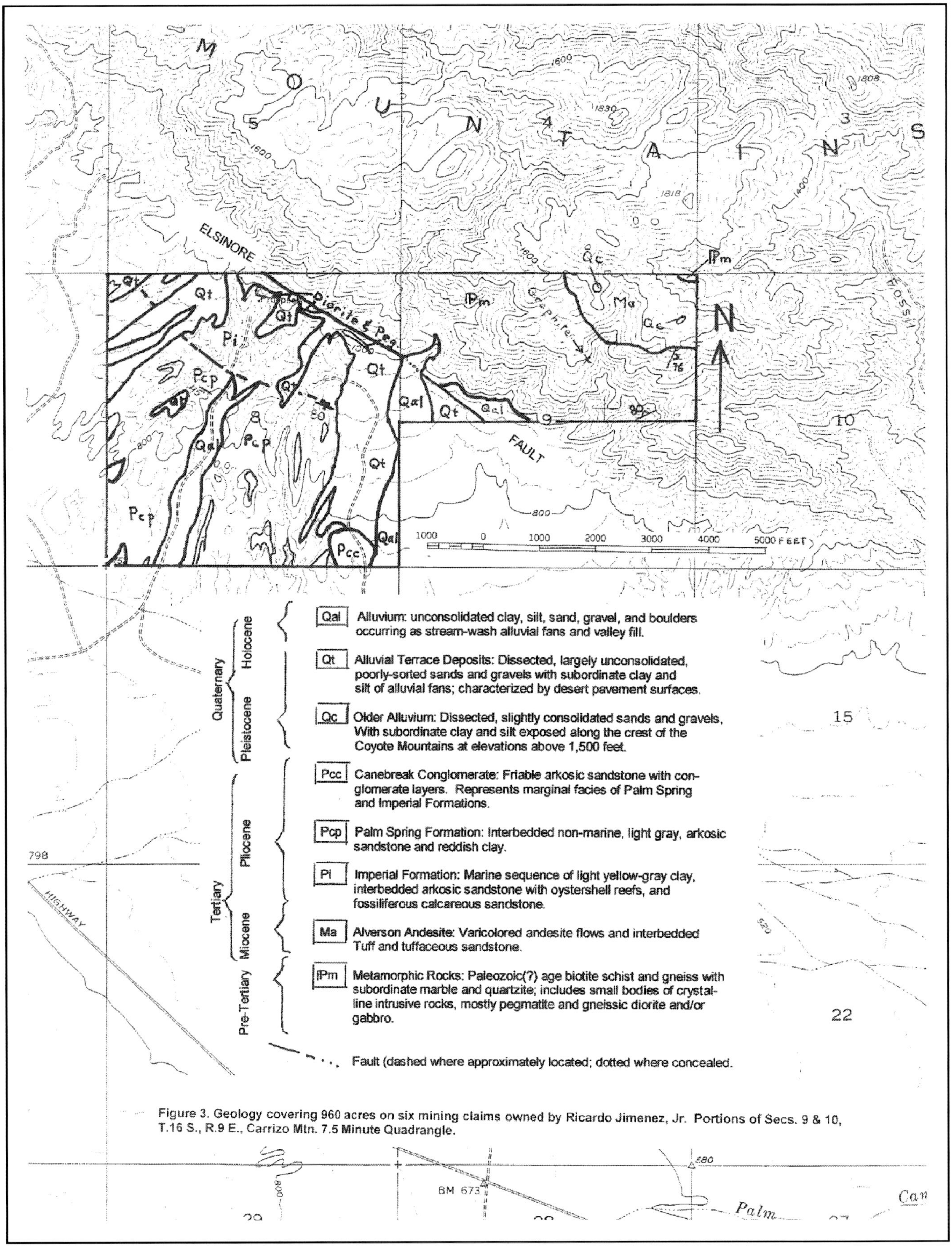

Figure 3. Geology covering 960 acres on six mining claims owned by Ricardo Jimenez, Jr. Portions of Secs. 9 & 10, T.16 S., R.9 E., Carrizo Mtn. 7.5 Minute Quadrangle.

Andesite is a sequence of mostly andesite flows of varied color and interbedded tuff, breccia, and nonmarine sandstone and conglomerate (Morton, 1977, p. 17). In the area northwest of Ocotillo, the andesite flow rock is mostly brown to orange-brown but locally has a maroon hue or even black on fresh surfaces. In places, the andesite is underlain by gray, tuffaceous sandstone which is underlain by pale green-gray, friable tuff.

The Imperial Formation is of marine origin and is composed predominantly of yellow and gray claystone with interbedded sandstone and oyster-shell reefs (Morton, 1977, p. 17). The Formation is as much as 2700 feet thick on the south side of Carrizo Creek in the Fish Creek Mountains (Dibblee, 1954, p. 22). Morton (1977, p. 34) states that the claystones attain a thickness of 450 feet. The Imperial Formation grades laterally and upward into the Palm Spring Formation which is composed of a thick, nonmarine sequence of interbedded light-gray arkosic sandstone and reddish lacustrine clay (Morton, 1977, p. 18). The Palm Spring Formation was deposited throughout the Imperial depression after marine waters of the Gulf were barred from it, probably by the damming action of the Colorado River delta (Dibblee, 1954, p. 23). Fragments of silicified wood are common throughout the Formation. According to Dibblee (1954, p. 22), the Palm Spring Formation is 4800 feet thick south of Carrizo Creek.

The Canebrake Conglomerate is the coarse, marginal conglomerate facies of the Palm Spring and Imperial Formations (Dibblee, 1954, p. 23). Typically, it is composed of conglomerate containing mostly granitic and metamorphic debris (Morton, 1977, p. 18). Where it occurs in the study area, the Formation consists of a 30-foot-thick exposure of friable, tan, subangular, fine- to medium-grained arkosic sandstone with layers of coarse-grained sandstone and conglomerate. The pebbles and cobbles are up to 6 inches in diameter and are comprised of rounded to subrounded clasts of mostly granitic rocks, gneiss, quartzite, and andesite.

Quaternary alluvium in the Coyote Mountains area can be divided into three main units. From oldest to youngest, these are continental (nonmarine) deposits (Qc) exposed along the crest of the Coyote Mountains, nonmarine terrace deposits (Qt) at the flanks of the Coyote Mountains, and valley fill (Qal).

Alluvial deposits (Qc) along the crest of the Coyote Mountains were designated as the "Garnet Formation" in a Masters thesis by Andrew Christensen (1957, p. 127). According to Christensen, the Formation probably represents the remnants of a fanglomerate deposited in the earliest part of the Pleistocene before tectonic activity along the Elsinore fault zone fully exposed the Alverson Andesite and older formations. Christensen (1957, p. 127) states that the Formation ranges in thickness from less than a foot to over 60 feet and is restricted to elevations greater than 1500 feet. In the exposures observed by Christensen (1957, p. 128) the Formation is comprised of a "stratified, buff to brown sandstone and cobble and pebble conglomerate with local shoestrings and lenses of siltstone." The cobbles and pebbles are composed predominately of subangular and subrounded schist and marble.

The nonmarine terrace deposits (Qt) are dissected remnants of a Pleistocene alluvial fan surrounding the Coyote Mountains. The terrace alluvium lies unconformably above the preceding Formations except the older alluvium resting on the crest of the Coyote Mountains. The terrace deposits are unconsolidated, crudely stratified sands and gravels with subordinate silt, clay and boulders. Christensen (1957, p. 131) described the pebble and cobble clasts as subangular to subrounded and composed of, in decreasing order of abundance, schist, gneiss and quartzite, marble, granitic rock and andesite. The deposits range in thickness from less than a foot to over 20 feet. The surface of the deposits is characterized by well-developed desert pavement. Most of these deposits are too small, isolated, and contain too much silt and clay to be of commercial value. The deposits in E1/2 E1/2 Section 8 and SW1/4 NE1/4 Section 8, however, have commercial potential.

The youngest alluvial deposits (Qal) are composed of unconsolidated clay, silt, sand, gravel, and boulders occurring primarily as alluvial fan and valley fill streamwash deposits. In the study area, the alluvium is mostly gravelly sand with subangular to subrounded pebbles and cobbles. The surface gravel is composed of an estimated 70 percent gneiss and granitic rocks, 20 percent schist, and 10 percent marble, andesite and quartzite.

MINERAL DEPOSITS

Pegmatite

The Snow White silica prospect, located in the NW1/4 Section 8 (T.16 S., R.9 E.), consists of three vertical, northwest trending pegmatite dikes in gneissic diorite or gabbro. The two upper dikes are a few feet thick and discontinuous for a total overall length of nearly 500 feet. The third and lowest pegmatite dike is about 100 feet long and 10 feet thick. The main quarry is at the west end of the lower pegmatite dike and is 30 feet long by 18 feet high in feldspar-rich pegmatite with local quartz veins up to 1 foot wide. The pegmatite becomes finer grained to the west and contains clay-bearing shear zones 1 to 3 feet wide. About

25 feet east of the quarry and slightly above it is a caved adit heading north at the base of a cut, 20 feet by 10 feet high exposing an 8-foot thickness of feldspar-rich pegmatite with quartz and muscovite mica. There is also some biotite mica present. A third set of workings is located at the east end of the prospect and is separated from the other pegmatite dikes by a gulley. These workings consist of two cuts. The upper one is 12 feet long by 4 feet high with about two-thirds feldspar and one-third quartz with about 5 percent muscovite mica. The lower cut is 10 feet long by 3 feet high exposing a 3-foot-wide quartz vein in intensely fractured pegmatite. The access trail to the prospect has been washed out in places, requiring about a third of a mile of walking to reach it. An estimated 200 to 400 tons of silica was mined from the property in 1928-29 (Morton, 1977, p. 91). The pegmatite dikes are probably too thin and irregular to be profitably mined for their quartz and feldspar content.

A 1939 location notice was found next to a 4-foot-diameter prospect pit in a pegmatite body 30 feet by 20 feet, enclosed in diorite located in the NE1/4 SE1/4 NE1/4 Section 8. The pegmatite contains about 10 percent muscovite mica with sheets up to one inch in diameter.

Graphite

A graphite prospect is located in schist about 1300 feet north 34 degrees east from the center of Section 9 (T.16 S., R.9 E.). Graphite schist along a shear zone has a maximum thickness of 5 feet and can be traced for 30 feet. The shear zone strikes north 65 to 85 degrees west, and dips 80 degrees to the northeast. Workings consist of two trenches and a cut. The main trench is 20 feet long by 4 to 56 feet wide and 3 to 8 feet high, and the second trench is 12 feet long by 6 feet wide by 2 to 5 feet high and follows the southeast end of the shear zone. The cut is about 12 feet long and follows the northeast part of the exposed shear zone. The deposit contains an estimated 60 short tons of graphite schist. Although there is no known domestic production, Mexican graphite was selling for $218 per short ton in 2002 (Taylor, 2003, p. 29). The deposit is probably too small and inaccessible to be mined profitably. There is no vehicular access to the prospect or to any part of the claim.

Montmorillonite

Tertiary claystone occurs in the Imperial Formation. An analysis of the Tertiary clay, once known as the Jumbo clay deposit in Section 6, 7, and 8 (T.16 S., R.9 E.), is given in the California Journal of Mines and Geology (Sampson and Tucker, 1942, p. 132) as follows:

Silica (SiO2)	53.50%
Aluminum oxide (Al2O3)	14.10%
Iron Oxide (Fe Al2O3)	3.90%
Magnesium oxide (MgO)	3.62%
Calcium oxide (CaO)	6.80%
Sodium & potassium oxides	3.98%
Loss on ignition	14.16%

The loss represents mostly loss of carbon dioxide gas upon heating the calcium and magnesium carbonates contained in the clay. According to Ryman (1964, p. 3), the high ignition loss precludes use as brick clay.

A BLM industrial minerals specialist, Andrew Regis (January 1978, p. 6), classified clay he sampled from the Imperial Formation in Section 6 (T.16 S., R.9 E.) as calcium bentonite. There are two basic definitions of bentonite. One is a geologic definition proposed by Ross and Shannon (1926): "Bentonite is a rock composed essentially of a crystalline clay-like mineral formed by devitrification and the accompanying chemical alteration of glassy igneous material, usually a tuff or volcanic ash; and it often contains variable proportions of accessory crystal grains that were originally phenocrysts in the volcanic glass."

A mineralogical definition for bentonite was later proposed by Grim and Guven (1977): "...a clay composed essentially of a smectite clay mineral, and whose physical properties (plasticity, drilling mud yield, water holding capacity, etc.) are a consequence of the smectite composition." According to Regis (January 1978, p. 5) montmorillonite is probably the best generic term for the layered clays, but it is "bentonite" that has become the most popular.

X-ray diffraction analysis of three samples from Section 6 indicated calcium montmorillonite and quartz as major constituents, and calcite, feldspar, and gypsum as minor phases (Regis, February 1978, p. 8). The same samples when tested (p. 9) had low viscosity values of 30 barrels per ton and high water loss values ranging from 55 to 70 milliliters. Regis stated (p. 29) that such a bentonite is somewhat restricted in usage to animal feed binders, pet adsorbents and other uses not requiring a swelling type bentonite. The claystones of the Imperial Formation sampled by Regis in Section 6 look very similar to the other deposits on the southwest slope of the Coyote Mountains, including the area investigated in the 1986 study by the author.

In a Department of Interior hearing — U.S. v. Lloyd O'Callaghan, Sr., et al. — clay deposits from the Coyote Clay Nos. 3 and 4 placer claims in Section 14 (T.16 S., R.9 E.) were found to be "common clays not subject to location...."

Sand and Gravel

When the investigation was made in 1986, aggregate produced in the Ocotillo area was used for road base (about 60%), asphalt concrete (about 25-30%), and concrete (about 10%) according to an employee of Massey Sand and Rock Co. (J. Gotti). An undetermined but relatively small percentage of the reject material passing the No. 4 sieve is sold as fill. During Massey's operation in SE1/4 Section 9 and NE1/4 Section 16, about 10-15 percent waste (minus No. 4 sieve) resulted when producing base aggregate for asphalt concrete, and about 50-55 percent waste resulted when producing aggregate for asphalt concrete.

In the Ocotillo area in general, the material is fairly clean at the surface but tends to become dirtier with depth. This is especially true in the older terrace alluvium. In Sections 14 and 15, near Fossil Canyon, the overall amount of material rejected is about 30 percent according to an employee of Val-Rock (R. Becker).

On November 20, 1985, I took a grab sample consisting of 228 pieces of crushed rock from the 3/4- to 1-inch pile at the Massey plant. The aggregate was mined from SE1/4 Section 9 and should be fairly random in composition and representative of the deposit after having gone through the crushers and screens. The sample contained the following percentages of various rock types:

Rock Types	Percent
Schist	48
Gneiss	20
Granitic Rock	10
Quartzite	9
Andesite	8
Marble	3
Sandstone	1.3
Quartz	0.4

The high percentage of schist contributes to a relatively poor showing when subjected to abrasion tests. The Quaternary alluvial fan material on Jemeniz' mining claims has a higher percentage of gneiss and granitic rocks which would make the material harder.

MINERAL PRODUCTION HISTORY

Introduction

The Coyote Mountains district takes in an area of about 50 square miles in Imperial County. The reader is referred to Weber (1963) for deposits in the west portion which is in San Diego County. The following paragraphs on production and mineral potential are from California Division of Mines and Geology County Report 7, "Geology and Mineral Resources of Imperial County, California (Morton, 1977, pp. 25, 26).

> A remarkable variety of mineral deposits have been found in the Coyote Mountains, although none of these has yielded significant tonnages of commercial material. Occurrences of beryllium, clay, limestone, marble, nickel, quartz, roof granules, sand and gravel, and silica sand are known to exist there. Of these, however, only clay, marble, quartz, roof granules, and sand and gravel have been sold. It is, nonetheless, an area of considerable interest in that it does contain deposits of potential economic significance. Large reserves of cement-grade limestone are present in the Carrizo Mountain area in addition to numerous smaller deposits (see Coyote Mountains limestone deposit). Several areas are underlain by deposits of clay that heretofore have been utilized only as a cattle feed supplement. Nickel prospects near Fossil Canyon have aroused considerable attention in recent years, although none has proved to be of economic grade. Beryl occurrence in pegmatite dikes of the western part of the district are of undetermined grade and extent.
>
> The mineral deposits of the Coyote Mountain district, as would be expected from the wide variety of commodities represented, are of widely different origins. The limestone and marble are of sedimentary origin and are to be found predominantly in the metamorphic sequence. A very pure detrital limestone (coquina) in the Imperial Formation unconformably overlies the older limestone near Carrizo Mountain, and may enhance the value of the underlying deposits (see Coyote Mountains limestone).
>
> The nickel occurrences near Fossil Canyon (Roark prospect) are associated with a pyroxenite body which has been intruded into gneiss. The garnierite-morenosite occurrences noted thus far have been adjacent to the southern intrusive contact of the pyroxenite with quartz diorite. Another nickel occurrence (Edwards prospect), 2 miles eastward, however, lies within foliated granitic rocks. Pegmatite dikes, intruded into the metamorphic rocks, are the host rock of the beryl occurrences and also for the quartz mined for silica. Alverson volcanics have provided the source material for a roofing

granule operation (Weaver deposit). Clay and silica sand occur in the Imperial Formation. Sand and gravel from Holocene fanglomerate at the mouth of Fossil Canyon is utilized for asphalt concrete.

Probably the best potentially economic mineral resource of the Coyote Mountain district lies in the development of the limestone deposits near and on Carrizo Mountain and elsewhere in the metamorphic series. Population growth in the Imperial Valley and San Diego areas may some day provide a market of sufficient size to support a cement plant utilizing these sources of limestone. Also of potential value are the extensive clay beds in the Imperial Formation which are well exposed south of Carrizo Wash. These are composed of low-grade clays or clay-size material that may be found useful in the future.

Clay

For about two years (1950-1951) P.M. "Rusty" Shaefer hauled 50 tons of montmorillonite clay every 3 to 6 months from the west half of Section 8 (T.16 S., R.9 E.) to the Coyote Wells where it was shipped by rail car to Denver, Colorado. The clay was presumably used as a binder or supplement for cattle feed. Morton (1957, p. 35) reported little or no production from the Jumbo clay deposit in Sections 6, 7, and 8, T.16 S., R.9 E. Extensive deposits, mostly in the Imperial Formation, of high-silica, low-alumina clays underlie much of the areas in, and flanking, the range.

Older clay removal areas lie along the lower slopes of the Coyote Mountains at northern portions of Sections, 14 and 15, representing old operations of various parties through the years (Ryman, 1964, p. 2). The material reportedly had small use in the late 1930s and early 1940s as filler and binder for aspirin tablets (Ryman, 1964, p. 3).

Morton (1977, p. 35) reported production as undetermined but small from the O'Callahan deposit in NE1/4 SE1/4 Section 14 (T.16 S, R.9 E.). The clay was mined by bulldozer from low-lying erosional remnants of the Imperial Formation, crushed, ground, and bagged on the property and sold for use as a livestock feed supplement. O'Callahan testified (U.S. v. Lloyd O'Callaghan, Sr., et al., 1971, p. 6) that from 1954 to 1956 he produced clay from the state pit area (Sec. 14) for cattle feed, and that Oscar Jones and a Mr. Hogue produced clay from the mining claims until 1968. During this period, approximately 20,000 tons of clay was produced from the Bolsa de Oro #5 mining claim (NE1/4 Sec. 15) and the Coyote Clay group in N1/2 Section 14 according to O'Callaghan's testimony. Also, a quantity of clay was given to the county for use as a sealant in Sunbeam Lake.

Sand and Gravel

The correlation between sand and gravel production and population growth in Imperial County is not particularly good. The sand and gravel industry in the county has relied heavily on government highway projects, although general concrete construction and ditch lining and filter media around drain tiles for agricultural drainage have also been significant. The late 1950s and early 1960s were particularly active for highway paving because more money was available then.

The first sand and gravel producer to work in the Ocotillo area was Rusty Schaefer et al. who produced 3000 to 6000 tons per year from the Minoca placer claim in SE1/4 Section 9, T.16 S., R.9 E. from 1955 to 1964. William H. Raley produced an estimated 200,000 to 300,000 cubic yards within the Coyote Clay Nos. 3 and 4 placer mining claims in NW1/4 Section 14 (T.16 S., R.9 E.) in the Fossil Canyon area under a lease from the O'Callaghans from 1959 to 1964.

After 1964, much of the production was by George (Bud) Conway of Valley Transit Cement Co. who leased from Raley. Under a BLM contract, 710,000 tons were produced from 1964 to 1972 (Ryman, 1973). The first period of major production from the state material site in S1/2 Section 14 (T.16 S., R.9 E.) was around 1960 to 1961. The material was used for a highway job at Mountain Springs Grade, west of Ocotillo.

Around 1973 Raley bought Conway's crusher and moved to land owned by Carroll in N1/2 Section 16 and operated there for 10 years until it was bought by Richard Becker. In early 1974, Massey bought out Cal-Western and took over their pit in NE1/4 Section 16 on the Bolsa de Oro #6 mining claim.

In 1983, R. Becker formed Val-Rock and obtained a BLM material sale contract for 100,000 cubic yards valued at $38,000 in SW1/4 Section 14 and SE1/4 NE1/4 Section 15. Total production through November 1985 was 48,000 cubic yards.

Current producers in the Ocotillo district are Granite Construction (Shell Canyon area, also known as Fossil Canyon, the Ammex and Jimenez; Val-Rock; Cal-Grade (formerly Masters); and Imperial County. Val-Rock and Granite produce between 300,000 and 500,000 cubic yards annually. All of the operations are within habitat for the threatened Peninsular Bighorn Sheep.

REFERENCES CITED

Becker, R., Dec. 4, 1985, personal communication.

Christensen, A.D., July 1957, Part of the geology of the Coyote Mountain area, Imperial County, California: California Univ., Los Angeles, M.A. thesis, 188 p.

Dibblee, T.W., Jr., 1954, Geology of the Imperial Valley region, California, in Jahns, R.H., ed., Geology of southern California: California Division Mines Bull. 170, ch. 2, cont. 2, pp. 21-28, pl. 2, 1:375,000.

Gotti, Joe, Dec. 11, 1985 and Jan. 23, 1986, personal communication.

Grim, R.E., and Guven, N., 1978, Bentonites — geology, mineralogy properties and uses: Elsevier Scientific Publishing Co., Amsterdam, Netherlands.

Morton, P.K., 1977, Geology and mineral resources of Imperial County, California: California Division of Mines and Geology, County Rpt 7, 104 p.

O'Callaghan, Lloyd, Sr., Dec. 10, 1985, personal communication.

Orfanos, H.S., Oct. 18, 1985, personal communication.

Regis, A.J., Jan. 1978, Correlation between physical properties and exchangeable chemistry of bentonites from the western United States: Bureau of Land Management technical note T/N 313, 56 p.

__________, Feb. 1978, Mineralogy, physical and exchangeable chemistry properties of bentonites from the western United States, exclusive of Montana and Wyoming, Bureau of Land Management technical note T/N 315, 35 p.

Ross, C., and Shannon, P.J., 1926, The minerals of bentonite and related clays and their physical properties: Jour. American Ceramic Soc., V. 9, No. 2, 79 p.

Ryman, M.E., 1964, Validity determination, Bolsa de Oro #5 Coyote Clay #3-#4 placer claims of Ross O'Callaghan, et al., involving mineral trespass, BLM mineral report, 3 p., map.

__________, Dec. 1972, Production table for sand and gravel removed by W.H. Raley under Contract R 05668 from NW1/4 Sec. 14, T.16 S., R.9 E. for the period Aug. 1964-Dec. 1972.

Sampson, R.J. and Tucker, W.B., 1942, Mineral resources of Imperial County, in Thirty-eighth Rept. Of the State Mineralogist: California Division of Mines, pp. 104-145, Plate.

Sharp, R.P., 1975, Field guide, southern California, K/H geology field guide series, Kendall/Hunt Publishing Co., Dubuque, Iowa, 208 p.

Strand, R.G., 1962, Geologic map of California, San Diego-El Centro sheet: California Division of Mines and Geology, 1:250,000.

Taylor, H.A., Jr., June 2003, Industrial minerals: Mining Engineering, pp. 15-56.

Weber, F.H., Jr., 1963, Geology and mineral resources of San Diego County Rept. 3, 309 p.

The Emplacement of the Pegmatite Fields of San Diego and Riverside Counties: A Regional Perspective

Jeffrey E. Patterson, *University of Calgary*

The gem-bearing, highly fractionated, granitic pegmatites of southern California are of the rare-element LCT (Lithium, Cesium, Tantalum) Complex classification (Cerny, 1991a), meaning that they can be considered as the end members of the fractionation trend for sub-alkalic rocks (Bowen, 1928). That the complete trend from gabbro to lithium pegmatite exists in southern California has been clearly established at the Little 3 Mine in Ramona (Patterson, 1997a,b) and will serve as the basis for observations during this field trip. Previous county field trips have visited the gem-bearing pegmatites southwest of the Elsinore Shear Zone (ESZ) in Pala, Mesa Grande, and Ramona, which are related to the I-type, oceanic affinity, granites (Jahns, 1954c, Jahns, 1961, Chappell and White, 1974, Jahns, 1979, Todd and Shaw, 1985, Shigley, et al., 1986, Foord, et al., 1991). This trip will visit two gem-bearing pegmatites, that are associated with the S-type, continental affinity, two mica granites. Characteristic of this type of granite is the La Posta pluton, parent to the Sacatone Springs pegmatites. These pluton-pegmatite pairs differ in their chemistry and internal structure from their counterparts to the southwest (Chappell and White, 1974, Todd and Shaw, 1985, Walawender, et al., 1990, Taylor and Wise, 1996, Kennedy and Morton, 2003). The purpose of this paper is to present a new approach to the emplacement of these pegmatite systems by incorporating the concepts of plate tectonics and fracture mechanics that were either not available to, or not incorporated by, the previous investigators (Waring, 1905, Jahns and Wright, 1951, Jahns, 1979).

The geographical relationship of the pegmatite districts with the fault systems of the Peninsular Ranges Batholith (PRB) has been observed and published since the heyday of gem tourmaline mining in the region (Fairbanks, 1893, Merrill, 1915, Figure 1). Yet, the tectonic relationships between the pegmatites and faults are still not well understood. In general, LCT pegmatites appear to have been emplaced systematically with relation to calc-alkaline granitic plutons found in continental margin subduction zones (Cerny, 1991b). Since subduction zones are located at island arcs, continental margins, or continental sutures. This has direct bearing on the regional stress regime at the time of emplacement, Faure (2001). The apparent relationships, shown in Figure 1, between most of the parent plutons and their pegmatite districts shows that most of the districts are and on the continental side rather than the trench, given an oblique subduction direction at the time of emplacement (Atwater, 1970). Therefore, consideration of the regional geologic history is critical to establish the parameters that affected the ascent and emplacement of the southern California pegmatite systems. Since LCT pegmatites are *generally not* found within their parent granites (Cerny, 1991b, Pezzotta, 2000), we are fortunate on this field trip to examine two dyke systems that *do* appear to be emplaced within the marginal facies of their respective parent plutons.

A simplified regional perspective on the strikes and dips of the pegmatite districts and generalized dating pattern for the plutons and their respective pegmatites of the PRB is shown on Figure 2. What is distinctly noticeable is the trend toward younger ages to the south and east, especially for the pluton-pegmatite pairs. Also clearly evident is the primarily north-striking/east-dipping trend of the major districts northeast of the ESZ. Since several methods have been used to date the various rock units, each unit cannot be compared directly, but the general trend serves well for this field trip (Bushee, et al., 1963; Earl, 1965; Clinkenbeard and Walawender, 1989; Snee and Foord, 1991). Table 1 provides a simplified timeline to follow the critical geologic events which are discussed below.

The scenario of pegmatite emplacement begins with a depiction, from a tectonic simulation, of Rodinia, the postulated Proterozoic landmass (Scotese, 2002). Within pegmatite studies, the provenance and protolith has been given much speculation. This scenario is presented, starting far enough back in geological history, as to allow for a sufficient number of events and processes to occur so that the provenance and protolith can be well established. Figure 3 provides a starting point when the west coast of North America was apparently connected to Australia and Antarctica (Doughty, et al., 1998).

This west coast of North America is postulated to have rifted in the late Proterozoic, as shown in Figure 4. It is quite reasonable to assume that a stable deep water continental margin was created, as these sediments contain limestone lenses with fossils indicative of a sulfide/boron-rich "black smoker" environment (Silberling, et al., 1961). Subsequently, the detrital materials that make up the Bedford Canyon formation and the Julian Schist were laid down

The Pegmatite Fields of Riverside and San Diego Counties

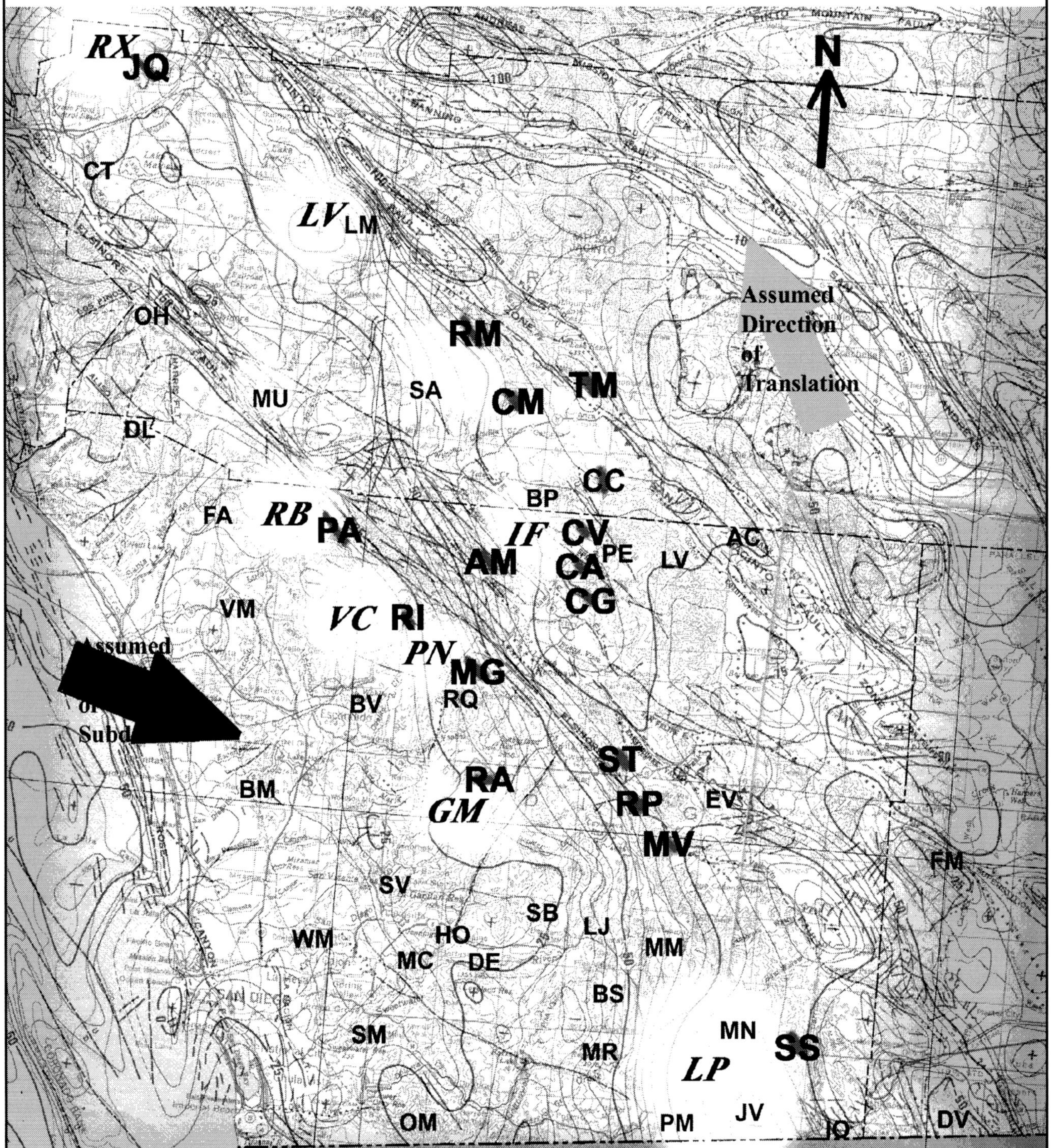

Legend: **CG** Gem Bearing Pegmatites

BV Boron, Feldspar, & Tin Pegmatites

IF Probable Parental Granite for Pegmatites

To Pegmatite Dykes of Baja California, Mexico

Scale: 1" ~ 12 miles (2.5 cm ~ 20 km)

Figure 1

The Pegmatite Fields of Riverside and San Diego Counties
Approximate Dates, Strikes, and Dips

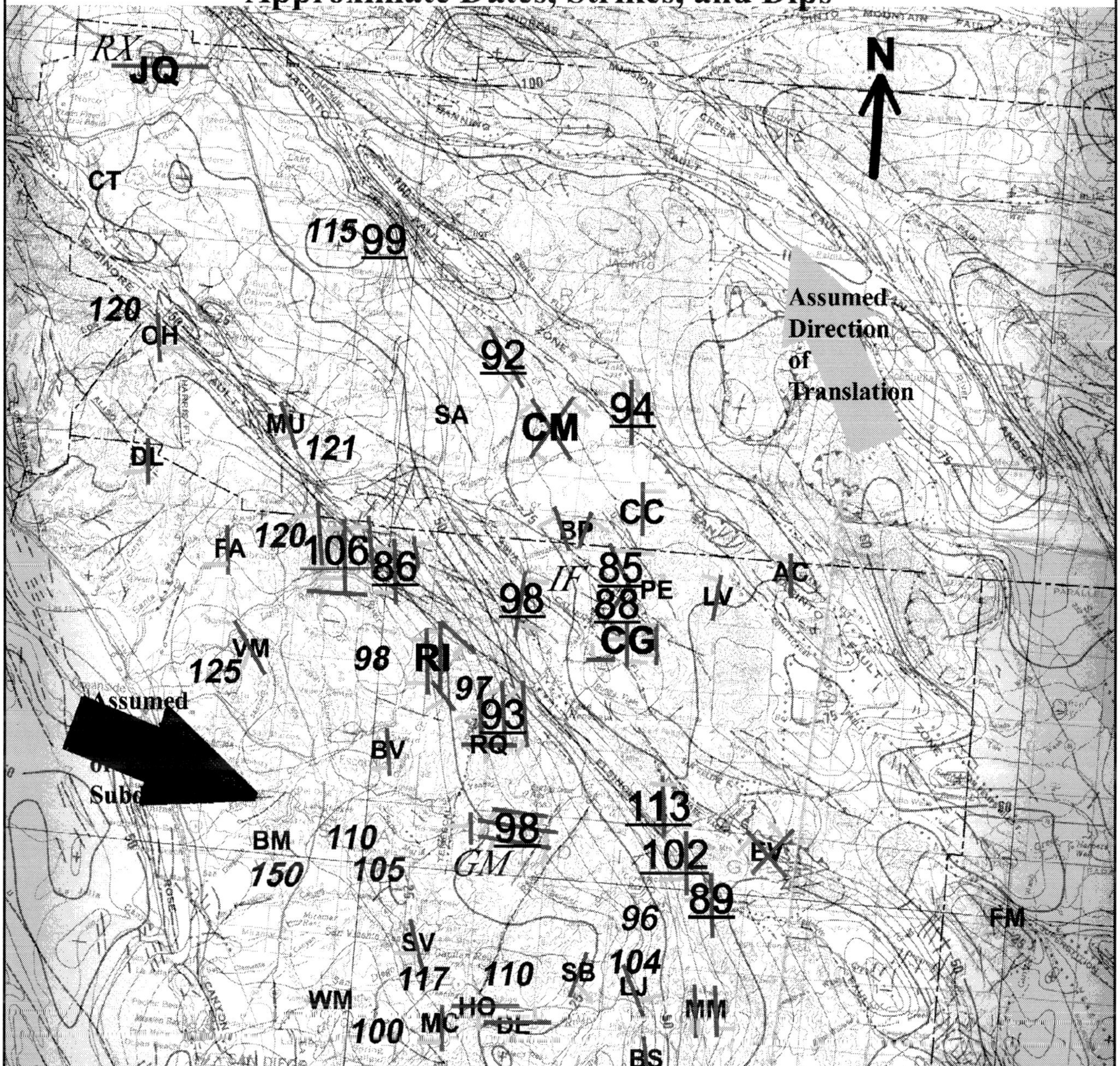

Legend: **106** ma Approximate Dates of Pegmatites

121 ma Approximate Dates of Related Plutons

| General Strike and —— General Dip of Pegmatites

Scale: 1" ~ 12 miles (2.5 cm ~ 20 km)

To Pegmatite Dykes of Baja California, Mexico

107 86

Figure 2

Index for Figures 1 & 2. **Gem-Bearing Pegmatite Districts**: AM = Aguanga Mountain; CA = Carmellita Mine; CC = Coyote Canyon; CG = Cryo-Genie Mine; CM = Cahuilla Mountain; CV = Chihuahua Valley; JQ = Jensen Quarry; MG = Mesa Grande; MV = Mason Valley; PA = Pala; RA = Ramona; RI = Rincon; RM = Red Mountain; RP = Royal Pandora Mine; SS = Sacatone Springs; ST = Starr Prospect; TM = Thomas Mountain. **Probable Parental Plutons**: GM = Gower Mountain; IF = Indian Flats; LV = Lakeview Mountains; LP = La Posta; PN = Pine Mountain; RB = Rainbow Granite; RX = Rubidoux Granite; VC = Valley Center. **Boron, Feldspar, and Tin Pegmatites**: AC = Alder Canyon; BP = Beauty Peak; BS = Buckman Springs; BV = Bear Valley; CT = Cajalco Tin Mine; DE = Dehesa; DL = De Luz; DV = Davies Valley; EV = Earthquake Valley; FA = Fallbrook; FM = Fish Creek Mountains; HO = Hoover Deposit; IO = In-ko-pa Overlook; JV = Jewel Valley; LJ = Laguna Junction; LM = Lakeview Mountains; LV = Lost Valley; MC = McGinty Quarry; MM = Metal Mountain; MR = Morena Reservoir; MU = Murietta; OH = Ortega Highway; OM = Otay Mountain; PE = Pearson Deposit; PM = Pacific Feldspar Mine , RQ = Rose Quartz; SA = Sage; SB = Spanish Bayonet; SM = San Miguel Mountain; SV = San Vincente; VM = Vista Moosa Canyon; WM = Cowles Mountain. (Base Map adapted from Oliver, 1980.)

Table 1- SIMPLIFIED GEOLOGIC TIME LINE: RIVERSIDE and SAN DIEGO COUNTIES

AGE ma	Geology	Reference
1200	RODINIA – Rifting of North America - Antarctica	Scotese, 2001
1000	Erosion of White Picacho Pegmatites – Lithium and Beryllium	Jahns, 1952
Ordovician	Limestone – Conodonts – San Marcos, Baja	Lothringer, 1984
Missippian	Limestone – Hemet – Riverside County	Schwarcz, 1969
Triassic	Bedford Canyon Formation (950 Zircon)	Larsen, 1948
Triassic	Julian Schist (Triassic Amminoid Imprint) (750 Zircon)	Hudson, 1922
234	Harper Creek Gneiss	Thomson & Girty, 1994
165 – 120	Santiago Peak Volcanics	Anderson, 1991
140	Closure of Back arc Basin (Model)	Gastil, 1983
130 – 105	Emplacement of Western PRB into BCF and JS	Gromet & Silver, 1967
106–95	**Emplacement of Pegmatites of Pala and Western PRB**	**Earl,1965, Snee&Foord,1991**
110 – 85	Emplacement of Eastern PRB into JS	Walawender, et.al., 1990
97-95-93	La Posta Pluton – Hornblende Biotite - Granodiorite	Clinkenbeard, 1987
90	**Emplacement of Packrat Pegmatite, Sacatone Springs**	**Snee & Foord,1991**
88	**Emplacement – Carmellita Pegmatite, Chihuahua Valley**	**Earl, 1965**
85	**Emplacement – Blue Lady Pegmatite, Chihuahua Valley**	**Earl, 1965**
85 – 75	Uplift and Erosion – Lusardi Formation	Nordstrom, 1970
65 - 40	Erosion – Point Loma Formation	Abbott, 1999
35 - 30	Peneplain Deposition – Poway Conglomerate	Bellemin & Merriam, 1958
30	Beginning of Rifting of Gulf of California	Atwater, 1970
18.7 +/- 1.3	Jacumba Andesite and Dacite	Hawkins, 1970
8.3 +/- .5	Emplacement of Murietta Basaltic Flows	Hawkins, 1970
5.5	**Formation and Transport of Elsinore Shear Zone**	**Atwater, 1970**
2.4	Uplift of Agua Tibia – Mt. Palomar Block	Magistrale & Rockwell, 1996

as deep water turbiditic sediments, with significant quartzitic layers, starting in the early Paleozoic as indicated by detrital zircons (Stewart, 1972, 1988, and Figure 5). The eroded sediments, enriched in lithium and beryllium, from the White Picacho Pegmatites would also have been part of this sequence, as shown in Figure 5.

By Permian time, Figure 6, the North American Trench had most probably been established, possibly due to the drastic change in plate movement caused by the continental flood basalts in Siberia. Whatever the reason, the Farallon Plate began to subduct and the series of hemipelagic clays that had been deposited off the western coast of North America were metamorphosed to the schists, quartzites, and calc-silicate lenses of the Bedford Canyon and Julian Schist formations, ranging from greenschist to sillimanite-amphibolite facies. The proposed enrichment of these

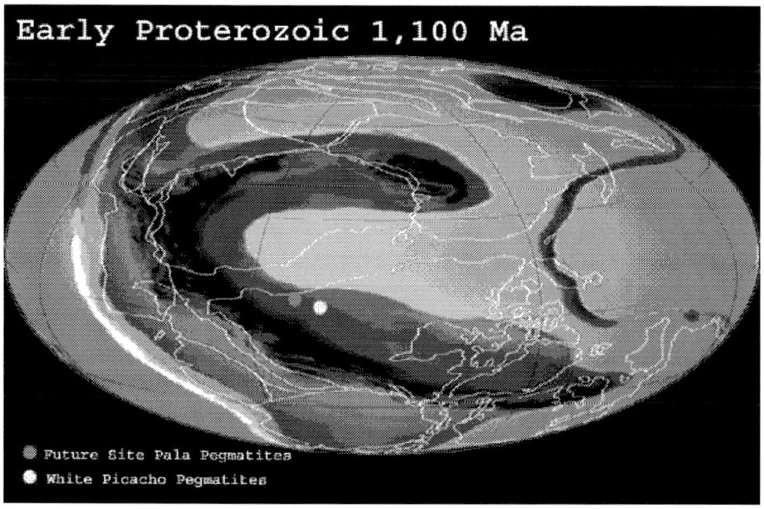

Figure 3. Simulation of Rodinia during the Early Proterozoic.(After a figure located at: http://www.scotese.com/precambr.htm.)

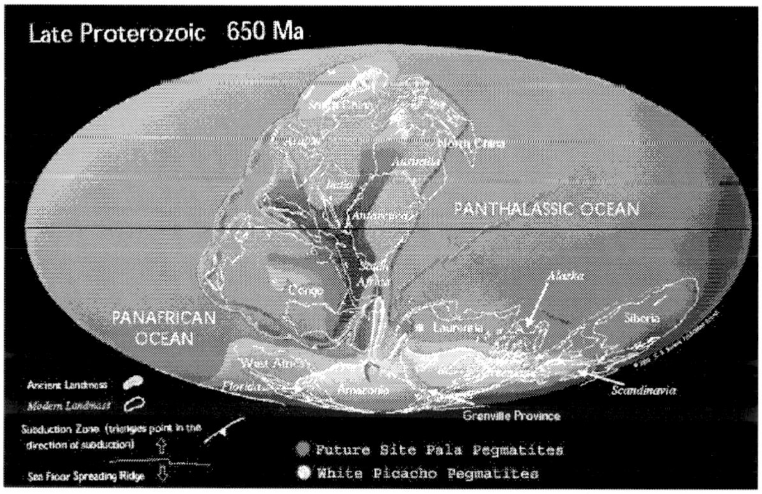

Figure 4. Simulation of the conditions during rifting of Laurentia (After a figure located at: http://www.scotese.com/precambr.htm.)

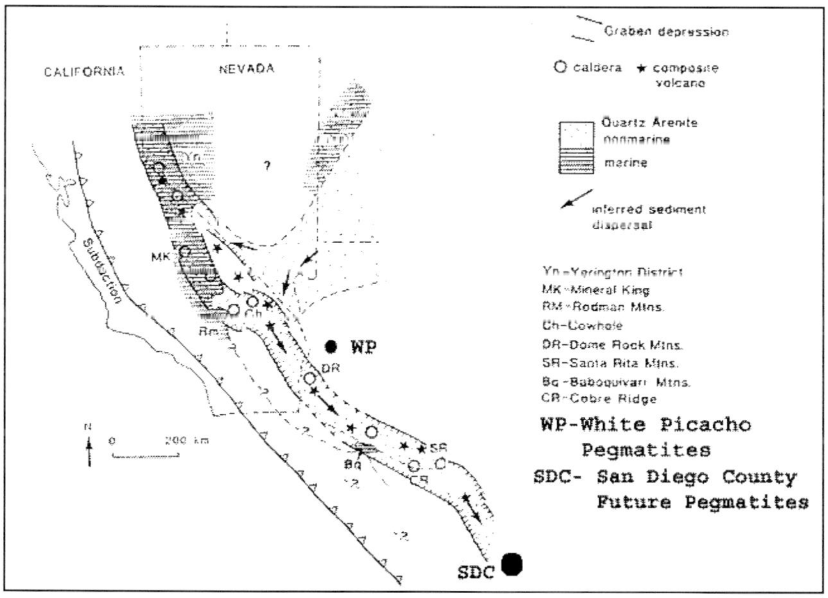

Figure 5. Possible path of Lithium and Beryllium bearing sediments along the coast of North America during the Pre-Cambrian to the Cretaceous. (After Busby-Spera, 1988).

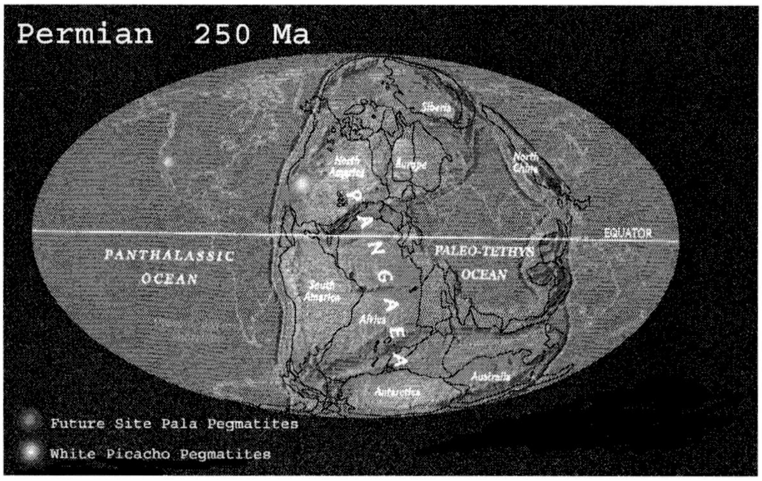

Figure 6. Simulation of the conditions assumed during the initiation of the North American Trench. (After: National Geographic, 2002.)

sediments in lithium, beryllium, or boron is apparently undocumented. However, the enrichment of the lithosphere above the subducting slab in this region is well documented (You, 1994). Thus, a plausible source protolith existed.

The emplacement of the overlying Santiago Peak Volcanics (SPV) sequence, followed by the plutons and pegmatites of the PRB during the late Cretaceous, is well documented (Hudson, 1922, Larsen, 1948, Earl, 1965, Fife, et al., 1967, Gastil, 1983, Clinkenbeard, 1987, Todd, et al., 1988, Anderson, 1991, Snee and Foord, 1991, Herzig, 1991, Walawender, et al., 1991, Todd, 1995). Figure 7 provides an assumed image of the global plate orientation at that time. Figure 8 provides a sketch of an assumed crustal cross-section of the PRB between 106 and 85 ma. Subsequently, the entire region was uplifted and began to shed detritus to the west (Nordstrom, 1970, Todd, 1995, Abbott, 1999). The pegmatites of the western PRB were on the surface by this time, as clasts of pegmatite can be found in the Paleocene Lusardi Formation (Nordstrom, 1970).

One of the concepts developed by plate tectonic theory is the sequence of geologic events during subduction. These include driving off of water and volatiles between 60 to 100 km depth, due to the collapse of hydrous minerals, like biotite, and the bringing in heat from the Asthenosphere by corner flow of the overlying mantle wedge (Figure 9). If a slab is subducting at a nominal 30 degree angle, then a band of volcanic activity would form about 120 to 200 km inboard from the trench with regularly spaced volcanic centers at about 70 km separation. Subsequently, a band of plutons would form slightly inboard of this volcanic chain. Should the subducting slab acquire an undulating or faulted surface, the orientation of the volcanic centers would follow this shape as well (Moores and Twiss, 1995). As a subducting environment matures, the sequence of emplacement shifts from volcanic-extrusive to plutonic-intrusive (D'Lemnos, 1992, Moores and Twiss, 1995). The additional water, volatiles, and heat from these processes aid in the anatectic melting of the overlying

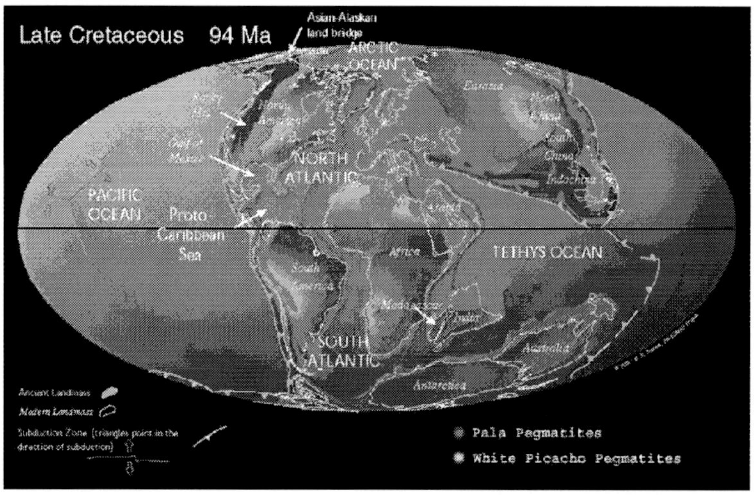

Figure 7. Simulation of plate boundary conditions assumed during the emplacement of the PRB. (After a figure located at: http://www.scotese.com/cretaceo.htm.)

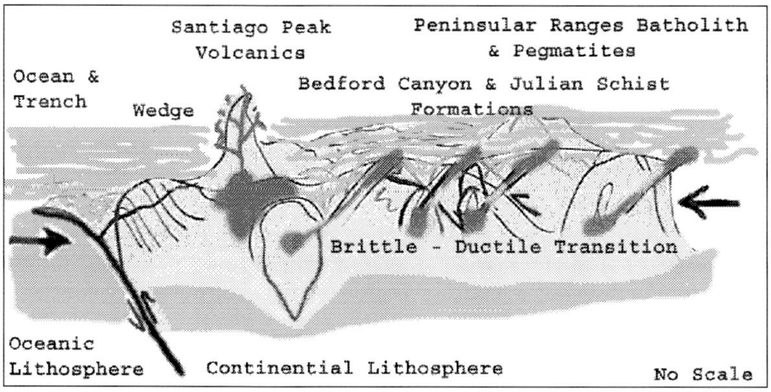

Figure 8. Cross-sectional sketch of the for a typical parent pluton – pegmatite field in the regional compressive environment assumed to be active during the late Cretaceous. The large Black arrows depict the assumed compressional condition at the time of pegmatite emplacement. The Pegmatites are represented in Red. Vertical scale is exaggerated.

sedimentary formations and produce the large range of fractionated melts that we see in the region.

During the Eocene, the entire region was apparently a vast peneplain, with an extensive covering of the Poway Conglomerate, a fluvial conglomerate of exotic clasts of rhyolite and quartzite from a source in Sonora, Mexico (Bellemin and Merriam, 1958, Woodford, et al., 1968, Abbott and Smith, 1989). Following the Eocene, the sequence of events leading to the rifting and opening of the Gulf of California began (Atwater, 1970, Abbott, 1999). Prior to this time, no major dextral northwest-trending faults existed (Atwater, 2003). For additional information, an animated depiction of the Post-Cretaceous plate movement can be found at: http://www.geol.ucsb.edu/faculty/atwater/.

During the Pliocene the ESZ, as a coupled shear zone with the San Andreas fault system, emerged (Magistrale and Rockwell, 1996, Atwater, 2003). The relationship between the ESZ and the pegmatite districts now becomes evident. Previous workers assumed that there was some inherent crustal weakness that allowed the ESZ and related faults to form and indeed there is. When the stress regime shifted from left-lateral compression to right-lateral extension, the coupled lateral motion produced a northwest directional transtension that pulled the blocks of southern California apart in shear (Dickinson, 1996). The failure occurred through the weakest zones, which in this case lies behind the line of major plutons and their respective pegmatite districts. The transpression of the Santa Ana, Perris, and Elsinore-Temecula blocks, during

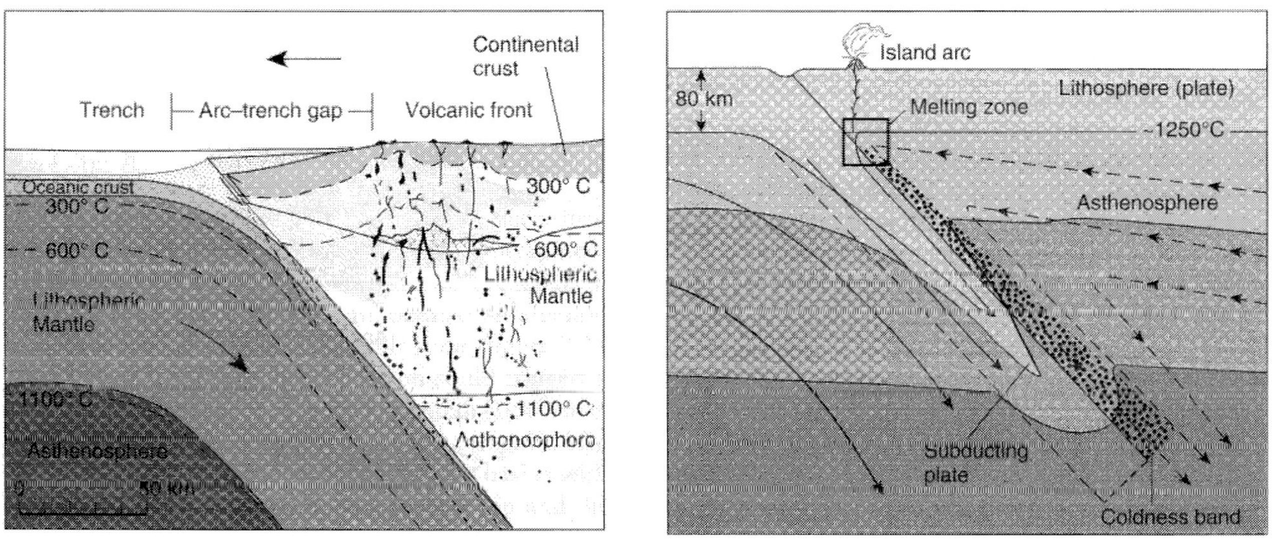

A: Volatile Ascent -
Assisting Anatectic Melting
after Ernst, et.al., 1970

B: Corner Flow -
after Marsh, 1979

Figure 9. Water, volatiles, and additional heat from the Asthenosphere create the proper conditions for anatectic melting of the overlying sediments.

their subsequent lateral translation northwestward, created a condition of thrusting, which led to the uplift of Palomar (Smiths) Mountain (Magistrale and Rockwell, 1996). This is essentially the topography that we see today.

For the past 100 years, the accepted thought has been that the southern California pegmatites were intruded into unique sets of preferentially oriented cooling cracks in the host gabbros (Waring, 1904; Anderson, 1931; Donnelly, 1936; Jahns, 1946, 1954a,b,c, 1961, 1979; Hanley, 1951; Jahns and Wright, 1951; Simpson, 1960, 1965; Shigley, 1982; Herbert, 1982; Torres, 1982; Marcusson, 1985a,b; Shigley, et al., 1986; Houser, 1986; Foord, 1976; Foord, et al. 1991; Laurs, 1991; Crother, 1992, Bloomfield, 1997; Webber, et al. 1999, and others) — which are apparently only exposed where filled by the pegmatites (Hanley, 1951). As recently as 1979, Jahns states:

> "The pegmatites, which represent the youngest igneous activity within the batholith complex, typically form very thin to fairly thick dikes with northerly trends and gentle to moderate westerly dips. With few exceptions, the gem-bearing and closely associated dikes are confined to plutons of gabbroic rocks and mafic tonalites. This intimate relationship with relatively early members of the batholith assemblage is ascribable to structural control.
>
> The pegmatites *plainly* were emplaced along well-developed sets of fractures that not only are independent of primary structural features in the enclosing rocks but also transect contacts between major rock units. These fractures evidently represent regional cooling and contraction within the batholith as a whole, and they appear to have been formed only in plutons that were essentially unjointed or unfractured during earlier stages in their respective emplacement and cooling histories."

It can be deduced that the intimate relationship of the pegmatites is to their parent granites, rather than to the batholith as a whole. Each parent-pegmatite pair follows the same regional age trends to the south and east, with the pegmatites being emplaced about 5 to 15 ma after the parent granite. The pegmatites are not the youngest igneous activity within the PRB, as the pegmatites of Pala, Ramona, and Aguanga Mountain are older than the La Posta and Indian Flats plutons. It is suggested that a new countywide dating program be conducted to provide a consistent method of comparing the emplacement dates of the parent pluton — pegmatite pairs.

The observation that the gem-bearing pegmatites are primarily emplaced in gabbroic host rocks (Jahns, 1979) holds true for the dykes on Thomas Mountain near Anza, Queen Mountain in Pala, and Gem Hill in Mesa Grade, but it does not hold for the remainder of southern California. The dykes on Cahuilla Mountain are emplaced in the overlying schist (Geffner and Fisher, 1997). The dykes at Rincon, Ramona, Banner, and on Pala Chief Mountain, Hiriart Mountain at Pala, are in granodiorite or in the contact zone of the granodiorite and an overlying gabbroic or metasedimentary screen (Jahns and Wright, 1951, Hanley, 1951, Patterson, 1997a,b). The dykes of Chihuahua Valley and Sacatone Springs are emplaced within their parent granites (Taylor and Wise, 1996).

Regionally, the strike of the pegmatites is to the north with westerly dips. This holds true for many of the dykes south of the ESZ, where they strike northerly (+/- 20°) with westerly dips ranging 20°-60° and averaging 30° (Figures 2 and 10). However, things change in the Ramona district, where the pegmatites primarily strike N 50°-70° W with southwesterly dips, and then abruptly flex to the north with westerly dips at the mouth of the Hatfield Creek Valley, as exposed along Highway 78 (Figures 12 and 13). Further, the gem pegmatites north of the ESZ on Aguanga Mountain, north of the Indian Flats Pluton, and on Cahuilla Mountain, and some scattered feldspar pegmatites in the Cuyamaca mountains dip easterly (Figure 11). Some

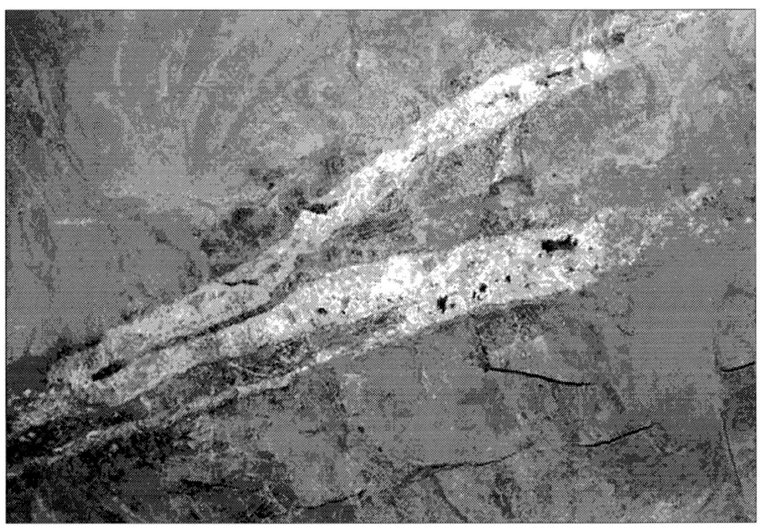

Figure 10. North-striking, west-dipping dyke juncture at the Himalaya Mine, western magazine drift, looking north. This is the typical strike and dip of the major gem-bearing pegmatites in the region. Note the dyke parallel thermal shock fracture below the dilating pegmatite zone.

Figure 11. North-striking, east dipping dyke at the Blue Lady Mine, Chihuahua Valley, looking northeast. The average dip of the pegmatites at this mine is 22° east. This property is currently under the authority of San Diego State University.

Figure 12. West-striking, southwest-dipping dyke at Highway 78, view looking northwest. This dyke is typical of those in the Ramona district. Note the rake of the fine lines of tourmaline on the center section beneath the geologic hammer handle. These lines represent the trend of intruding melt during hydraulic fracture.

Figure 13. North striking, west dipping continuation of the dyke at Highway 78, looking north. Small crystals of Aquamarine were recovered from the pocket next to the hammer. Same pegmatite as shown in Figure 12. Note the sharp upper and lower contacts, the zone of tourmaline at the center, and the thin aplite filled vein about the dyke.

exceptional stringers have also been observed to strike westerly and dip north or south.

That the pegmatites were "*plainly emplaced along well-developed sets of fractures*" does not bear the scrutiny of current regional observations. Field observations do not show any regional north-striking/west-dipping fracture systems on the surface or underground within the workings of the various gem mines (Hawk, 1980, Patterson, 1996). Rather, a second joint set is present. This set is related to the stress regime imposed on the batholith during the northwest transport during the opening of the Gulf of California (Walker, 1958, Figures 14 and 15). As the regional strike and dip of the major pegmatites is essentially unchanged across the batholith, from Pala to Sacatone Springs, there must have been a consistent uniform regional stress condition during the late Cretaceous.

Figure 14. West-dipping listric fracture pattern in granodiorite along Highway 67, Pala, looking north. These fractures formed during the Pliocene as a result of the rifting of the San Luis Rey river valley. The ESZ is about 10 km northeast.

Figure 15. Multi-directional joint sets in weathered boulders of the La Posta Pluton on the north side of Interstate 8 between Boulevard and Jacumba, looking northeast. No 30° west-dipping sets or mineral fillings are evident.

In general the principal stress direction in the crust at the continental margin of North America is horizontal (Turcotte, et al., 1977, Zoback and Zoback, 1980), which would hold closed any open asperities at the depth of emplacement (5–10 km, 1.5-3 kbar) of the LCT pegmatites (Pollard, 1976, London and Burt, 1982, Brown and Scholz, 1986). Under the compressional regime in the subduction zone, any significant seismic event could have exerted sufficient force upon the hardened carapace of the parental pluton to cause it to fail by hydraulic fracture (Anderson and Grew, 1977, Knapp and Norton, 1981, and Hibbard and Watters, 1985). Alternatively, chemical stress corrosion by water, boron, or fluorine could initiate fracture. Once the exterior carapace failed, the interior melt would quickly leave the parent pluton, perhaps by a filter-press mechanism, and flow in the surrounding country rock with sufficient over-pressure to dilate the pegmatites.

The *Quickfield* finite element analysis, shown in Figure 16, gives a close 2-D simulation of the pegmatite extrusion process, showing the external forces and resulting stress regime and deformation. The boundary conditions for this analysis assumes a compressional environment with 2 kbar vertical, 3 kbar left horizontal, and 2.5 kbar right horizontal. Either a magnitude 7 earthquake or stress corrosion by water, boron, and fluorine is assumed to initiate the hydraulic fracture of the

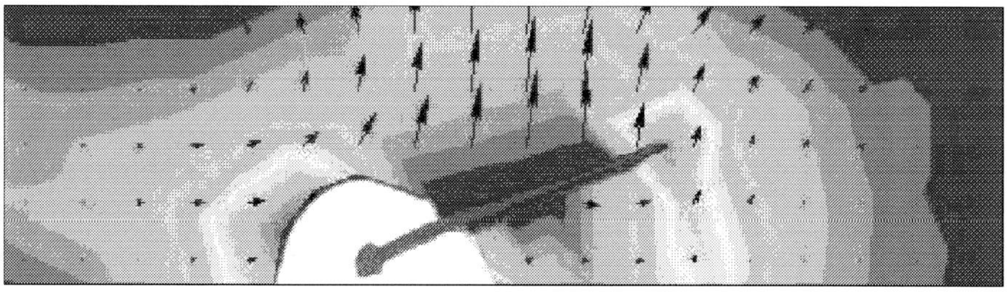

Figure 16. Quickfield 2-D finite element analysis image of a parental pluton, shown in white, under compression at 2 kbar vertical, 3 kbar left horizontal, and 2.5 kbar right horizontal, with an additional magnitude 7 earthquake to initiate fracture. Pegmatite melt is shown in red.

surrounding host rocks. Note that the departing angle is roughly 23°, which is slightly less than the average dip of the regional pegmatites. The average dip angle of the pegmatites at the Blue Lady Mine is 22° to the east.

The nature of the dilation of the very high aspect ratio dykes, with bridges, spurs, straps, horns, cusps, and dyke parallel fractures, suggests that these pegmatite systems were emplaced in hydraulically fractured en-echelon arrays and supports the contention that there were no pre-existing fractures (Gorlov, 1973, Delaney, et al., 1986, Engelder, 1987, Engelder, et al., 1993, Van Horn, et al., 1999, Van Horn and Stephan, 2001). Examples of these dilatational features are shown in Figures 17, 18, and 19. The dyke parallel and perpendicular fractures, interpreted

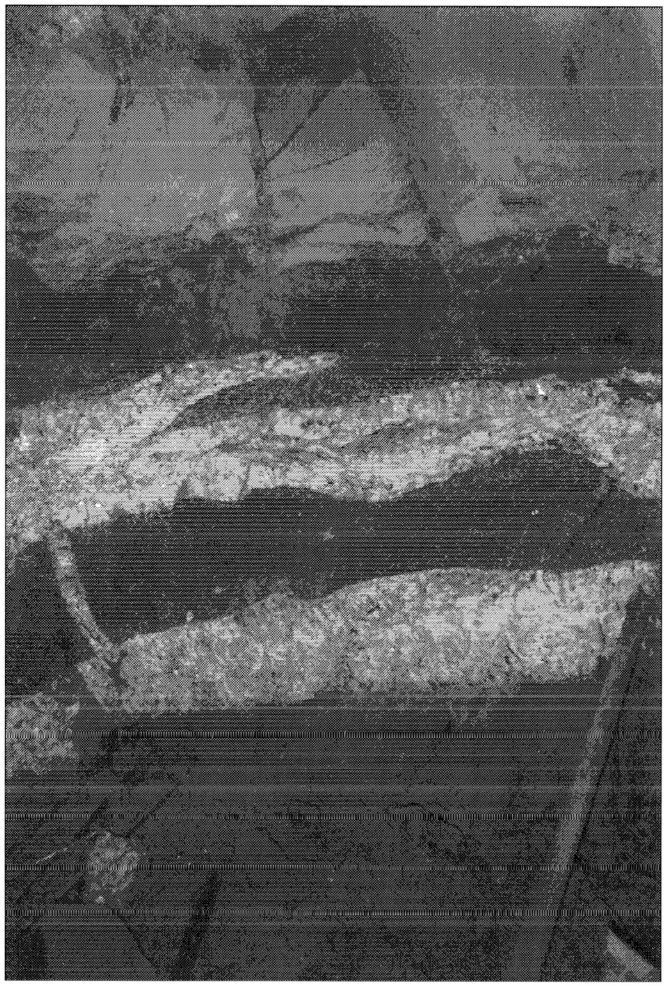

Figures 17 and 18. North-striking, west-dipping, multiple dyke segments of the main gem-bearing pegmatite at the Himalaya Mine, looking north. Both figures show dilational features found in hydraulically fractured rock, a common condition at the Himalaya Mine. The central zone of the upper dykes, which pinch-out updip, contain flow-banded tourmaline-rich albite segregations that flow into the continuing lower dyke. Gem tourmaline pockets were found, by GPR, on the north and updip sides of both locations.

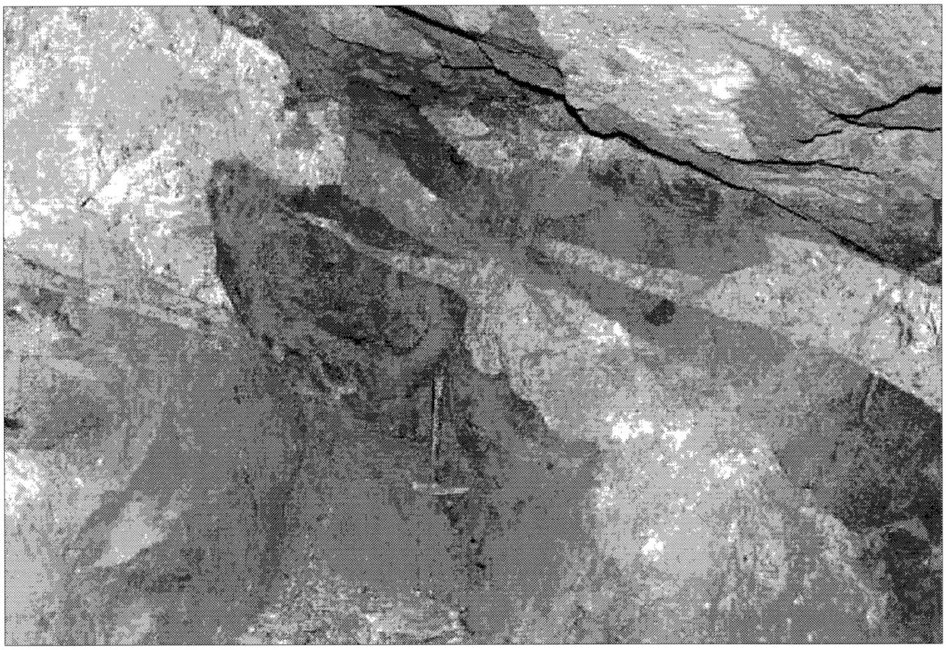

Figure 19. Multiple dilational features of the Himalaya dyke, downdip of the major gem tourmaline bearing President's Day Pocket zone, looking south. Note the dilating horns and straps, and the cuspate embayed lower contact near the geologic hammer. At this point, two dykes join from downdip to form the 1.7 m thick continuation visible on the left.

by others as nonfilled regional joints and the cause of pocket formation, are more likely thermal shock fractures or process zone fractures created during the initial hydraulic fracturing and melt emplacement (Marcusson, 1985a,b, Houser, 1986, Pollard, 1987).

The concepts presented in this paper provide a brief review of current geologic thinking with regard to the emplacement of the pegmatite systems in southern California and their relationships to the Elsinore Shear Zone and the Peninsular Ranges Batholith as a whole. The geologic timeline has demonstrated that a plausible lithium, boron, beryllium enriched protolith existed. The plate tectonic and fracture mechanics approach has provided a reasonable scenario for the hydraulic fracturing of the host rocks along with extrusion of the pegmatite melts. Subsequent uplift and rifting of the batholith during the Pliocene, and lateral translation created the ESZ as we see it today. Thus demonstrating that the relationship of the pegmatite districts to the regional faults is due to the inherent crustal weakness created behind the line of obliquely emplaced parent granite-pegmatite pairs.

ACKNOWLEDGMENTS

Financial support was provided by the Natural Sciences and Engineering Research Council of Canada in the form of a Research Grant to my advisor at the University of Calgary, Frederick A. Cook, Ph.D., Department Geology and Geophysics, and by a grant from the E.E. Patterson Family Trust. Jim Means and John Watson provided invaluable assistance in review.

REFERENCES

Abbott, P.L., 1999, *The Rise and Fall of San Diego: 150 million years of history recorded in sedimentary rocks*, Sunbelt Publications, Inc., San Diego, California, 231 p.

Abbott, P.L., and Smith, T.E., 1989, *Sonora, Mexico, source for the Eocene Poway Conglomerate of Southern California*, Geology, 17, 329-332.

Anderson, C.L., 1991, *Zircon Uranium-Lead Isotopic Ages of the Santiago Peak Volcanics and Spatially Related Plutons of the Peninsular Ranges Batholith, Southern California*, unpublished M.Sc. Thesis, San Diego State University, 103 p.

Anderson, O., 1931, *Discussions of certain phases of the genesis of pegmatites*, Norsk Geologisk Tidsskrift, 12, 1-56.

Anderson, O.L. and Grew, P.C., 1977, *Stress corrosion theory of crack propagation with applications to geophysics*, Reviews of Geophysics and Space Physics, 15, 77-104.

Atwater, T., 1970, *Implications of Plate Tectonics for the Cenozoic Evolution of Western North America*, Geological Society of America Bulletin, 81, 3513-3535.

Bellemin, G.J., and Merriam, R.H., 1958, *Petrology and Origin of the Poway Conglomerate*, Geological Society of America Bulletin, 69, 199-220.

Bloomfield, M.J., 1997, *Gem tourmaline pegmatite deposits. A chemical comparison between tourmalines from two LCT-family pegmatites and their role as petrogenetic indicators*, unpublished M.Sc. Thesis, University of Leicester, England, 149 p.

Bowen, N.L., 1928, *The Evolution of Igneous Rocks*, New York, Princeton University Press, 334 p.

Brown, S.R., and Scholz, C.H., 1986, *Closure of Rock Joints*, Journal of Geophysical Research, 91, B5, 4939-4948.

Busby-Spera, C.J., 1988, *Speculative tectonic model for the early Mesozoic arc of the southwest Cordilleran United States*, Geology, 16, 1121-1125.

Bushee J., Holden, J., Geyer, B., and Gastil, G., 1963, *Lead-Alpha Dates for some Basement Rocks of Southwestern California*, Geological Society of America Bulletin, 74, 803-806.

Cerny, P., 1991a, *Rare-element Granitic Pegmatites, Part I: Anatomy and Internal Evolution of Pegmatitic Deposits*, Geoscience Canada, 18, no. 2, 49-67.

Cerny, P., 1991b, *Rare-element Granitic Pegmatites, Part II: Regional to Global Environments and Petrogenesis*, Geoscience Canada, 18, no. 2, 68-82.

Chappell, B.W., and White, A.J.R., 1974, *Two Contrasting Granite Types*, Pacific Geology, 8, 173-174.

Clinkenbeard, J.P., 1987, *The mineralogy, geochemistry, and geochronology of the La Posta pluton, San Diego and Imperial Counties, California*, unpublished M.Sc. Thesis, San Diego State University, 219 p.

Clinkenbeard, J.P. and Walawender, M.J., 1989, *Mineralogy of the La Posta Pluton: Implications for the Origin of Zoned Plutons in the Eastern Peninsular Ranges Batholith, Southern California*, American Mineralogist, 74, 1258-1269.

Crother, R.A., 1992, *Host Rock Lithologies and Structural Constraints on Complex Pegmatites in the San Jacinto Mountains of California*, unpublished B.Sc. Thesis, California State Polytechnic University, Pomona, 40 p.

Delaney, P.T., Pollard, D.D., Ziony, J.I., and McKee, E.H., 1986, *Field Relations Between Dikes and Joints: Emplacement Processes and Paleostress Analysis*, Journal of Geophysical Research, 91, B5, 4920-4938.

Dickinson, W.R., 1996, *Kinematics of Transrotational Tectonism in the California Transverse Ranges and Its Contribution to Cumulative Slip Along the San Andreas Transform Fault System*, Geological Society of America, Special Paper 305, 46 p.

D'Lemos, R.S., Brown, M. and Strachan, R.A., 1992, *Granite magma generation, ascent, and emplacement within a transpressional orogen*, Journal of the Geological Society, London, 149, 487-490.

Donnelly, M.G., 1935, *The lithia pegmatites of Pala and Mesa Grande, San Diego County, California*, unpublished Ph.D. Thesis, California Institute of Technology, Pasadena, California, 235 p.

Doughty, P.T., Price, R.A., and Parrish, R.R., 1998, *Geology and U-Pb geochronology of Archean basement and Proterozoic cover in the Priest River Complex, northwestern United States, and their implication for Cordilleran structure and Precambrian continent reconstructions*, Canadian Journal of Earth Sciences, 35, 1, 39-54.

Earl, J.L., 1965, *X-Ray Fluorescence Rubidium : Strontium Age Determination of Minerals from the Southern California Batholith*, unpublished M.Sc. Thesis, San Diego State University, 115 p.

Engelder, T., 1987, *Joints and Shear Fractures in Rocks*, in Atkinson, B.K., ed., Fracture Mechanics of Rock, Academic Press, Inc., Orlando, Florida, Chapter 2, 27-69.

Engelder, T., Fischer, M.P., and Gross, M.R., 1993, *Geological Aspects of Fracture Mechanics*, Geological Society of America Short Course Notes, Annual Meeting, Boston, 281 p.

Ernst, W.G., Seki, Y., Onuki, H., and Guilbert, M.C., 1970, *Comparative study of low-grade metamorphism in the California Coast Ranges and outer metamorphic belt of Japan*, Geologic Society of America, Memoir 124, 276 p.

Fairbanks, H.W., 1893, *Geology of San Diego County, also of portions of Orange and San Bernardino counties*, California State Mining Bureau, 11th Annual Report of the State Mineralogist, 76-120.

Fife, D.L., Minch, J.A., and Crampton, P.J., 1967, *Late Jurassic Age of the Santiago Peak Volcanics, California*, Geological Society of America Bulletin, 78, 299-304.

Foord, E.E. 1976, *Mineralogy and Petrogenesis of layered Pegmatite-aplite Dikes in the Mesa Grande District, San Diego County, California*, unpublished Ph.D. Thesis, Stanford University, Stanford, California, 326 p.

Foord, E.E., London, D., Kampf, A.R., Shigley, J.E., and Snee, L.W., 1991, *Gem-Bearing Pegmatites of San Diego County, California*, in Walawender, M.J. and Hanan, B.B. (eds.), Geological excursions in Southern California and Mexico, San Diego State University, Department of Geological Sciences, 128-146.

Gastil, R.G., 1983, *Mesozoic and Cenozoic granitic Rocks of Southern California and Western Mexico*, Geological Society of America, Memoir 159, 265-275.

Geffner, P., and Fisher, J., 1997, *Recent Work at the Lithia Dike Claim, Cahuilla Mountain, Riverside County, California*, Rocks and Minerals, 72, 156-161.

George, P.G., 1993, *Tectonic implications of fission-track thermochronology and amphibole thermobarometry studies of the northern Peninsular Ranges batholith, Southern California*, unpublished Ph.D. dissertation, Louisiana State University, 195 p.

Gorlov, N.V., 1973, *Structural control of pegmatites in the White Sea Region*, Akademiya Nauk SSSR, Leningrad, 93 p.; in Russian, translated, Anisimov, I., University of St. Petersburg.

Hanley, J.B., 1951, *Economic Geology of the Rincon Pegmatites, San Diego County, California*; California Division of Mines Special Report 7B, 24 p.

Hawk, R.N., 1980, *A Geologic Reconnaissance and Stability Investigation of the Himalaya Mining Company Claim*, unpublished B.Sc. Thesis, San Diego State University, 21 p.

Herbert, E.L., 1982, *Clay Mineralogy of Granitic Pegmatites in the Pala District, San Diego County, California*, unpublished M.Sc. Thesis, San Diego State University, 73 p.

Herzig, C.T., 1991, *Petrogenetic and tectonic development of the Santiago Peak Volcanics, northern Santa Ana Mountains, California*, unpublished Ph.D. Dissertation, University of Riverside, Riverside, California, 376 p.

Hibbard, M.J., and Watters, R.J., 1985, *Fracturing and Diking in Incompletely Crystallized Granitic Plutons*, Lithos, 18, 1-12.

Houser, C., 1986, *Structural Controls and Mineralogical Indicators for the Formation and Distribution of Pockets in the Elizabeth R Mine, Pala, San Diego County, California*, unpublished Senior Thesis, San Diego State University, 43 p.

Hudson, F.S., 1922, *Geology of the Cuyamaca region of California with special reference to the origin of the nickeliferous pyrrhotite*, University of California, Department of Geology Science Bulletin, 13, 175-252.

Jahns, R.H., 1954a, *Pegmatites of Southern California*, in Geology of Southern California, California Division of Mines Bulletin 170, Chapter 7, Contribution 5, 37-50.

Jahns, R.H., 1954b, *Geology of the Peninsular Range Province*, in Geology of Southern California, California Division of Mines Bulletin 170, Chapter 2, Contribution 3, 29-52.

Jahns, R.H., 1954c, *Geologic Guide for the Northern Part of the Peninsular Range Province*, in Geology of Southern California, California Division of Mines Bulletin 170, Geologic Guide No. 5, 59 p.

Jahns, R.H., 1961, *Notes on the Pala and Mesa Grande Pegmatite Districts, San Diego County, California, Field Trip No. 3*, in Blakemore, E.T., ed., Guidebook for Field Trips, Geological Society of America, Cordilleran Section, Annual Meeting 1961, San Diego State College, 44-53.

Jahns, R.H., 1979, *Gem-bearing pegmatites in San Diego County, California: The Stewart Mine, Pala district and the Himalaya Mine, Mesa Grande district*, in Abbott, P. L., and Todd, V.R., eds., Mesozoic Crystalline Rocks: Peninsular Ranges batholith and Pegmatites, Sal Point Ophiolite, San Diego, California, San Diego State University, pp. 151-162.

Jahns, R.H. and Wright, L.A., 1951, *Gem and Lithium Bearing Pegmatites of the Pala District, San Diego County, California*; California Division of Mines Special Report 7A, 72 p.

Kampf, A.R., Gochenour, K., and Clanin, J., 2003, *Tourmaline Discovery at the Cryo-Genie Mine, San Diego County, California*, Rocks and Minerals, 78, 3, 156-168.

Kennedy, M.P., and Morton, D.M., 2003, *Preliminary Geologic Map of the Murrieta 7.5' Quadrangle, Riverside County, California*, U.S. Geological Survey, Open File Report 03-189, http://geopubs.wr.usgs.gov/open-file/of03-189/ .

Kimbrough, D.L., Smith, D.P., Mahoney, J.B., Moore, T.E., Grove, M., Gastil, R.G., Ortega-Rivera, A., and Fanning, C.M., 2001, *Forearc-basin sedimentary response to rapid late-Cretaceous batholith emplacement in the Peninsula Ranges of Southern and Baja California*, Geology, 29, 6, 491-494.

Knapp, R.B., and Norton, D.L., 1981, *Preliminary numerical analysis of processes related to magma crystallization and stress evolution in cooling pluton environments*, American Journal of Science, 281, 35-68.

Kunz, G.F., 1905, *Gems, jeweler's materials, and ornamental stones of California*; California State Mining Bureau Bulletin 37, 155 p.

Larsen, E.S., Jr., 1948, *Batholith and Associated Rocks of Corona, Elsinore, and San Luis Rey Quadrangles, Southern California*, Geological Society of America Memoir 29, 182 p.

Laurs, B.M., 1991, *The Metasomatic effects of the Emplacement and Crystallization of the Himalaya Pegmatite-Aplite Dike System on the Gabbronorite Wallrock, San Diego County, California*, in Discovery, University of California at Santa Barbara Journal of Undergraduate Research, 14, 138 p.

Lothringer, C.J., 1984, *Geology of a Lower Ordovician Allochthon, Rancho San Marcos, Baja California, Mexico*, in Frizzell, V.A., Jr., ed., Geology of the Baja California Peninsula, Pacific Section Society of Economic Paleontologists and Mineralogists, 39, 17-22.

Magistrale, H., and Rockwell, T., 1996, *The Central and Southern Elsinore Fault Zone, Southern California*, Bulletin of the Seismological Society of America, 86, 6, 1793-1803.

Marcusson, C.R., 1985a, *Horizontal and Vertical Zonations within the Himalaya dike system, Himalaya Mine, Mesa Grande, California*, unpublished B.Sc. thesis, San Diego State University, 66 p.

Marcusson, C.R., 1985b, *Recent Work at the Himalaya Mine*, Mineralogical Record, 16, 419-424.

Marsh, B.D., 1979, *Island Arc volcanism*, American Journal of Science, 67, 161-172.

Merrill, F.J.H., 1915, *The Counties of San Diego and Imperial*, in Report XIV of the State Mineralogist Biennial Period, 1913-1914, Mines and Mineral Resources, San Francisco, California State Mining Bureau, 14, 635-722.

Moores, E.M., and Twiss, R.J., 1995, *Tectonics*, W.H. Freeman and Company, New York, 415 p.

Oliver, H.W., 1980, *Interpretation of the gravity map of California and its continental region*, California Division of Mines and Geology, Bulletin 205, 52 p.

Patterson, J.E., 1996, *Modeling of Layered Aplitic Pegmatite Dikes Using Ground Penetrating Radar, Little Three Mine, Ramona District, San Diego County, California*. The Spirit of Inquiry, University of Arizona Undergraduate Research Program, 17.

Patterson, J.E., 1997a, *A Geological Structural Analysis of the Hatfield Creek Valley, Ramona District, San Diego County, California*. The Spirit of Inquiry, University of Arizona Undergraduate Research Program, 15.

Patterson, J.E., 1997b, *The Little 3 Mine "Plug"; Evidence for a Zoned Plutonic Intrusion and a Plausible Parent Source of the Ramona Pegmatite District, Ramona, San Diego County, California*, University of Arizona, unpublished B.Sc. Honors thesis, 152 p.

Patterson, J.E., and Cook, F.A., 2002, *Successful application of ground-penetrating radar in the exploration of gem tourmaline pegmatites of Southern California*, Geophysical Prospecting, 50, 2, 107-117.

Pezzotta, F., 2000, *Internal structures, paragenesis and classification of the miarolitic Li-bearing complex pegmatites of Elba Island (Italy)*, in Pezzotta, F., ed., Mineralogy and petrology of shallow depth pegmatites; papers from the First international workshop, Memorie della Societa Italiana di Scienze Naturali e del Museo Civico di Storia Naturale di Milano, 30, 1, 29-43.

Pollard, D.D., 1976, *On the form and stability of open hydraulic fractures in the earth's crust*, Geophysical Research Letters, 3, 9, 513-516.

Pollard, D.D., 1987, *Elementary fracture mechanics applied to the structural interpretation of dykes*, in Halls, H.C., and Fahig, W.F., eds., Mafic Dyke Swarms, Geological Association of Canada Special Paper 34, 5-24.

Shigley, J.E., 1982, *Phosphate minerals in granitic pegmatites: A study of primary and secondary phosphates from the Stewart pegmatite, Pala, California*, unpublished Ph.D. Thesis, Stanford University, Stanford, California, 485 p.

Shigley, J.E., Kampf, A.R., Foord, E.E., and London, D., 1986, *Fieldtrip guidebook to Gem Pegmatites of Southern California*, 14th General Meeting of the International Mineralogical Association, July, 1986, Stanford University, Stanford, CA, 39 p.

Silberling, N.J., Schoellhamer, J.E., Gray, C.H., Jr., Imlay, R.W., 1961, *Upper Jurassic fossils from Bedford Canyon Formation, Southern California*, Bulletin of the American Association of Petroleum Geologists, 45, 10, 1746-1748.

Simpson, D.R., 1960, *Geology of the Central Part of the Ramona Pegmatite District, San Diego County California*, unpublished Ph.D. Dissertation, California Institute of Technology.

Simpson, D.R., 1965, *Geology of the Central Part of the Ramona Pegmatite District, San Diego County California*; California Division of Mines and Geology Special Report 86, 3-23.

Snee, L.W., and Foord, E.E., 1991, *$^{40}Ar/^{39}Ar$ Thermochronology of Granitic Pegmatites and Host Rocks, San Diego County, California*, abstract, Geological Society of America, Abstracts with Programs, A479.

Stewart, J.H., 1972, *Initial Deposits in the Cordilleran Geosyncline: Evidence of a Late Precambrian (<850 m.y.) Continental Separation*, Geological Society of America Bulletin, 83, 1345-1360.

Stewart, J.H., 1988, *Latest Proterozoic and Paleozoic southern margin of North America and the accretion of Mexico*, Geology, 16, 186-189.

Taylor, M.C., and Wise, M.A., 1996, *The Indian Flats Pluton and its pegmatite aureole: A fertile granite rare-element pegmatite system within the peninsular ranges batholith, Southern California*, abstract, Geological Association of Canada Mineralogical Association of Canada Annual Meeting, Programs with Abstracts, 21, A-94.

Todd, V.R., 1995, *Geology of the Mount Laguna Quadrangle, San Diego County, California*, U.S. Geological Survey, Open File Report 95-522, 34 p.

Todd, V.R., and Shaw, S.E., 1979, *Structural, metamorphic, and intrusive framework of the Peninsular Ranges batholith in southern San Diego County, California*, in Abbott, P.L., and Todd, V.R., eds., Mesozoic crystalline rocks: Peninsular Ranges batholith and Pegmatites, Sal Point Ophiolite, San Diego, California, San Diego State University, Department of Geological Sciences, 177-231.

Todd, V.R., and Shaw, S.E, 1985, *S-type granitoids and an I-S line in the Peninsular Ranges batholith, Southern California*, Geology, 13, 231-233.

Torres, S.G., 1982, *The Geology of the Blue Lady Pegmatite Dike of Chihuahua Valley, San Diego County, California*, unpublished B.Sc. Thesis, San Diego State University, 60 p.

Turcotte, D.L., Ahern, J.L., and Bird, J.M., 1977, *The State of Stress at Continental Margins*, in Denham, D., ed., Stress Environment and Displacements at Continental Margins, Tectonophysics, 42, 1-28.

Van Horn, S.R., Jones, A., Stansfield, W, and Markey, Z., 1999, *Emplacement mechanism of small lamprophyric dikes in north-central Vermont: Implications for their use as Paleostress indicators*, Geological Society of America Abstracts with Programs, Northeastern Section, 31, A75.

Van Horn, S.R., and Stephan, H.J., 2001, *Emplacement mechanism of Mesozoic dikes in the Champlain Valley: Examples from Charlotte Beach and the Redstone Quarry*, Geological Society of America Abstracts with Programs, Northeastern Section, 33, A2672.

Walawender, M.J., Gastil, R.G., Clinkenbeard, J.P., McCormick, W.V., Eastman, B.G., Wernicke, R.S., Wardlaw, M.S., Gunn, S.H., and Smith, B.M., 1990, *Origin and evolution of the zoned La Posta-type plutons, eastern Peninsular Ranges batholith, Southern and Baja California*, in Anderson, J.L., ed., The nature and origin of Cordilleran magmatism, Boulder, Colorado, Geological Society of America Memoir 174, 1-18.

Walker, G.E., 1958, *Intrusive Relations of the Batholith of Southern California near Bonsall, California*, unpublished M.Sc. Thesis, University of Arizona, Tucson, 137 p.

Waring, G.A., 1905, *The pegmatite veins of Pala, San Diego County, California*, American Geologist, 35, 356-369.

Webber, K.L., Falster, A.U., Simmons, W.B., and Foord, E.E., 1997, *The Role of Diffusion-Controlled Oscillatory Nucleation in the Formation of Line Rock in Pegmatite-Aplite Dikes*, Journal of Petrology, 38, 12, 1777-1791.

Webber, K.L., Simmons, W.B., Falster, A.U., and Foord, E.E., 1999, *Cooling rates and crystallization dynamics of shallow level pegmatite-aplite dikes, San Diego County, California*, American Mineralogist, 84, 708-717.

Weber, F.H., Jr., 1963, *Geology and Mineral Resources of San Diego County, California*; California Division of Mines and Geology County Report, no. 3, 309 p.

Woodford, A.O., Welday, E.E., and Merriam., R., 1968, *Siliceous tuff clasts in the upper Paleocene of Southern California*, Geological Society of America Bulletin, 79, 1461-1486.

You, C.F., 1994, *Lithium, beryllium, and boron isotope geochemistry: Implications for fluid processes in convergent margins*, unpublished Ph.D. dissertation, University of California, San Diego, 237 p.

Zoback, M.L., and Zoback, M., 1980, *State of Stress in the Conterminous United States*, Journal of Geophysical Research, 85, 11, 6113-6156.

Geomorphic Evidence for Ancient Bedrock Landsliding at Buckman Springs, San Diego County, California

William J. Elliott, *Consulting Engineering Geologist*
P.O. Box 541, Solana Beach, California 92075

INTRODUCTION

While returning to San Diego during the summer of 1988 on Interstate Freeway 8, lost in some childhood memories about warm summer days and camping just south of Buckman Springs at the old U.S. Forest Service Boulder Oaks Campground (Figure 1, Photo 3), I was gently returned to the present by late afternoon shadows cast across what appeared to be the back-scarp and hummocky toe area of an ancient bedrock landslide (Photos 1 and 2). Somewhat surprised to see this geomorphic signature in San Diego's back country (an otherwise familiar sight in coastal San Diego County sedimentary terrane), I made a mental note to check further into this errant anomaly.

Buckman Springs, the site of two old burned out road houses is situated in southeastern San Diego County at the intersection of Old U.S. Highway 80 and Old Buckman Springs Road. It is located about four miles south-southeast from Pine Valley, near the head of Cottonwood Valley (Section 20, T16S, R5E, SBBM).

Cottonwood Valley is a broad, flat, elongated, north-south trending alluviated highland valley, bounded by rugged, steep-sided mountains which are underlain by metamorphic and late Mesozoic granitic rocks of the Peninsular Ranges batholith. A long history of tectonic activity is recorded in fractures, faults and photo/topo lineaments in this crystalline bedrock terrane. These planes of weakness, when combined with other key ingredients for slope instability, such as over-steepened slopes and a likely wetter paleo-climate, could have tripped the scales in favor of gravity, and produced bedrock landsliding.

A review of black-and-white and color stereographic aerial photographs, and both large- and small-scale topographic maps, showed that indeed, there was good geomorphic expression, suggestive of an ancient landslide. Typical landslide features included: a steep arcuate back-scarp, hummocky topography, and an irregular toe area which could be "stretched" out to, and possibly even across a portion of Cottonwood Creek, about as far east as Old U.S. Highway 80 (Figure 1).

Maximum dimensions of this apparently west-to-east displaced mass of probably mostly granitic bedrock, are: width (north to south) approximately 7400 feet, and length (east to west) approximately 3100 feet.

Based upon a general knowledge of the relative degree of geomorphic expression with respect to the ages of other landslides in the southern California area, it is estimated that this event would likely have occurred during the last Pleistocene glacial period, on the order of about 10,000 to 30,000+ years ago.

Field checking remains to be done — during the cooler winter months when the wiggly critters are less active. Further work may show that this striking geomorphic expression results from nothing more than differential erosion, masquerading in the shape of an ancient landslide. Or, perhaps this, along with other similar anomalies in this backcountry region, will turn out to be heretofore generally unrecognized ancient granitic/metamorphic bedrock landslides.

This previously unpublished text was prepared during late summer 1988.

SEPTEMBER 2003 UPDATE

Since 1988, new geologic data has become available. For example, Todd (1995) published a geologic map of the Mount Laguna 15' quadrangle. And, newer stereographic aerial photographs have become generally available for photo-geomorphic interpretation.

A couple of years ago, at 30,000 feet, on an early morning commercial flight east, I was treated to a fleeting low-sun-angle view of Cottonwood Valley from the south. Again, I became immersed in the photo- and topo-geomorphic signatures so frequently associated with the artifacts of landsliding. Quick mental notes were made of several other probable bedrock landslides along the sides of Cottonwood Valley — especially the east side.

Review the topography shown in Figure 1, and if time permits, carefully review both sides of Cottonwood Valley from the ground. How many locations can you find where unstable hillsides may have succumbed to the forces of gravity? Are these features artifacts of differential erosion — or of landsliding? You decide.

As a footnote, the Boulder Oaks store and gas station (Photo 3), once located on the west side of Old

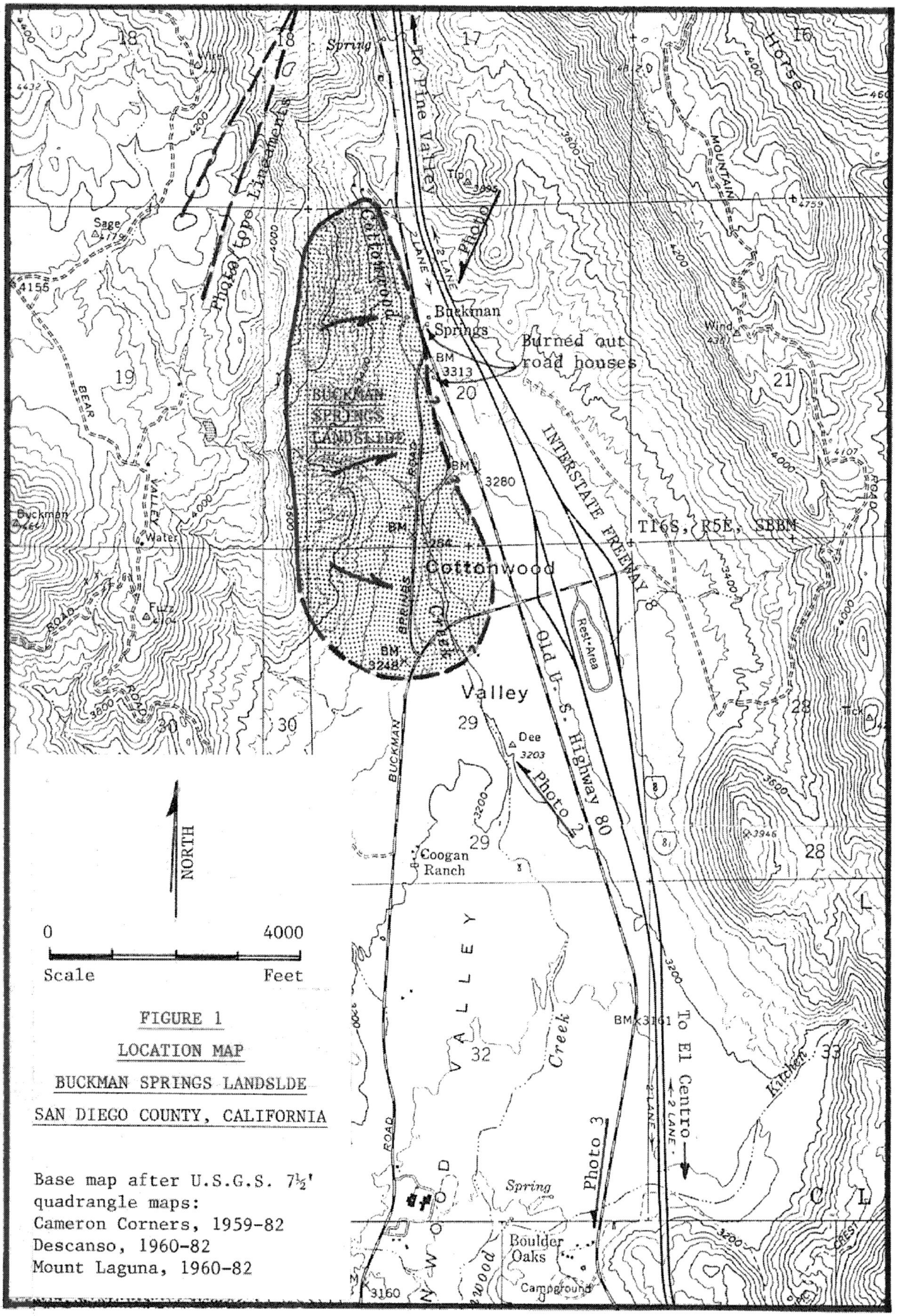

FIGURE 1

LOCATION MAP

BUCKMAN SPRINGS LANDSLDE

SAN DIEGO COUNTY, CALIFORNIA

Base map after U.S.G.S. 7½' quadrangle maps:
Cameron Corners, 1959-82
Descanso, 1960-82
Mount Laguna, 1960-82

Photo 1. Buckman Springs Landslide, view to southwest. Structure on the right is part of an old burned out roadhouse east of Old U. S. Highway 80. Photo taken August 1988. Compare with Photos 5 and 6.

U.S. Highway 80, one-tenth mile south of Kitchen Creek, were razed in about 1998.

The remains, however, of concrete and rock structures at Buckman Springs (Figure 1, Photos 1 and 6) yet stand as silent and soulful testament to the hopes and dreams of days long past. An historical account of Buckman Springs, beginning in the 1870s follows, and is on display at the nearby Buckman Springs Safety Rest Area along Interstate 8 freeway.

HISTORY OF BUCKMAN SPRINGS

The text that follows can be found on a public display poster at the Interstate Freeway 8 Buckman Springs Safety Rest Area, located approximately one mile south-southeast from Buckman Springs (Figure 1). With two minor bracketed, [ ], modifications (based upon original text by Davidson, 1937), it is reproduced herein with permission from CalTrans (2001). A July 1912, black and white photograph showing the bottling plant in operation at Buckman Springs is reproduced herein as Photo 4 (San Diego Historical Society, 1912a), and can also be viewed at the rest stop poster.

"Buckman Springs is an isolated group of small carbonated springs located along Interstate 8 in the Pine Valley Region of San Diego County, some 54 miles east of the city of San Diego. The pioneering Buckman family first developed the springs as a source of natural soda water in the 1870s. The site was a popular watering stop for travelers en route from Yuma, Arizona to San Diego for more than 50 years. Remains of this bottling plant and springs can be located on the site to this day.

"Buckman Springs is comprised of six small carbonated springs, about 20-30 yards apart on a gentle slope on the eastern side of Cottonwood Creek. The flow rate of this highly carbonated water is about one gallon per minute through limestone outcroppings and it has a high concentration of silica, iron and salt.

"Amos Buckman settled and developed the springs in the area then known [to the Indians] as Agua de Soda or Soda Springs. Buckman, native of Vermont, journeyed overland from Massachusetts during the Gold Rush. His family, wife Frances and daughters Hattie and Emma, followed shortly thereafter by ship travelling the treacherous Cape Horn route. Two other daughters, Winnifred and Mary, were born after the family had settled in California.

"Arriving in San Diego in the 1860s, Amos Buckman opened a carpenter's shop...[near what is now Kettner

Photo 2. Buckman Springs Landslide, view to the northwest from Old U.S. Highway 80. Photo taken August 1988.

Photo 3. Store and gas station at Boulder Oaks Campground — razed in about 1998. On V-J Day, during the late summer of 1945, I remember people honking their car horns and lining up to buy gas in celebration of the end of gas rationing, the end of hostilities with Japan, and the end of world War II! Photo taken August 1988.

Boulevard]. In the early 1870s, he began prospecting for gold in the eastern portion of San Diego County and heard about natural soda springs located near Campo. Shortly afterward he discovered the mineral springs and staked a homestead.

"President James A. Garfield on Aug. 20, 1881, signed a homestead grant of 160 acres in the name of Amos Buckman, formalizing his title to the property around the springs. By this time the Amos family had settled on the land, improved the spring as well as started a vineyard and apple orchard using the soda water as irrigation. As early as the spring of 1880, attempts were made to bottle the spring water and market it in San Diego.

"By the next spring, Amos Buckman was providing camp sites for visitors who wanted to stay while consuming and bathing in the effervescent waters. Throughout the 1880s Buckman's naturally carbonated soda water experienced some popularity as a tonic and elixir throughout portions of the county. Supplies, however, were intermittent as the process and bottling plant had not been perfected. Amos Buckman in 1887 hired a San Diego tinsmith to improve the facility and fountains. Throughout his life, Amos Buckman dreamed of developing a resort and spa around his bubbling mineral waters, however, by the 1890s his health and finances were lagging. He died March 28, 1898 at the age of 78.

"Many factors operated against the success of the spring's mineral waters; the fact that the springs were cold springs rather than thermally heated, the remoteness and barren nature of the location and the high iron content of the waters, which would coat the bottles a metallic orange color. The mineral water was marketed in Southern California under a variety of trade names, most often as 'California Club Water' and 'Buckman Springs Lithia Water.'

"The depression of the 1930s affected the business like many struggling industries of the era. It wasn't until January 1940 that the bottling of the spring water jumped back into business when Amos Buckman's daughter, Winnifred Casebeer (Buckman) leased the property to a Los Angeles syndicate. The sale of the lithia water to hospitals and patients on doctor's recommendation was unsuccessful in the long run.

Photo 4. "Historical photo shows workers bottling water from the small carbonated springs that made Buckman Springs a popular stop for travelers during the 1880s. Photo courtesy of San Diego Historical Society Photograph Collection." (CalTrans, 2001.) This July 1912 photograph is reprinted with permission from the San Diego Historical Society (1912a).

Photo 5. Easterly view of the two-story wood-construction Buckman Springs restaurant and store, and adjacent one-story stone-masonry house. Photo taken July 1912, and reprinted with permission from the San Diego Historical Society (1912b). The sign under the roof, in the middle of the wood building reads:

EVERYTHING
TO EAT
AT THE SPRINGS
STORE

GASOLINE 30¢
HAY GRAIN

Photo 6. Same view as Photo 5. Only the walls of the burned out stonemasonry house and some remnants of the stone-masonry foundation for the two-story wood building remain.

These sentinels are located approximately one-tenth mile north from the intersection of Old U.S. Highway 80 and Old Buckman Springs Road.

Photo taken September 6, 2003.

"Meanwhile, the area had become an important stop on U.S. 80, along with Jacumba, as the only two spots between El Centro and San Diego where a supply of water was available for filling radiators. Even this function faded away when Interstate 8 was opened in July 1975."

A chemical analysis by Smith, Emery & Company (Tucker, 1925) of the 60° to 65° Fahrenheit Buckman Springs water follows:

	Grains per U.S. Gallon
Silica (SiO_2)	5.55
Alumina (Al_2O_3)	0.43
Iron bicarbonate ($Fe(HCO_3)_2$)	2.45
Calcium bicarbonate ($Ca(HCO_3)_2$)	61.90
Magnesium bicarbonate ($Mg(HCO_3)_2$)	10.18
Magnesium sulphate ($MgSO_4$)	4.75
Calcium sulphate ($CaSO_4$)	0.86
Magnesium chloride ($MgCl_2$)	2.09
Manganese perchloride ($MnCl_4$)	Trace
Potassium chloride (KCl)	1.21
Lithium chloride ($LiCl$)	14.93
Sodium chloride ($NaCl$)	47.65
Sodium nitrate ($NaNO_2$)	Trace
Sodium bicarbonate ($NaHCO_3$)	6.03
TOTAL SOLIDS PER GALLON	158.03

REFERENCES

CalTrans, 2001, History of Buckman Springs: Poster display at the Buckman Springs Safety Rest Area, Interstate Freeway 8, approximately four miles south-southeast from Pine Valley. Text used with permission from CalTrans. July 1912, photo (see Photo 4, herein) used with permission from the San Diego Historical Society (1912a).

Davidson, John, 1937, Buckman Springs named after pioneer owner: San Diego Tribune, April 23, 1937. Article on file at the San Diego Historical Society Research Library, Subject File No. 581-1, Springs A-H.

Hart, M.W., 1991, Landslides in the Peninsular Ranges, Southern California, in Walawender, M.J., and Hanan, B.B., eds., Geological excursions in Southern California and Mexico: Guidebook for the 1991 Annual Meeting, Geological Society of America, San Diego, California, pp. 349-365.

Hoggatt, W.C., and Todd, V.R., 1977, Geologic map of the of Descanso [7.5'] quadrangle, San Diego County, California: U.S. Geological Survey, Open-File Report 77-406, 22 p., approximate map scale 1"=2000'.

San Diego, County of, 1970a, Stereographic, color aerial photographs, Nos. 24-12, 24-13, approximate scale 1"=2000', flown October 16, 1970.

San Diego, County of, 1970b, Topographic survey, map sheet Nos. 214-1923, 218-1923, and 222-1923, scale 1"=200', contour interval 5'.

San Diego Historical Society, 1912a, Black and white photograph (see Photo 4, herein). View of the bottling operation at Buckman Springs. Photo No. 91:18564-712, taken July 1912.

San Diego Historical Society, 1912b, Black and white photograph (see Photo 5 herein). Easterly view of the two-story Buckman Springs store/restaurant and adjacent stone-masonry house. Photo No. 91:18564-708, taken July 1912.

Strand, R.G., compiler, 1962, Geologic map of California, San Diego — El Centro sheet: California Division of Mines and Geology map sheet, approximate scale 1"=4 miles.

Todd, V.R., 1979, Geologic map of Mount Laguna [7.5'] quadrangle, San Diego County, California: U.S. Geological Survey, Open-File Report 79-862, 48p., approximate map scale 1"=2000'.

Todd, V. R., 1995, Geology of the Mount Laguna [15'] quadrangle, San Diego County, California: U.S. Geological Survey, Open-File Report 95-522, 34 p., approximate map scale 1"=1 mile.

Tucker, W.B., 1925, Report XXI of the State Mineralogist (Lloyd L. Root), Los Angeles Division, San Diego County: California State Mining Bureau, Volume 21, Number 1, p. 381.

U.S. Department of Agriculture, 1953, Stereographic, black-and-white aerial photographs, Nos. AXN-11M-206, 207, 208, and 209, approximate scale 1"=1667', flown April 24, 1953.

U.S. Department of Agriculture, 1980, Stereographic, false-color infrared aerial photographs, Nos. 24-635020-1079-99, 100, 101, 102, and 103, approximate scale 1"=2130', flown April 9, 1980.

U.S. Geological Survey, 1959-82 and 1997, Cameron Corners, California, 7.5' topographic quadrangle map, scale 1"=2000', contour interval 40'.

U.S. Geological Survey, 1960-82 and 1997, Descanso, California, 7.5' topographic quadrangle map, scale 1"=2000', contour interval 40'.

U.S. Geological Survey, 1960 and 1997, Morena Reservoir, California, 7.5' topographic quadrangle map, scale 1"=2000', contour interval 40'.

U.S. Geological Survey, 1960-82 and 1997, Mount Laguna, California, 7.5' topographic quadrangle map, scale 1"=2000', contour interval 40'.

Weber, F.H., Jr., 1963, Geology and mineral resources of San Diego County, California: California Division of Mines and Geology, County Report 3, 309 p., approximate map scale 1"=2 miles.

Geology of the Coyote Mountains, Imperial and San Diego Counties, California

Thomas W. Dibblee, Jr., *Dibblee Geological Foundation*

ABSTRACT

The Coyote Mountains are in the western Imperial basin about 22 km (13 mi.) north of the USA/Mexico border. They form a low southeast trending range about 24 km (15 mi.) long, uplifted on the northeast side of the Elsinore fault as this fault dies out to the southeast.

The Coyote range is composed of a core of Peninsular Range crystalline basement, mostly adjacent to the Elsinore fault. The basement ranges from east to west from biotite schist through gneiss to gneissoid quartz diorite of Mesozoic age or older. All these rocks include marble layers. In adjacent areas the basement is mostly quartz diorite.

This basement is overlain by the sedimentary series of the Imperial basin, in this range as thick as 3000 m, (10,000 ft.); this series consists of a local basal volcanic unit (Miocene), the shallow marine Imperial Formation (Pliocene) and the terrestrial Palm Spring Formation and Canebrake Conglomerate (Pleistocene) in ascending order. This series is well exposed on the north flank of the range where it thins greatly and coarsens westward onto the east sloping basement platform.

Foliation structure of the basement rocks is vertical and strikes normally northeast in the central area. Westward near the Elsinore fault it strikes anomalously nearly west. This suggests clockwise rotation from right lateral drag on the fault, prior to deposition of the sedimentary series.

The sedimentary series dips northwest from the basement core into a major syncline. East of this core it is anticlinally folded plunging east toward the Imperial Valley. Many minor faults striking northwest and northeast are related to dextral movement on the Elsinore fault.

The Elsinore fault dips steeply northeast in at least one place. About 10 km (6 mi.) of right slip may have occurred on this fault in late Pleistocene time.

INTRODUCTION

Previous Investigations

Fossiliferous marine strata and underlying volcanic rocks exposed in Fossil (formerly Alverson) Canyon in the Coyote Mountains and in other areas in the Imperial basin attracted the attention of early investigators (Hanna, 1926; Woodring, 1931; Tarbet and Holman, 1944; Tarbet, 1951).

In the 1920s and more intensely in the 1940s, the marine beds (Imperial Formation of Woodring, 1931) of the Imperial basin became the target of intensive oil exploration by several oil companies. That was when I mapped the basic geology of the entire Imperial basin in 1943 and 1944, as part of an oil exploration project (Dibblee, unpublished geologic maps, scale 1:62500 (1944); Dibblee, 1954; 1984, 1996, 1997).

Physiography

The Coyote Mountains are in the western part of the Imperial basin or Salton Trough, about 22 km (13 mi.) north of the USA/Mexico border. As shown on Figure 1, the Coyote Mountains form a low southeast trending range about 24 km (15 mi.) long and as wide as 10 km (6 mi.). The southwest boundary is sharp and straight along the Elsinore fault. The north and east boundaries are vague and gradual with Carrizo Creek wash separating the range from the Vallecitos Badlands to the north.

The main and highest part of the Coyote Mountains is essentially a southeast trending low mountain range of crystalline basement rocks about 16 km (10 mi.) long extending south of east from the low northeastern margin of the basement rocks of the Peninsular Range terrane. The highest part of the Coyote Mountain basement terrane is in the eastern part, from where this ridge turns west of north as a spur to form an overall horseshoe shaped ridge of basement rock. This spur includes Carrizo Mountain, the high point of the Coyote Mountains at about 2400 ft. above sea level.

The southeast trending ridge of basement rock of the Coyote Mountains rises abruptly northeastward from Coyote Wells Valley. This straight steep southwest flank, about 15 km (9 mi.) long is the escarpment of the Elsinore fault. The north flank of this ridge slopes gradually into a badland terrane (Carrizo Badlands) of sedimentary rocks, which surrounds the high north spur of basement rocks of the Coyote Mountains.

The Coyote Mountains are flanked on the northwest by Vallecitos Valley, a structural depression northeast of the Peninsular Range basement. This valley is transected by Carrizo Creek that drains from the Peninsular Mountains. Eastward from Vallecitos Valley the hill terrane of the northern Coyote Mountains extends northward into the Vallecitos Badlands.

Unlike the mountains north of the Vallecitos Badlands the Coyote Mountain uplift within the Imperial basin is almost isolated from the Peninsular Range terrane. To the south the Coyote Mountains are separated from that terrane by Coyote Wells Valley of the Imperial basin.

GEOLOGIC SETTING

Geologically the Coyote Mountains are within the Imperial basin at its western margin (Figure 2) where the crystalline basement of the Peninsular Range terrane slopes eastward or northeastward under this basin filled with a clastic sedimentary series of Miocene to Pleistocene age. Under the central part of the Imperial Valley this series is as much as 6200 m (20,000 ft.) thick, it thinning out and coarsening westward against the Peninsular Range basement platform. This condition is superbly exposed on the north flank of the Coyote Mountains, as well as in the terrane to the northwest.

From this condition it is evident that in this part of the Imperial basin the Peninsular Range basement platform was being tilted downward to the northeast as it became buried by the sedimentary series of the basin. This basement platform was contemporaneously elevated in the Peninsular Range terrane where it shed detritus.

However, this northeastward tilt of the basement platform was disrupted by tectonic turbulence related to dextral transpressive movements on several active high angle faults, such as the Elsinore fault and others to the north, that strike east-southeast from the Peninsular Range uplift into the Imperial basin (Figure 2). Tectonic movements on these faults and related northeast striking cross faults prevailed while this sedimentary series accumulated and accelerated after in late Quaternary time when the sedimentary series became severely deformed, as shown on Figures 5 and 6. During this diastrophism the Coyote Mountains were elevated from the Imperial basin by dextral transpression movement on the northeast side of the Elsinore fault as it dies out to the southeast into this basin and evolved by uplift of the once buried basement platform under the sedimentary series to the surface on the crest of this uplift.

ROCK UNITS

Crystalline Basement Rocks

The crystalline basement rocks exposed in the Peninsular mountains west of the Imperial basin and the Coyote and other mountains within it (Figure 2), are composed primarily of plutonic igneous rocks of the southern California batholith, generally considered to be of late Mesozoic (Cretaceous) age. In places these plutonic rocks engulf small pendant remnants of older metasedimentary rocks.

Metasedimentary rocks. The oldest rocks of the crystalline basement exposed in the Coyote Mountains are metasedimentary rocks of pre-late Mesozoic age. In the Carrizo Mountain area these rocks are composed of dark gray, thin bedded biotite schist-phyllite metamorphosed from micaceous shale. These dark colored rocks include many layers of white marble (Figure 6), some as thick as 100 m.

Metasedimentary rocks comprise nearly all the basement rocks exposed in the Coyote Mountains, but the grade of metamorphism varies. In the eastern part, including Carrizo Mountain, these rocks are the least metamorphosed. Westward the grade of metamorphism increases to gneiss and gneissoid quartz diorite, so that in the westernmost exposures the predominant rock is gneissoid quartz diorite. The grade of metamorphism from east to west is so gradual that the schistose rocks appear to have been "granitized" to quartz diorite. The structural grain of these rocks is consistent (Figures 5 and 6).

Despite this increasing east to west metamorphic grade, the marble layers or lenses persist throughout these exposures. They are more crystalline in western exposures. They seem to have resisted being "granitized" with respect to the schistose rocks presumably due to their carbonate composition.

Plutonic rocks. In the Peninsular Mountains southwest of the Coyote Mountains nearly all the basement is of quartz diorite, massive to gneissoid, with a small strip of gneiss in Carrizo Canyon. In the Fish Creek Mountains to the north, all of the basement rock is of massive quartz diorite.

In the vicinity of the Coyote Mountains nearly all of the plutonic rocks are of quartz diorite, or tonalite composition. They are light gray, medium-grained, holocrystalline, massive to gneissoid, composed mostly of sodic plagioclase feldspar and lesser amounts of quartz, potassic feldspar and biotite. In some areas in the Peninsular mountains and in the Coyote Mountains this rock contains dark gray finer grained discoid inclusions or xenoliths like that of the Bonzal Tonalite (of Larsen, 1948), oriented parallel to the gneissoid structure of the rock.

In the Coyote Mountains the basement rocks are in places cut by small pegmatite dikes, less than a meter wide, trending mostly north-south, some of which contain black tourmaline. These are the youngest of the basement rocks.

Sedimentary Series

In the western Imperial basin the deeply eroded surface of the crystalline basement under the Imperial basin is overlain by the clastic sedimentary series of Miocene through Pleistocene age that fills this subsided basin. In the Vallecitos Badlands north of Carrizo Wash this series is as much as 4300 m (13,900 ft.) thick in exposed thickness. In the badlands of the Coyote Mountains to the south it is estimated to be as thick as 3000 m (10,000 ft.). A nearly complete section is exposed on the northwest flank of this range.

The sedimentary series of these exposures is weakly lithified and ranges from fine claystone through sandstone into coarse conglomerates. The claystone was deposited as fine mud, probably in large part from the Colorado River. However, the coarse sediments to the west were derived from the Peninsular Mountains as that terrane was elevated contemporaneously when the Imperial basin subsided. This is evident by the westward coarsening, thinning and buttressing of this clastic series against the basement in the Coyote Mountains. In the eastern Coyote Mountains, as well as in the Fish Creek Mountains to the north, a thin volcanic unit called Alverson Volcanics, is at or near the base of this series.

The stratigraphic sequence of this series in the Coyote Mountains is as shown on Figure 3, and described briefly below in ascending order.

Redrock Formation. The Redrock or Anza Redbed Formation is best exposed in the Fish Creek Mountains, where it overlies basement and is overlain by volcanics. In the Coyote Mountains this redbed unit occurs only as one exposure just west of Fossil Canyon, where it rests on basement and is overlain by a thin layer of Alverson Volcanics (Figure 5). In other areas north and south of the Coyote Mountains, this unit is represented by conglomerate.

Alverson Volcanics. In the Coyote Mountains the volcanic rocks were named by Tarbet and Holman, 1944, as the Alverson Volcanics, after Alverson Canyon (now Fossil Canyon), the type locality. This unit is exposed in the eastern Coyote Mountains where it overlies the basement platform and is not present in the western Coyote Mountains. This volcanic unit is also exposed south of the Coyote Mountains about 10 km (6 mi.) west of the Coyote Mountain exposures, where it likewise overlies the basement (Figure 5). The Alverson Volcanics are as thick as 250 m (800 ft.) in the Coyote Mountains and composed of dark brown basaltic andesite as flows and flow breccias. It was erupted onto the eroded basement surface from an unknown source.

Samples from the Alverson Volcanics of the Coyote Mountains were radiometrically age dated (K-Ar) at about 16 MA (Gastil and others, 1979) or late early Miocene.

Imperial Formation. Marine clastic sediments of the Imperial basin were named Imperial Formation by Woodring, 1931. This marine unit is as thick as 1000 m (3300 ft.) in Fish Creek Canyon of the Vallecitos Badlands, the suggested type locality (Dibblee, 1996). On the north flank of the Coyote Mountains the folded Imperial Formation is estimated to be as much as 600 m (2000 ft.) in eastern exposures, thinning out in western exposures against basement (Figure 5).

The Imperial Formation overlies the Alverson Volcanics and on basement elsewhere in the Coyote Mountains. In and west of Fossil Canyon the basal unit of the Imperial Formation is a transgressive shallow marine sandstone and conglomerate unit as thick as 100 m (330 ft.) of mostly basement detritus. At Fossil Canyon this unit is very fossiliferous with coral reefs and many large sand dollars (Clypeaster).

The Imperial Formation is a weakly lithified shallow marine clastic unit of claystone and interbedded sandstone deposited in the Imperial basin when it was submerged under shallow water of the Gulf of California. The sandstones are tan to rusty brown, fine to medium grained and well stratified. The claystone is light olive gray, bedded, concoidally fractured and weathered to a distinctive light tan residual soil. The claystone includes hard thin gray reef-beds of small oyster shells in many places. The Imperial Formation contains no biogenic organic material such as diatomite. The Imperial Formation is interpreted to be of late Miocene-early Pliocene age.

Palm Spring Formation. The marine Imperial Formation grades upward and in part intertongues laterally westward into the overlying stream-laid Palm Spring Formation, named by Woodring, (1931) for Mountain Palm Spring west of the Coyote Mountains (Figure 1).

In the western Coyote Mountains the Palm Spring Formation is about 1480 m (4800 ft.) thick at the type locality in a northwest-dipping section (Figure 5) but thins rapidly westward and all but its lowest part intertongues westward into the coarse Canebrake conglomerate. In that area the Palm Spring Formation overlies basement where the Imperial Formation thins out.

The Palm Spring Formation is a weakly lithified unit of well stratified light gray to tan sandstone and light red claystone. It rarely contains conglomerate except where it grades westward through pebbly sandstone into the coarse Canebrake conglomerate facies. In many places the finer grained sandstones contain calcareous concretions of various shapes. It is rarely fossiliferous except in a few places where it contains silicified wood fragments and verterbrate remains.

The age of the Palm Spring Formation was thought to be Pliocene from its enormous thickness and severe deformation similar to that of thick Pliocene clastic formations of coastal basins. But in recent years the age of this formation has been revised upward to Pleistocene on the basis of vertebrate remains found at some places in this formation (Downs and White, 1967, Downs and Woodard, 1961). Frost et al., 1996, 1997).

Canebrake Conglomerate. The coarse western marginal facies of the Palm Spring Formation was named Canebrake Conglomerate (Dibblee, 1954) after Canebrake Wash on the southwest side of Vallecitos Valley. The type section was designated in the nearby Vallecitos Badlands to the north of that wash in a southwest dipping section (Dibblee, 1954) where this conglomerate attains its maximum thickness of 2100 m (7000 ft.).

At the west end of the Coyote Mountains the Canebrake Conglomerate in a northwest-dipping section (Figure 5) is thinner, 700 m (2000 ft.). Its lowest part intertongues eastward into sandstone and reddish clays of the Palm Spring Formation. In this area this conglomerate is on both sides of the Elsinore fault. Southwest of the fault it is thinner, grades eastward into the Palm Spring Formation with both formations together overlying the basement.

The Canebrake Conglomerate is composed of coarse gray to brown crudely bedded fanglomerate of quartz diorite detritus that grades eastward into finer bedded conglomerate and coarse sandstone. The Canebrake Conglomerate is the same age as the upper part of the Palm Spring Formation in which it intertongues into and overlies.

Alluvial Deposits. Throughout the elevated areas of the southwestern Imperial basin adjacent to the Peninsular Range uplift, the beveled erosion surface of the severely deformed sedimentary series and parts of the adjacent crystalline basement are covered by remnants of flat-lying late Pleistocene alluvial deposits.

These are slightly elevated, much dissected veneers a few meters to tens of meters thick of alluvial pebbly gravel and sand. They are extensive in the Vallecitos Badlands, and in the northwest and eastern margins of the Coyote Mountains where they slope slightly away from the mountains. This condition must be the effect of the latest surge of uplift of these mountains and suggests that these alluvial deposits were more extensive.

TECTONIC STRUCTURE

Since the sedimentary series accumulated in the Imperial basin in the late Cenozoic era, the basin has been severely deformed especially on its western margin, by the latest surge of uplift of the Peninsular Range, which encroached eastward into the Imperial basin. During this diastrophism this area was affected by severe tectonic turbulence, probably in large part related to dextral transpressive movements on several high angle faults striking east-southeast from the Peninsular Range uplift into the Imperial basin (Figure 4). The most southerly of these faults is the Elsinore fault along which the Coyote Mountains was elevated.

Because the severely deformed sedimentary series is in large part of Pleistocene age, this crustal turbulence must be very young and is ongoing. It is the effect of the late Coast Range orogeny that prevailed throughout the plate margins of California.

The structural setting of the Coyote Mountains is as shown on Figure 7.

Structure of the Basement of the Coyote Mountains

The structure of the basement of the Coyote Mountains is well defined but unusual with respect to that of adjacent areas, especially because the basement of this range is composed of layered metamorphic rocks such as schist to gneissoid rocks with strata of marble. In the Coyote Mountains these rocks are upended and strike nearly east west north of the Elsinore fault (Figure 5), but in the eastern part this strike curves abruptly northeast then northwest in the vicinity of Carrizo Mountain (Figure 6). Overall, this defines a horseshoe like structure, concave to the northwest. The structure may be that of a near vertical northwest-plunging syncline, or could be that of a vertical southeast plunging anticline

In the Peninsular Range region the grain of the basement rocks normally strikes north to northwest. In the Coyote Mountains this is the strike in the Carrizo Mountain area, as well as in the quartz-diorite-gneiss of the basement southwest of the Coyote Mountains (Figure 5). The east-northeast strike north of the Elsinore fault in the Coyote Mountains is therefore anomalous. This unusual condition may be the effect of right lateral drag on this segment of the Elsinore fault. If so, this lateral drag movement on the fault must pre-date deposition of the sedimentary series and Miocene volcanic rocks, suggesting that the earliest dextral movement on this fault may be pre-Miocene, and that this fault may have extended farther southeast, possibly into the aligned Laguna Salada fault to the southeast (Dibblee, 1996).

Structure of the Sedimentary Series

The structure of the sedimentary series that once covered the horseshoe shape structure of the Coyote Mountains basement is related to movement on the Elsinore fault and structure of the sedimentary series of the Vallecitos Badlands.

In the Coyote Mountains the sedimentary series dips essentially north from the basement elevated on the Elsinore fault (Figure 5). Throughout the Vallecitos Badlands this series dip southwest from the basement of the mountains to the north. This structure is that of a major syncline with an axis along Carrizo Wash, plunging west under Vallecitos Valley toward the Peninsular Range basement elevated southwest of the Elsinore fault (Figure 7; Dibblee, unpublished map, 1944). This major structure is referred to as the Vallecitos syncline. Vallecitos Valley is essentially a structural depression, adjacent to the Peninsular Range basement southwest of the Elsinore fault.

In the eastern Coyote Mountains the structure of the sedimentary series is essentially anticlinal plunging eastward from the core of basement of Carrizo Mountain, but is complexly disrupted by faults striking northwest and northeast (Figure 6).

In the Coyote Wells Valley southwest of the Coyote Mountains the structure of the sedimentary series, largely covered by alluvium, is that of an asymmetric syncline plunging southeast. The northeast limb is in part upended due to drag uplift on the Elsinore fault; the axis is close to the fault (Figure 5). To the southwest the northeast-dipping sedimentary series and volcanic unit thin out (Figure 5).

FAULTS

The only major fault in the vicinity of the Coyote Mountains is the Elsinore fault along their southwest border, the most southerly of several high angle right slip faults shown on Figure 2.

Many minor faults prevail in the Coyote Mountains (Figures 5 and 6), as well as in terrane to the north (Figure 2), as high angle faults up to about 5 km (3 mi.) long. These occur as two sets, northwest striking longitudinal faults and northeast striking cross faults. They are probably the effect of regional dextral transpressive shear stress.

Elsinore Fault

The Elsinore fault is a major high angle southeast striking transcurrent fault zone that extends some 150 km (95 mi.) through the Peninsular range terrane and dies out to the southeast in the western Imperial basin.

Within about 20 km (32 mi.) northwest of Carrizo Wash the Elsinore fault zone is somewhat irregular but was mapped along the northeast margin of the Peninsular Range basement where it was uplifted on the southwest side of the fault and bulges northeast into Vallecitos Valley (Figures 2 and 7). Along this segment the fault zone is marked by shear zones of crushed basement and several springs in the shear zones in the basement hills southwest of Vallecitos Valley. These shear zones are irregular but in places are as wide as several hundred meters. These shear zones and the northeastward bulge of the basement uplift into the Vallecitos Valley suggests this fault segment dips southwest.

Along this fault segment the basement on its southwest side was uplifted against the deepest part of the Vallecitos syncline. This condition indicates a very large amount of vertical displacement on this segment of the Elsinore fault zone that is not apparent from the low relief of the basement hills.

At Carrizo Wash the Elsinore fault apparently disappears where vertical movement reverses as the Coyote Mountains were uplifted on the northeast side of this fault, The dip of the fault also reverses. Where the fault is exposed 6 km (4 mi.) southeast of Carrizo Wash, it dips about 70 degrees NE. Similar attitudes were seen on a splay 1 km (.6 mi.) southeast of Carrizo Wash.

The Elsinore fault is traceable for 20 km (13 mi.) along the southwestern baseline of the Coyote Mountains which is remarkably straight (Figures 5 and 6). The greatest amount of vertical uplift in late Quaternary time is in the central part of this range near Fossil Canyon, and is probably equal to the present height of this range with respect to Coyote Wells Valley or more. Quaternary vertical movement decreases to zero near Carrizo Wash, and at its southeast end near Coyote Wash.

As previously mentioned evidence of right lateral drag movement on the Elsinore fault in the Coyote Mountains prior to deposition of the sedimentary series is suggested by the westward bending or clockwise rotation of the upended metamorphic rocks adjacent to the fault from their normal northwest trend. If so, strike slip on this fault segment was greatly rejuvenated in late Quaternary time since deposition of the sedimentary series. If most of the clockwise rotation of the metamorphic basement occurred before deposition of the sedimentary series, an additional amount may have occurred since. Possibly the sedimentary series of this uplift may also have been rotated clockwise during deformation.

The magnitude of right slip component on the Coyote Mountain segment of the Elsinore fault is not definitely known but may be several kilometers. A possible clue may be in the localized areal extent of the Alverson Volcanics on opposite sides of this segment. On the northeast side of the fault these rocks are exposed at the western margin of Coyote Wells Valley (Figure 5), about 10 km (6 mi.) northwest of the exposures in the Coyote Mountains northeast of the fault. If those rocks were originally adjacent, they may have been separated by that much right lateral shift on this segment of the Elsinore fault in Quaternary time.

The Vallecitos syncline in the sedimentary series that plunges westward against the Elsinore fault (Figure 7) may be the effect of that much right lateral drag movement on the northeast side of this fault since deposition of this series in the western part of the Imperial basin. This syncline has no counterpart southwest of the fault, so it formed only on the northeast side. The Coyote Mountains may be an anticlinal uplift of the sedimentary series on the Elsinore fault, south of the Vallecitos syncline.

Although the Elsinore fault apparently dies out near Coyote Wash, the sedimentary series to the southeast is compressed into several folds with northeast trending axes in the vicinity of Yuha Buttes on the northeast side of the Laguna Salada fault a right slip fault aligned with the Elsinore fault (Figure 6) and which increases in dextral slip magnitude into Mexico (Dibblee, 1979).

Faults within Coyote Mountains

The Coyote Mountains are disrupted by many minor local faults northeast of the Elsinore fault (Figures 5 and 6). All are high angle faults that strike northwest and northeast, and few north-south, east west. Several extend into the Vallecitos Badlands (Figure 7). None extend more that 4 km (3 mi.). Movements on these faults are mostly dip-slip, but some are strike-slip.

Several northwest striking faults occur in the central and eastern Coyote Mountains. Two of these faults in the central area strike southeast into the Elsinore fault with uplift on the northeast side of both (Figure 5). Another occurs west of Carrizo Mountain with uplift of basement on the east side. In the eastern area, near Painted Gorge, the volcanic rocks and Imperial formation are displaced by minor right slip on several close northwest striking faults. Aligned with these faults to the northwest is a narrow fold in the Imperial Formation and another fault striking northwest across Carrizo Wash into the Vallecitos Badlands, with uplift on the northeast side.

Faults of northeast strike are most prominent in the western Coyote Mountains, in both the basement and sedimentary series. In the basement these faults strike east-northeast along the structural foliation of the basement rock, and veer northeast into the northwest-dipping sedimentary series (Figure 5). Movements on these faults is small, mostly left slip. These faults must be the effect of slippage along foliation planes in the basement related to clockwise rotation northeast of the Elsinore fault.

GENESIS OF THE COYOTE MOUNTAINS

Prior to deposition of the sedimentary series since Miocene time the area now the Coyote Mountains and vicinity was part of the Peninsular Range batholith of late Mesozoic age, originally overlain by a thick roof of older metamorphic rocks. Since it was emplaced this batholith was uplifted from great depth and exposed extensively since removal of most of the overlying roof by long erosion. The deeply eroded surface must have extended eastward under what evolved into the Imperial basin as this basement surface subsided under it since Miocene time to become the platform of the sedimentary series of this basin.

The western margin of this subsided basin extend through what is now the Coyote Mountains and vicinity. Along this vicinity the basement platform was tilted eastward toward the basin as it subsided, from the Peninsular Range terrane to the west where uplift continued and shed detritus, as previously mentioned.

In the vicinity of the Coyote Mountains there is no evidence of detachment of the sedimentary series from the basement platform on a listric normal fault as commonly interpreted (Frost et al., 1996, 1997).

Very soon after the sedimentary series accumulated with accelerating rapidity in Pleistocene time in the Imperial basin, it was severely deformed in late Pleistocene time by folding, probably related to dextral transpressive movements on the high angle southeast striking faults that extend south of east from the Peninsular Range terrane into the Imperial basin. These include the Elsinore fault and others to the north shown on Figure 7. This extensive deformation must have occurred within the last 1 or 2 million years, and is probably still ongoing. This late deformation was the effect of the late Coast Range Orogeny as described in Dibblee, 1995. The Plio-Pleistocene sediments derived from uplift of the present Peninsular Range terrane occurred during the early Coast Range orogeny.

During the late Coast Range orogeny the terrane north of Vallecitos Valley and that to the southeast were uplifted to form the Vallecitos Badlands and Coyote Mountains, respectively, in which the sedimentary series is extensively deformed and exposed. This uplift and deformation must be the effect of dextral transpressive contraction of the crust on the northeast side of the Elsinore fault as it dies out southeastward, relative to the more stable less deformed terrane southwest of this fault.

REFERENCES CITED

Dibblee, T.W.J., 1944, unpublished geologic map (AMS 15-minute quadrangles, scale 1:62500) of the Imperial basin.

Dibblee, T.W.J., 1954, Geology of the Imperial Valley region, California, in Jahns, R.H., ed., Geology of the natural provinces, California Division of Mines Bulletin 170, pp. 21-28.

Dibblee, T.W.J., 1984, Stratigraphy and tectonics of the San Felipe Hills, Borrego Badlands, Superstition Hills, and vicinity, in Rigsby, C.A., ed., Imperial Basin — tectonics sedimentation and thermal aspects; Pacific Section Society of Economic Paleontologists and Mineralogists, pp. 31-34 (reprinted 1996, in Abbott, P.L., and Seymore, D.C., 1996, eds., Sturzstroms and detachment faults, Anza-Borrego Desert State Park, California, South Coast Geological Society Annual Field Trip Guidebook No. 21, pp. 45-58.

Dibblee, T.W.J., 1995, Tectonics and depositional environment of the middle and upper cenozoic sequences of the coastal southern California region, in Fritsch, A.E., ed., Cenozoic Paleogeography of the western United States, Pacific Section SERM (Society for Sedimentary Geology Book 75, pp. 212-242.

Dibblee, T.W.J., 1996, Stratigraphy and tectonics of the Vallecitos-Fish Creek Mountains, Vallecitos Badlands, Coyote Mountains and Yuha Buttes, southwestern Imperial basin, in Abbott, P.L. and Seymore, D.C., eds., Sturzstroms and detachment faults, Anza-Borrego State Park, California; South Coast Geological Society Annual Field Trip Guide Book No. 21, pp. 59-79.

Dibblee, T.W.J. 1997, Geology of the southeastern San Andreas fault zone in the Coachella Valley area, southern California, in Baldwin, J., Levis, L., Payne, M., and Roquemare, G., eds., Southern San Andreas fault, Whitewater to Bombay Beach, Salton Trough, California; South Coast Geological Society Annual Field Trip Guide Book No. 25, 1997, pp. 35-56.

Downs, T., and White, J.A., 1967, A vertebrate funnel succession in superimposed sediments from late Pleistocene to middle Pleistocene in California xxiii International Geologic Congress v. 10, pp. 41-47.

Downs, T., and Woodward, G.D., 1961, Middle Pleistocene extension of the Gulf of California into the Imperial Valley; Geological Society of America Special Paper 68, p. 21.

Frost, E.G., Suitt, S.C. and Fattahipour, M., 1996, Emerging prospectives of the Salton Trough region with an emphasis on extension faulting and its implications for later San Andreas deformation, in Abbott, P.L., and Seymore, D.C., eds., Sturzstroms and detachment faults, Anza-Borrego Desert State Park, California, South Coast Geological Society annual Field Trip Guide Book No. 21, 1996, pp. 81-121, (reprinted 1997, in South Coast Geological Society Annual Field Trip Guide Book no. 25, 1997, pp. 57-97).

Gastil, G., Krummacher, D., and Minch, J.A., 1979, The record of Cenozoic Volcanism around the Gulf of California; Geological Society of America Bulletin v. 90, pp. 839-857.

Hanna, G.D., 1926, Paleontology of the Coyote Mountains, Imperial County, California; California Academy of Science Proceedings v. 14, no. 18, pp. 427-503.

Hoggatt, Wendy C., 1979, Geologic map of the Sweeney Pass Quadrangle, San Diego County, California, U.S. Geological Survey Open File Report, 79-754, scale 1:24000.

Tarbet, L.A., 1951, Imperial Valley in possible future petroleum provinces of North America; American Association of Petroleum Geologist Bulletin, v. 35, pp. 260-263.

Tarbet, L.A., and Holman, W.H., 1944, Stratigraphy and micropaleontology of the west side of the Imperial Valley, California [Abstract] American Association of Petroleum Geologists Bulletin, v. 28, pp. 1781-1782.

Woodring, W.P., 1931, Distribution and age of the Tertiary deposits of the Colorado Desert; Carnegie Institute of Washington Publication 148, pp. 1-25.

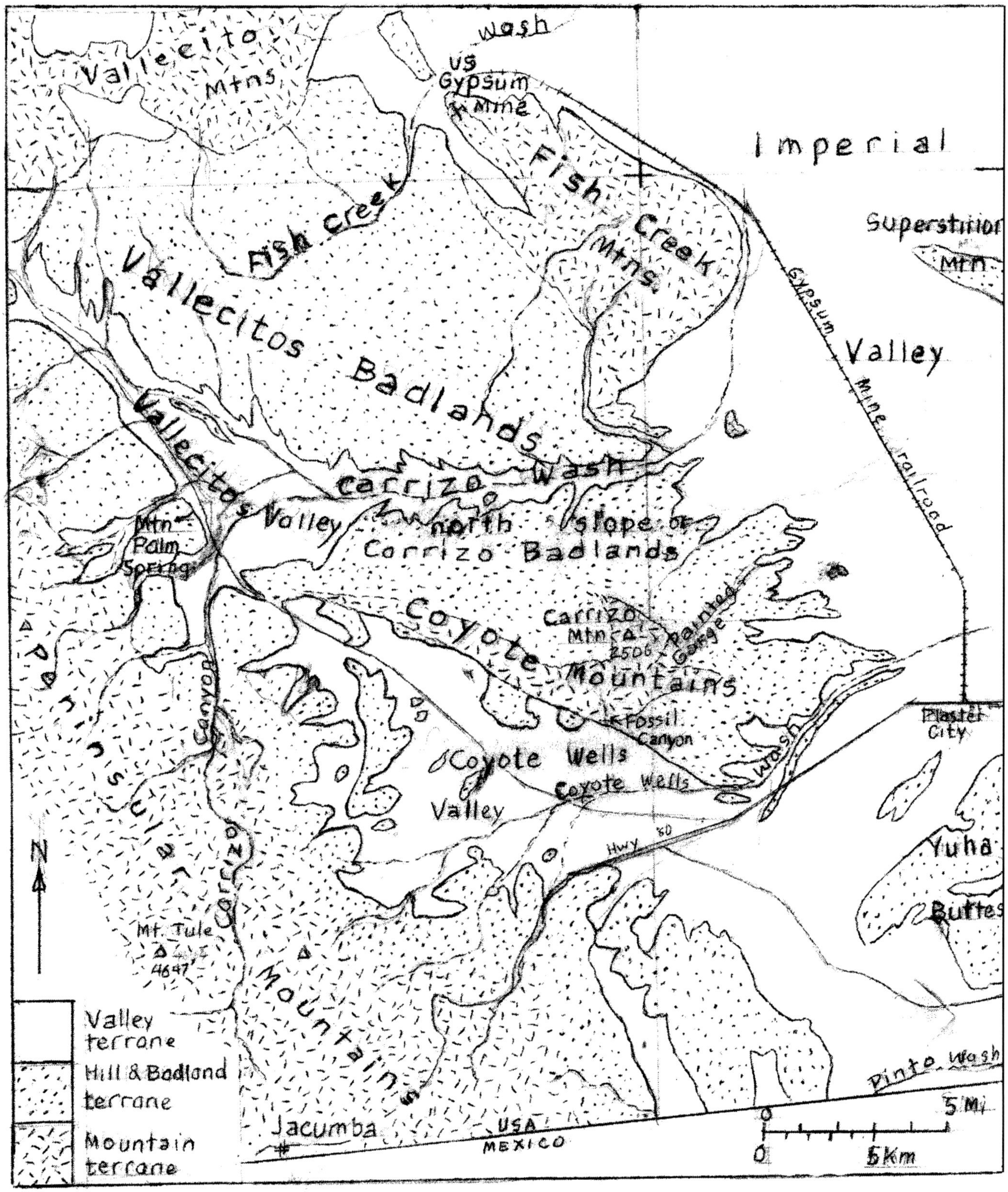

Figure 1. Physiographic setting of Coyote Mountains.

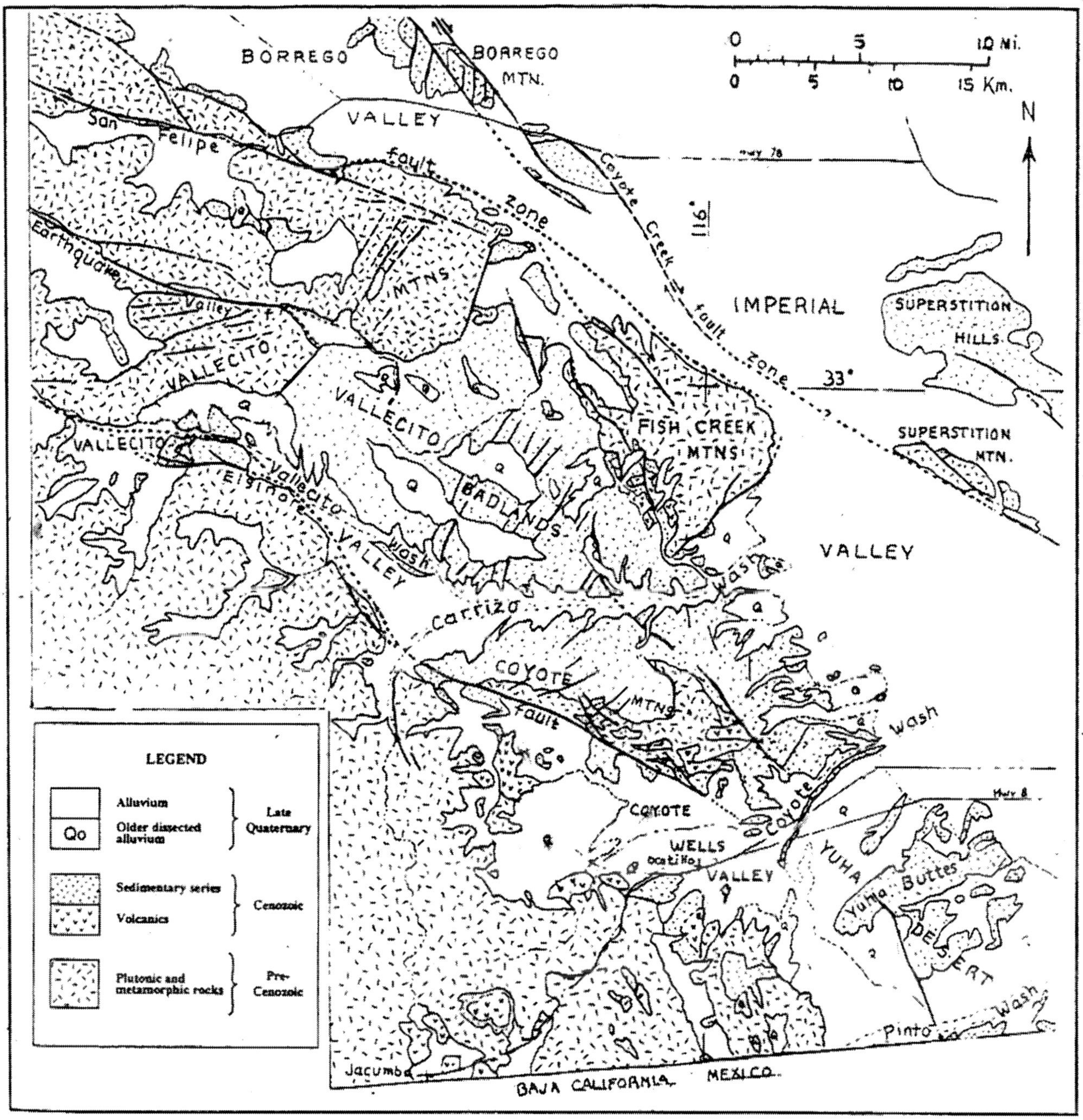

Figure 2. Generalized geologic map of southwestern Imperial basin and adjacent parts of Peninsular Range Mountains, showing physiographic and geologic features. (From Dibblee, 1997.)

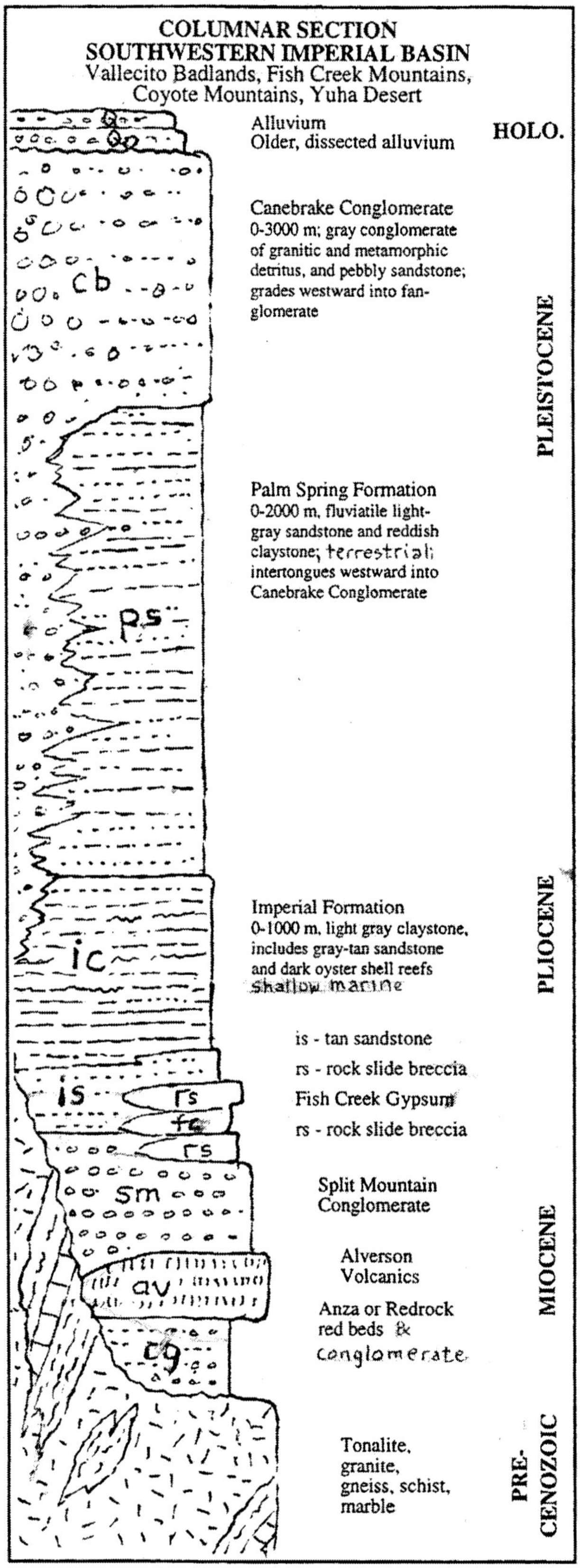

Figure 3. Stratigraphy of Cenozoic sedimentary series of southwestern Imperial basin.

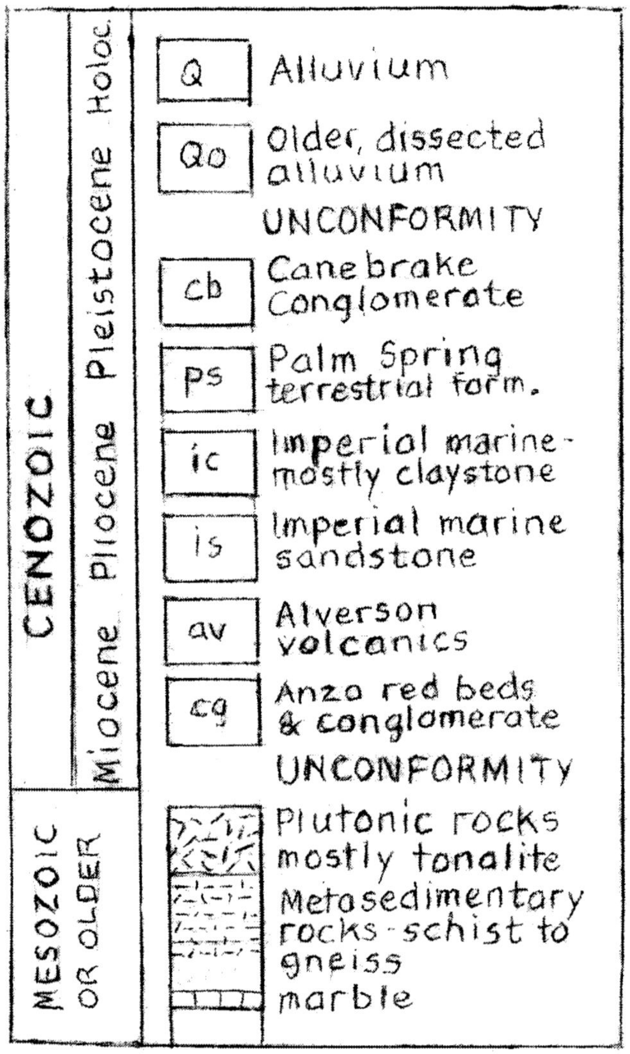

Figure 4. Legend for Figures 5 through 8.

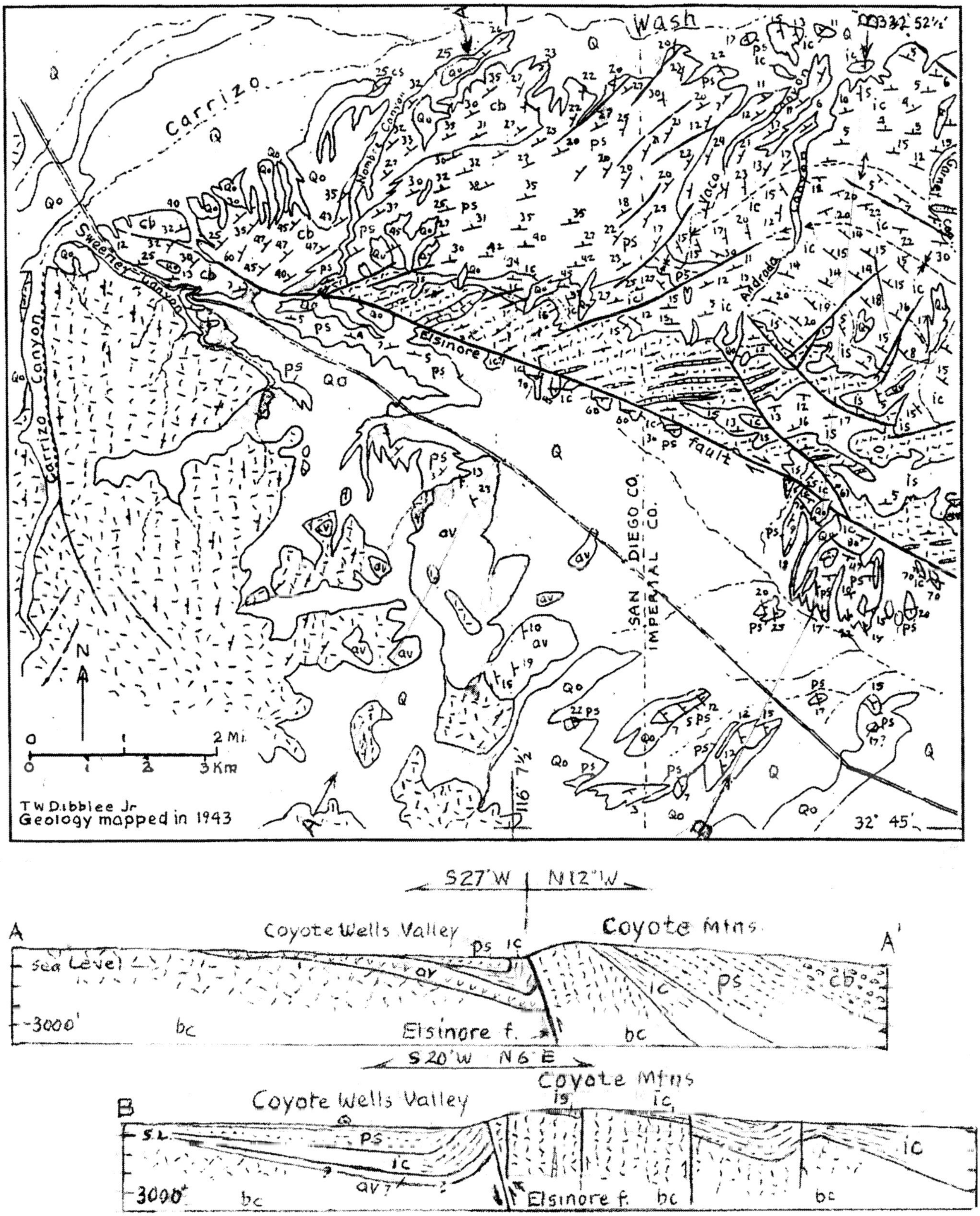

Figure 5. Geologic map of western Coyote Mountains. (From Dibblee, 1997.)

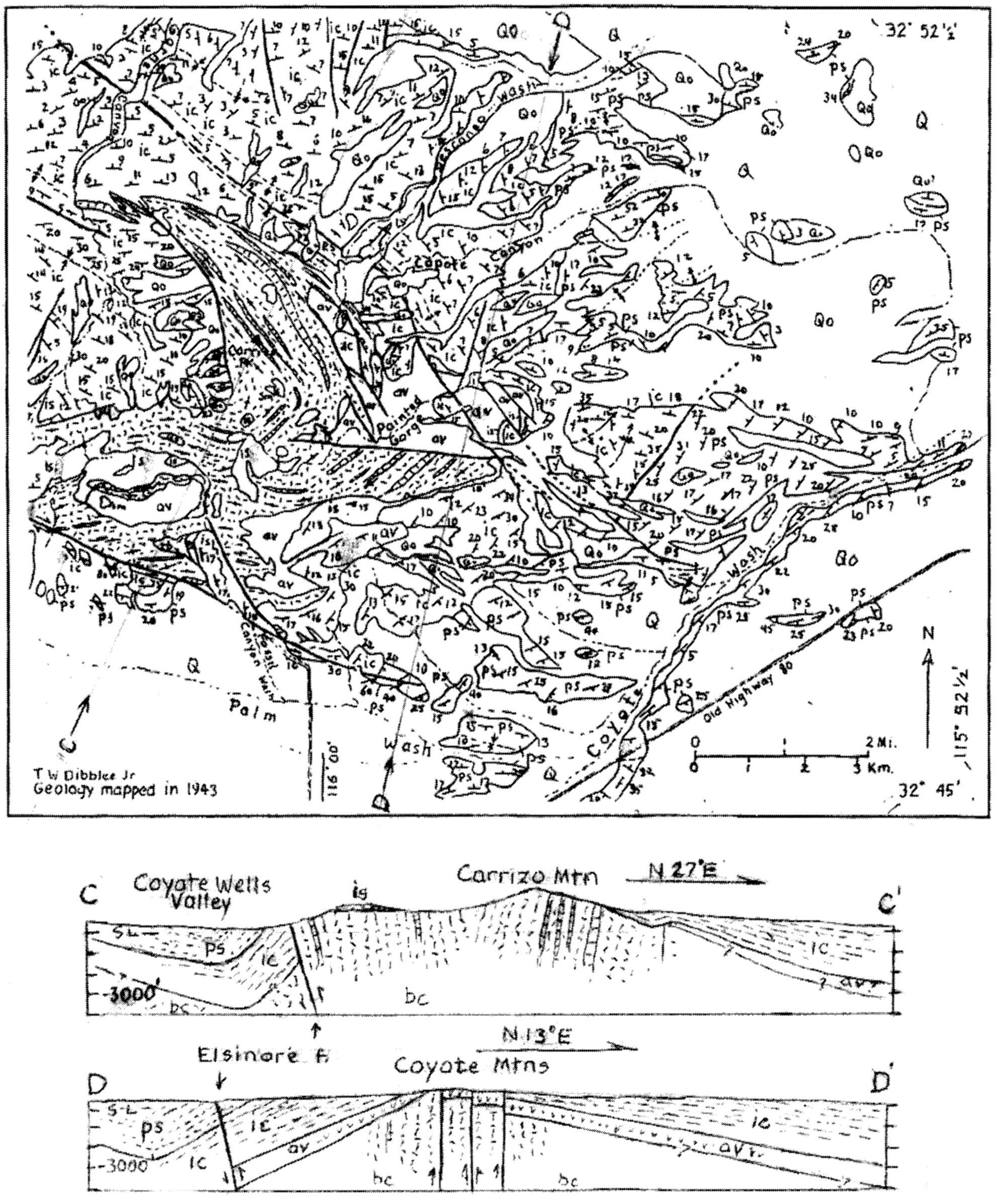

Figure 6. Geologic map of eastern Coyote Mountains. (From Dibblee, 1997.)

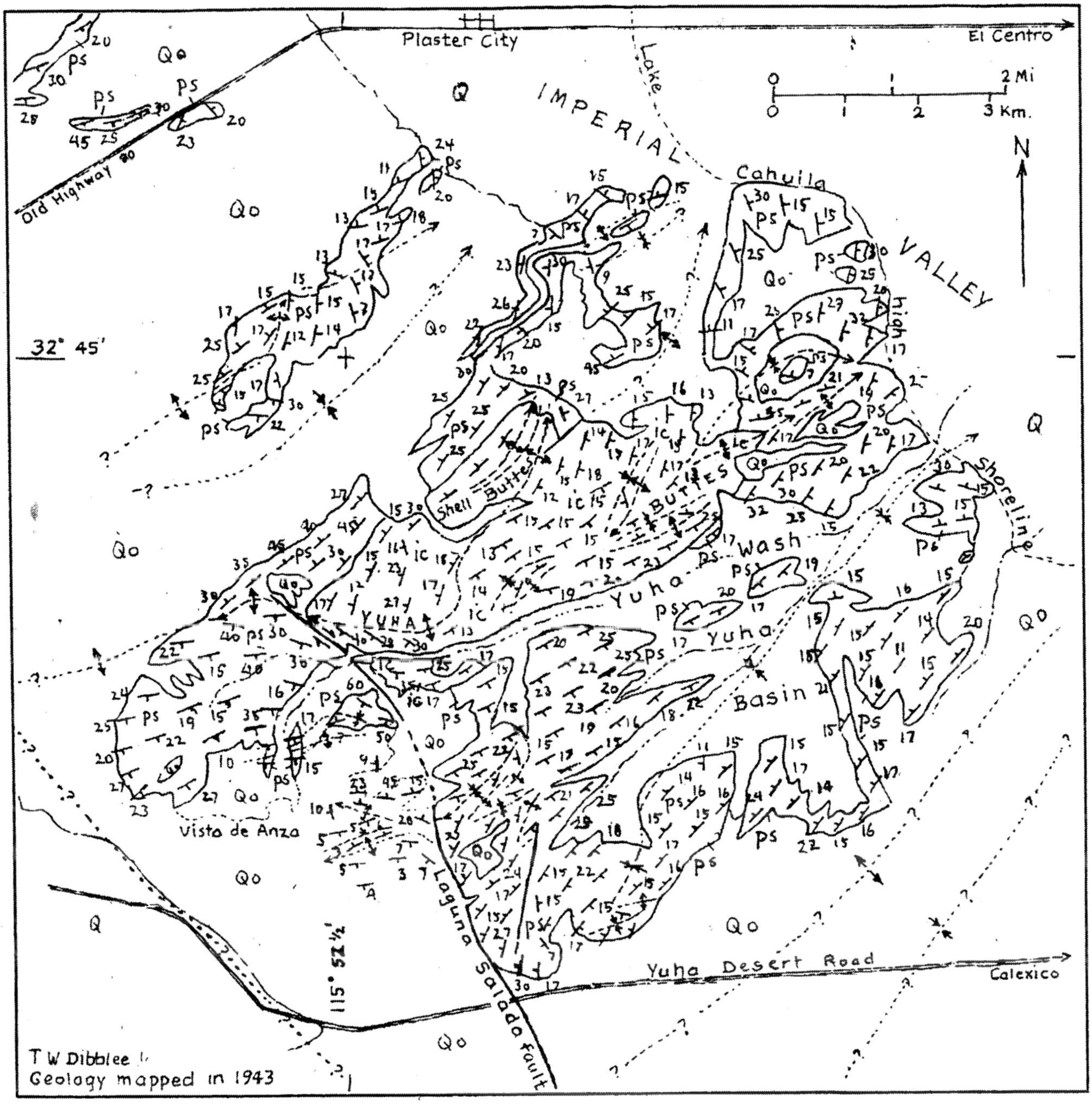

Figure 7. Geologic map of Yuha Buttes and vicinity, Yuha Desert. (From Dibblee, 1997.)

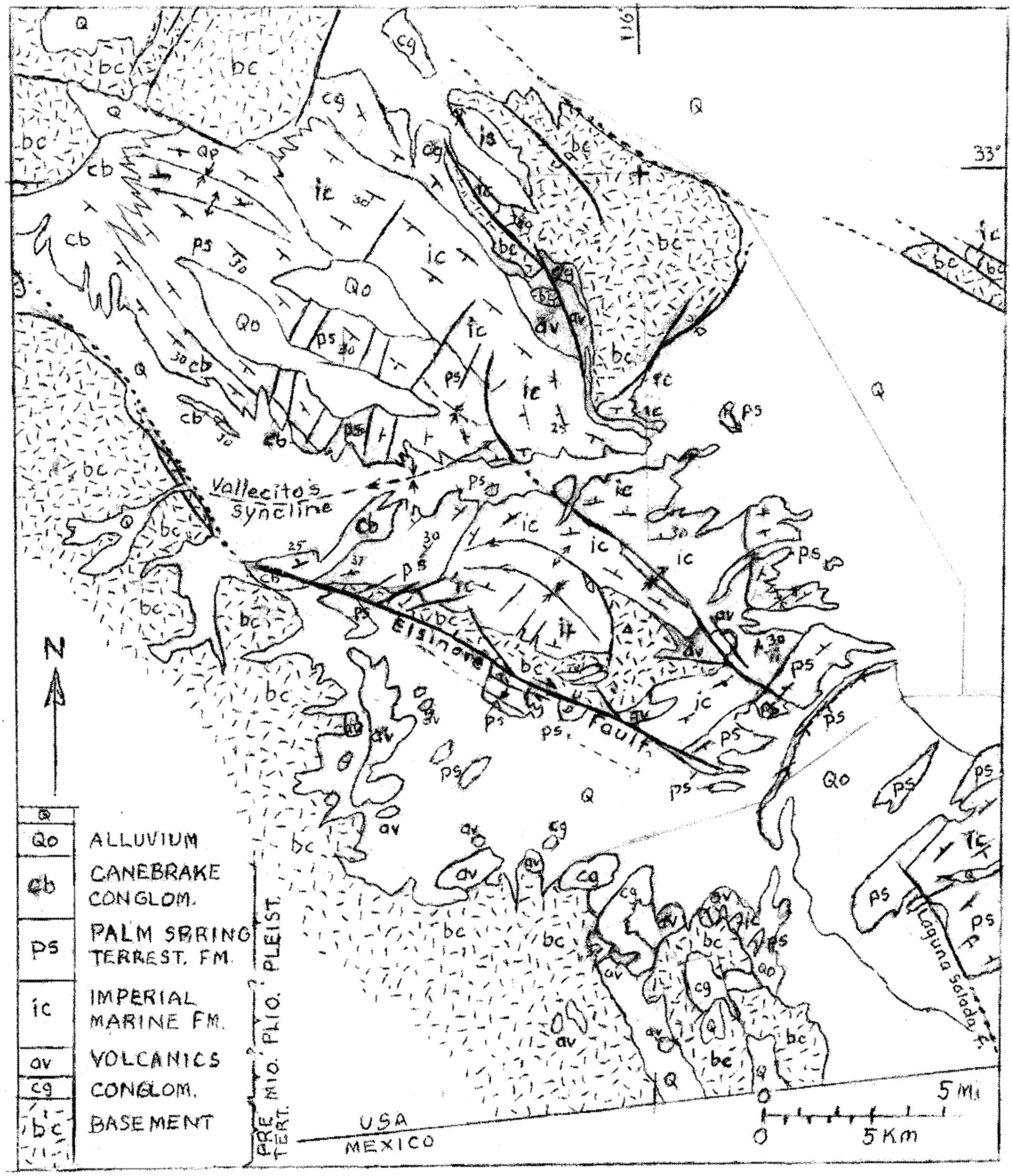

Figure 8. Structural setting of the Coyote Mountains.

IV.

Associated Papers

GEOLOGY OF THE JULIAN AREA

Joan Miller
Thomas Schar
Tanya Musselwhite
Gordon Gastil
Department of Geological Sciences
San Diego State University
San Diego CA 92182

ABSTRACT

Early studies of the Julian area of San Diego County divided the rocks into Julian Schist, Cuyamaca gabbro, and "mixed rock" (Merriam, 1958). In this study the "mixed rock" has been subdivided into seven informal units. The Julian Schist, and all but one or two of the newly recognized units have shared the regional deformation, but at least one unit, the tonalite of Wynola, is synchronous with or post-dates the deformation.

Recognizing units within the so-called "mixed" assemblage makes it possible to measure approximately three miles of right-lateral separation along the Elsinore fault.

INTRODUCTION

The area mapped for this study covers about 20 km^2 in the Santa Ysabel 15 minute quadrangle (Plate 1, Figure 1) of east-central San Diego County. The field work was done by three students from San Diego State University as their advanced field-mapping projects in 1992 Petrographic modal analyses were done by Joan Miller.

Previous mapping of this area was by Merriam (1958) and Germinario (1982). The vegetation of the area is Colter pine/black oak forest, open meadows, and local areas of brush. Outcrops are sparse, especially for the foliated rocks.

IDENTIFICATION OF ROCK UNITS

Twenty six samples from the new informal units were examined and point-counted in stained thin-sections. Plutonic rock names were assigned using the I.U.G.S classification of Streckeisen (1973).

JULIAN SCHIST

This unit appears to have been largely fine-grained clastic strata, with local sandstone and rare conglomerate. Relict bedding is observable locally. Throughout this study area the common aluminosilicate is andalusite. All contacts with granitic rocks appear to be intrusive.

CUYAMACA GABBRO

Earlier maps have shown the gabbro at Pine Hills to be continuous with the gabbro occurring on the Cuyamaca peaks (Merriam, 1958). This map adds several additional small bodies. Their textures and mineralogies resemble those described for the Cuyamaca peaks (Todd,1977,1979). Locally the gabbro is spatially associated with quartz and pegmatite dikes. Gabbro tends to outcrop as hills, and contacts with other units appear to be sharp.

The fabric of the gabbro shows little deformation. Although the gabbro contacts in Pine Hills are generally concordant with the foliation of adjacent rocks, intrusion may have followed existing foliation

(Plate I). There seems little cause to believe that the gabbro predates regional deformation.

TONALITE OF HOSKINGS RANCH

The tonalite of Hoskings Ranch is a fine- to coarse-grained pyroxene-hornblende-biotite tonalite which outcrops to the south of the Julian Schist (Plate 1, Table 1, Figure 2). Metamorphic foliation is general but varies widely in intensity. Foliation is more pronounced adjacent to the Julian Schist and Cuyamaca gabbro near Pine Hills. Biotite occurs in aggregate clumps. The rock contains no magnetite.

GRANODIORITE OF QUARTZ HILL

This granodiorite was mapped only in one locality northwest of Kanaka Flat which we refer to as Quartz Hill. It is an even-textured, fine-grained, gray rock with bouldery outcrops. The rock contains no magnetite. Foliation is not prominent, and the body may be post- or syn-deformation.

GRANODIORITE OF JULIAN ORCHARDS

This is a medium-grained hypideomorphic-granular, two mica, quartz-rich granodiorite which outcrops to the north of the Julian Schist. The fresh rock is gray, but it is buff to reddish brown in weathered exposures. The rock outcrops very poorly. Where exposed it can be seen to include numerous xenoliths of schist, quartzite, calc-silicate rock, and banded amphibolite. A distinguishing feature of this unit is the regolith lag deposit of weathering-resistant xenoliths accumulated on the surface. The rock contains no magnetite.

The xenoliths of schist are in places so ubiquitous, as to give a strong gneissic appearance to the rock. A place to observe this "gneissic" character is along Farmers Road between Wynola Road and the northern contact of the Julian Schist.

Contacts against the Julian Schist tend to be broad and diffuse. The unit is cut by numerous schorl-rich pegmatite dikes which appear to have shared the brittle regional deformation.

THE PLUTONIC COMPLEX OF KANAKA FLAT

This map unit, mapped only in the Kanaka Flat area, consists of several strongly foliated and mineralogically variable bodies of granite to tonalite. The rocks commonly weather out in large exfoliation cores. The degree of rock exposure and the detail of mapping was not sufficient to determine if the rock variation is that of a single zoned body or parts of unrelated plutons. Textures of solid state deformation are well developed. The rocks contain no magnetite.

GRANODIORITE OF SANTA YSABEL CREEK

This is a medium- to coarse-grained, strongly foliated, pyroxene granodiorite that characteristically weathers to a yellowish tan or reddish-brown color and is poorly exposed. Solid state deformation textures are well developed. Pyroxene has been largely replaced by biotite and chlorite. It was only mapped on the north side of the Elsinore Fault and the tonalite of Wynola. Its extent to the north of the mapped area is unknown.

GRANITE OF RUTHERFORD RANCH

This is a variably foliated, medium-grained biotite granite. It weathers gray to yellowish brown, forming large bouldery outcrops. In the one sample examined, potassium feldspar is

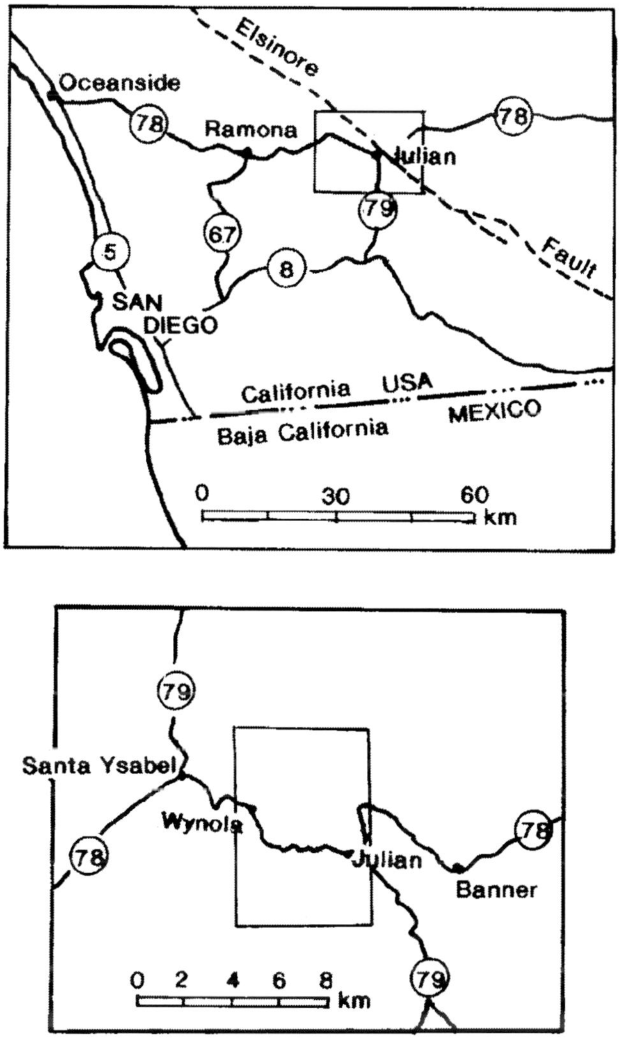

Figure 1. Index Map of the study area.

Legend and symbols for Plate 1

U	unidentified granitic rock
W	tonalite of Wynola
QH	granodiorite of Quartz Hill
CG	Cuyamaca Gabbro
HR	tonalite of Hoskings Ranch
JO	granodiorite of Julian Orchards
R	granite of Rutherford Ranch
SYC	granodiorite of Santa Ysabel Creek
KF	plutonic complex of Kanaka Flats
J	Julian Schist
	contact, dashed where approximate
	steep foliation trends in schist
	planar orientation in granitic rocks
	Elsinore fault

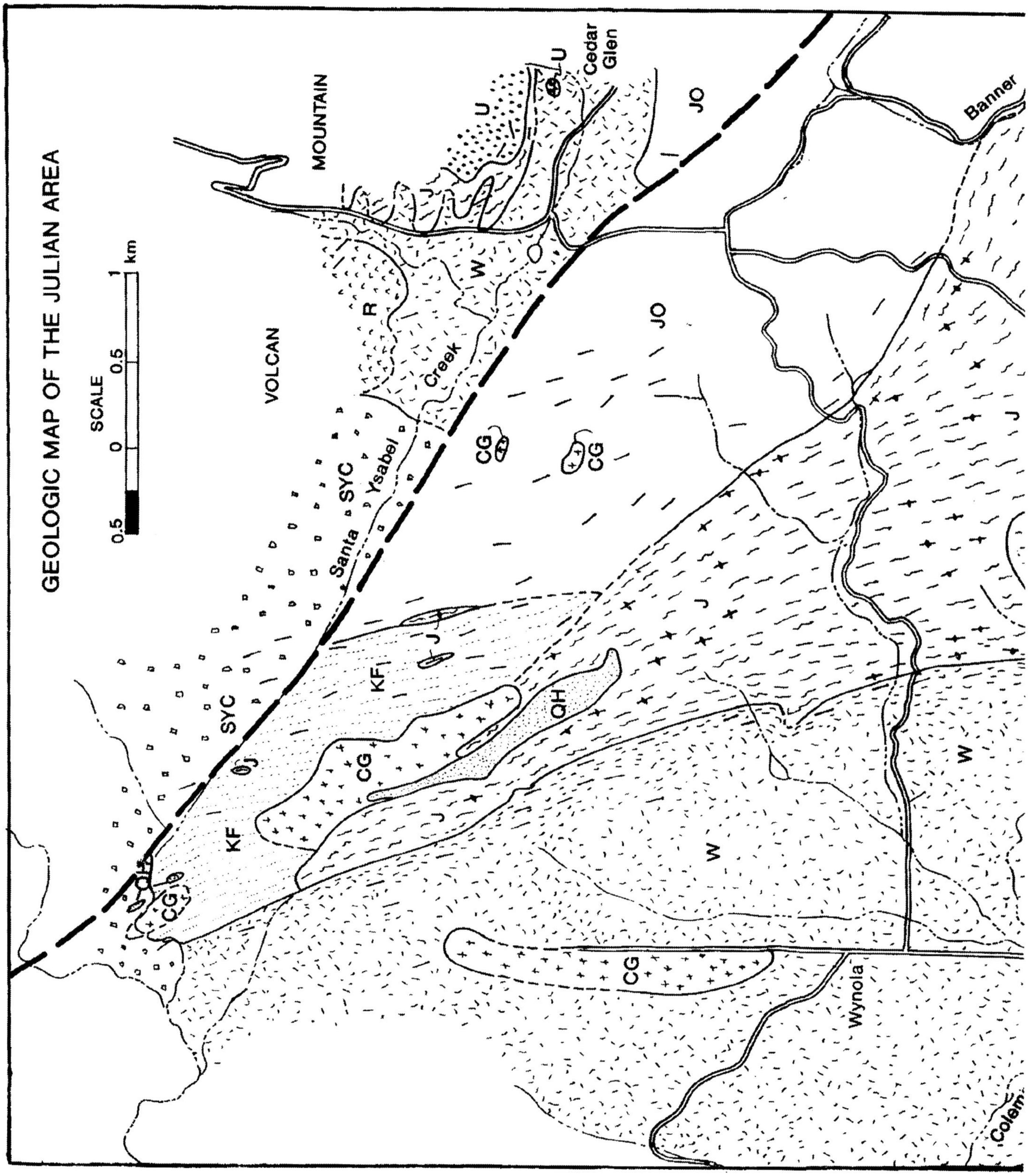
GEOLOGIC MAP OF THE JULIAN AREA
SCALE
0.5 0 0.5 1 km
VOLCAN
MOUNTAIN
Santa Ysabel Creek
SYC
KF
CG
QH
J
W
JO
R
U
Cedar Glen
Banner
Wynola

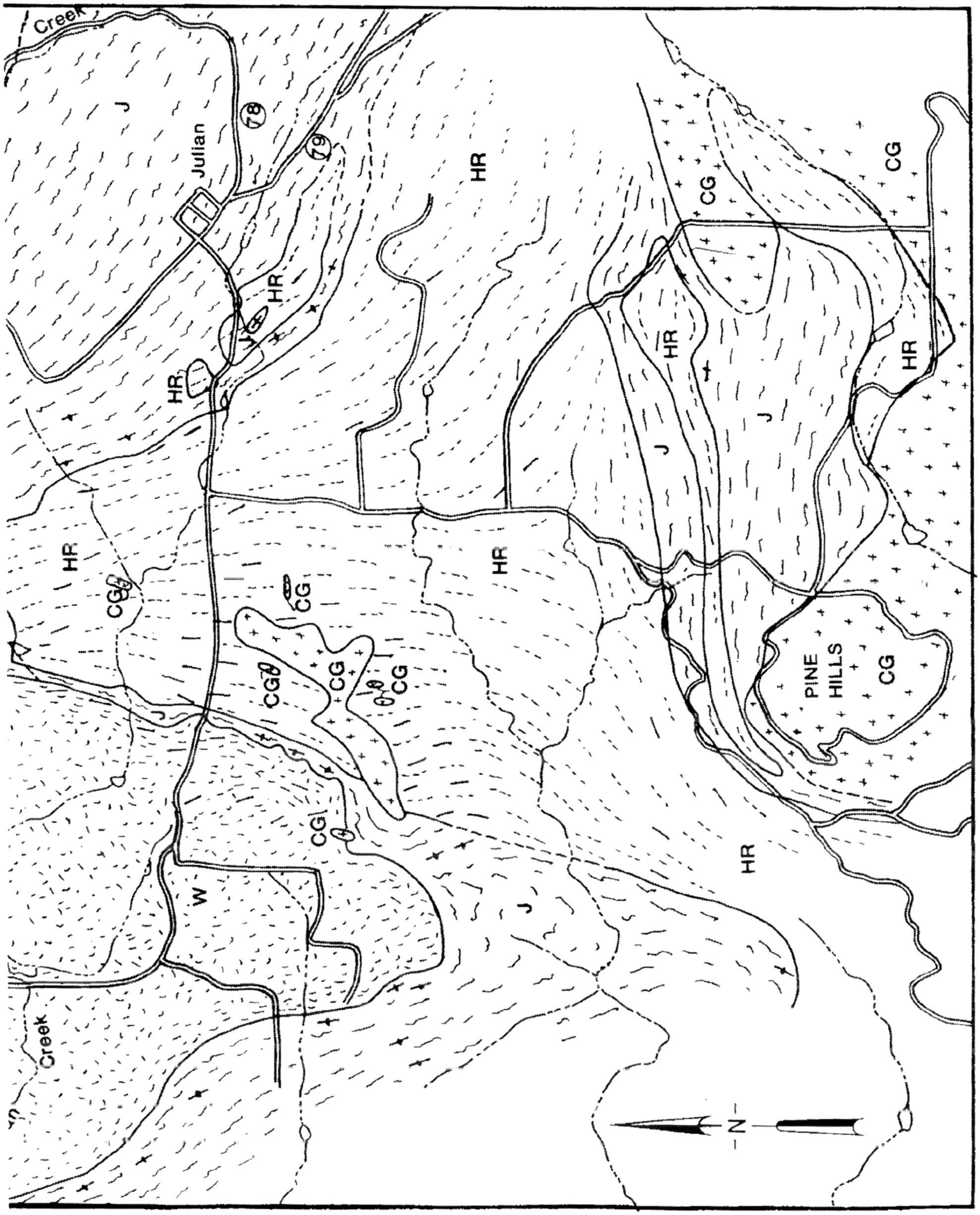
Creek
J
Julian
78
79
HR
CG
W
PINE HILLS
Creek
N

	granodiorite of Julian Orchards			plutonic complex of Kanaka Flats				granodiorite of Quartz Hill	granodiorite of Santa Ysabel Cr	granite of Rutherford Ranch	tonalite of Hoskings Ranch				tonalite of Wynola					
	J1	J02	J04	Q4	Q1	Q3	Q2	K	SYC	R4	EP6	H7	E4	H3	W0	W1	T6	W7	Wx	W23*
Quartz	38.0	45.0	42.5	43.5	37.5	34.5	22.0	29.5	35.5	28.0	27.0	31.0	27.5	27.0	20.0	20.5	24.5	26.5	15.5	9.0
Plagioclase	28.0	27.0	35.0	26.0	20.5	41.0	45.0	47.0	44.5	28.0	43.0	38.0	43.0	34.0	52.5	52.0	49.0	46.0	52.0	65.0
Potassium Feldspar	9.0	5.5	7.5	15.5	19.5	tr	—	5.0	10.5	33.0	2.1	2.1	6.6	6.9	—	—	3.3	1.4	—	—
Biotite	21.5	16.0	15.0	11.0	17.0	22.5	21.0	17.0	7.0	8.5	21.0	25.0	16.0	22.5	9.5	12.5	15.0	15.5	16.0	6.5
Muscovite	0.5	3.0	0.5	4.0	4.0	—	—	—	—	—	—	—	—	—	—	—	—	—	—	—
Amphibole	—	—	—	—	—	1.5	9.0	1.0	—	—	2.5	2.0	5.0	8.0	16.0	15.0	8.5	10.5	15.5	18.0
Pyroxene	—	—	—	—	—	—	—	0.5	tr	—	3.0	1.0	1.0	2.0	—	tr	tr	tr	tr	—
Sphene	—	—	—	—	—	0.5	0.5	0.5	2.5	1.0	0.5	1.5	2.0	1.5	tr	tr	tr	0.5	tr	tr
Opaques	4.0	3.5	tr	tr	1.5	tr	2.5	—	tr	1.0	1.5	tr	0.5	0.5	3.0	tr	tr	tr	1.0	tr

Table 1. Modal mineralogy expressed in % composition.

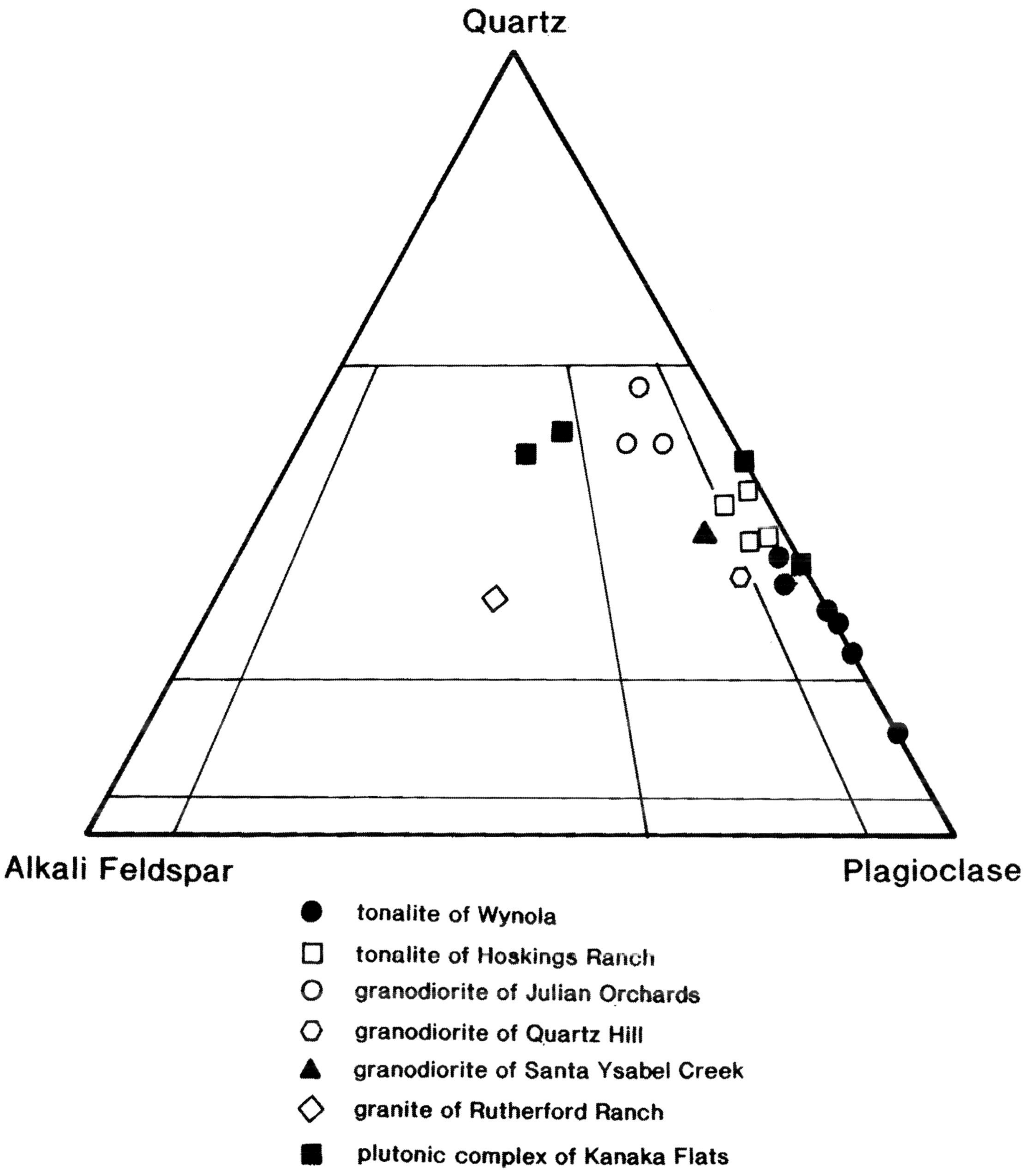

Figure 2. Modal classification of rock samples from the study area (after Streckiesen, 1973).

the most abundant mineral (33%). Sphene and opaque minerals are abundant, but magnetite was not detected. This unit, only mapped north of the Elsinore fault, is at the edge of the area studied and it has not been determined how far the unit extends to the north and east.

TONALITE OF WYNOLA

This is a medium- to coarse-grained grey tonalite (Table 1, Figure 2), commonly exposed in exfoliation cores. It appears in an elongate body which trends north through the village of Wynola to the Elsinore fault, where it appears to be offset right laterally. Its full extent north of the fault has not been determined.

The marginal portion shows strong magmatic foliation and does not contain magnetite. The interior part of the body shows only weak mineral orientation, idiomorphic hornblende up to one centimeter diameter and abundant magnetite (up to 0.7% volume) as calculated from readings on a susceptability meter. Dark enclaves are abundant, and locally it encloses xenoliths of Julian Schist.

The one sample selected north of the Elsinore Fault is actually a quartz diorite (sample W23 in Table 1, Figure 2), thus raising some question as to whether the rocks mapped as tonalite of Wynola north of the Elsinore fault are indeed part of that unit.

DISCUSSION AND CONCLUSIONS

The tonalite of Hoskings Ranch may have been continuous with the rocks mapped by Todd as the granodiorite of Cuyamaca Lake (1977) and the two-mica granodiorite of Julian Orchards probably correlates with the gneiss of Harper Creek (Todd, oral communication, 1991). The contacts and foliation of the Julian Schist and the enclosing foliated granitic rocks are nearly everywhere concordant. These rocks are magnetite-free and are part of what Todd and Shaw (1985) refer to as S-type granitic rocks.

The tonalite of Wynola appears to post-date all or most of the deformation that affected the Julian Schist and its enclosing granitic rocks. And, although the idiomorphic hornblende crystals are similar to those characterizing the hornblende facies of La Posta-type plutons, the pyroxene and magnetite content of the tonalite of Wynola indicate that it belongs with the plutons of the western part of the batholith.

The use of the eastern contact of the tonalite of Wynola as a piercing point to measure the total separation along the Elsinore fault is not without problems. In the field we first thought that the eastern edge of the complex of Kanaka Flat was the eastern edge of the tonalite of Wynola and thus showed us that a second strand of the fault existed about 1 km to the southwest of the main strand. However, our tentative conclusion here is that the rocks of Kanaka Flat are not part of the Wynola body and that despite unresolved complexities near Santa Ysabel Creek in the northwest corner of Section 14 (Santa Ysabel quadrangle), the eastern contact of the tonalite of Wynola extends northwest to within approximately half a km of the Elsinore fault. On the northeast side of the Elsinore fault the southern edge of the tonalite of Wynola here occurs in the northwest corner of section 29 against the granodiorite of Julian Orchards. Using Plate 1, and assuming that the tonalite of Wynola is correctly identified north of the fault, we measure a separation of approximately 5.6 km (3.5 miles).

REFERENCES CITED

Germinario, M.P., 1982, The depositional and tectonic environments of the Julian Schist, Julian, California: Masters thesis, San Diego State University, San Diego, CA, 95 p.

Merriam, R.H., 1958, Geology of the Santa Ysabel quadrngle, San Diego County, CA: California Division of Mines Bulletin, v. 177, p. 7-20.

Streckeisen, A.L., 1973, Plutonic rocks: classification and nomenclature recommended by the I.U.G.S. subcommission on the systematics of igneous rocks: Geotimes, v. 18, p. 26-30.

Todd, V.R., 1977, Geologic map of the Cuyamaca Peak quadrangle, San Diego County, CA: U.S. Geological Survey Open File Report 77-405, p. 1-13.

Todd, V.R. and Shaw, S.E., 1979, Structural, metamorphic, and intrusive framework of the Peninsular Ranges batholith in southern San Diego County, CA, in Abbott, P.L. and Todd, V.R., eds., Mesozoic Crystalline Rocks - Peninsular Ranges batholith and pegmatites, Point Sal ophiolite: Guidebook for Geological Society of America Annual Meeting, San Diego State University, San Diego, CA, p.178-231.

Todd, V.R. and Shaw, S.E., 1985, S-type granitoids and an I-S line in the Peninsular Ranges batholith, southern California: Geology, v. 13, p. 231-233.

Reprinted from *Geologic Investigations in Baja California, Mexico*, 1993, Annual Field Trip Guide Book No. 21, South Coast Geological Society, Inc., San Diego, California.

5. PEGMATITES OF SOUTHERN CALIFORNIA*

By Richard H. Jahns †

INTRODUCTION

For many years the southern California region has attracted the attention of geologists interested in pegmatites and pegmatite deposits, largely because of the well-known gem and lithium occurrences in San Diego and Riverside Counties. These world-famous pegmatites have been so often noted or described in the literature that they commonly are regarded as typical of the pegmatites in the region, even though this actually is far from the case. More than 90 percent of all published contributions on California pegmatites deal with the gem-bearing dikes of San Diego County alone!

It is the main purpose of this brief paper to summarize the distribution, occurrence, composition, and structure of all the known pegmatites in southern California, and to discuss several aspects of their geologic and economic significance. Much of the information has been obtained from the published record, a sampling of which is included in the list of references at the end of the paper. In larger part, however, the writer has found it necessary to draw from the results of his own observations, many of which were made in reconnaissance and hence are not wholly satisfactory as a background for generalizations. This qualification with respect to basic data plainly underlies the summary and discussions that follow.

The pegmatites of southern California can be divided into three major categories in terms of their general form and mode of occurrence:

1. Very small stringers, lenses, and pods of gabbroic and, in a few areas, more granitic composition, that generally appear to have been formed as segregations within masses of genetically related igneous rock.
2. Small, irregular masses, of tonalitic to granitic composition, that form layers, lenses, pods, branching swarms, and stockworks in *lit-par-lit* gneisses and other hybrid rocks, as well as in contact zones between masses of igneous rocks and older rocks that flank or surround them.
3. Small to very large individual masses, of tabular to pod-like form and tonalitic to granitic composition, that ordinarily occur in igneous and metamorphic rocks.

For the sake of brevity, the pegmatites of these three categories are hereinafter referred to as segregation masses, complexes, and discrete pegmatite bodies, respectively. These terms are of necessity much generalized, as each category includes pegmatite bodies of various forms. Further, the terms are intended to be mainly descriptive, and hence to involve a minimum of genetic implication. As might be expected, nearly all gradations can be observed among these three major types of pegmatites.

The distribution and occurence of pegmatites in southern California are summarized in table 1. Those masses that form the complexes are both widespread and locally abundant. They constitute by far the largest part of the pegmatite material in the region, but they have received little detailed attention from petrologists and other investigators. The segregation masses are considerably less abundant, although they are more widespread than published accounts of southern California geology might suggest.

By far the best known pegmatite bodies in the region represent the third category, which includes most of those with complex composition and spectacular mineralogy, and all of those that have yielded commercial quantities of economically desirable minerals. Some of the discrete bodies are very large, and many are structurally complex. These and other features of unusual interest are reflected in the attention given to the pegmatites of this category in the discussions that follow.

SEGREGATION MASSES OF PEGMATITE

Pegmatite masses of the segregation type are locally abundant in the Peninsular Range province (fig. 1), where they occur mainly in gabbroic rocks of the southern California batholith. They also are present in the western San Gabriel Mountains, in a few other, much smaller areas in the Transverse Range province, and in parts of the southern Sierra Nevada. Several scattered occurrences are known from the Mojave Desert and Basin-Range regions.

All of these pegmatites occur in igneous rocks to which they appear to be closely related in composition, age, and genesis. They appear as stringers (fig. 2), dikes, and pod-like masses (fig. 3) that are simple to highly complex in form. They rarely are more than 15 feet long, and most are less than 5 feet in maximum dimension. Contacts with the host rock are sharp to gradational, and the most blended contacts are typical of the most irregular masses of pegmatite.

Most of the segregation pegmatites are gabbroic in composition, and consist chiefly of hornblende and calcic to intermediate plagioclase in various proportions. Other mineral constituents include biotite, pyroxenes, apatite, magnetite, chlorite, epidote, and, rarely, quartz and alkali feldspars. Segregation masses of more granitic composition also have been observed, mainly in coarse-grained, felsic intrusives in parts of the Sierra Nevada and several mountain ranges of the eastern Mojave Desert. These pegmatites consist chiefly of perthite, quartz, and sodic plagioclase, with or without subordinate

* Contribution No. 653, Division of the Geological Sciences, California Institute of Technology.
† Professor of Geology, California Institute of Technology.

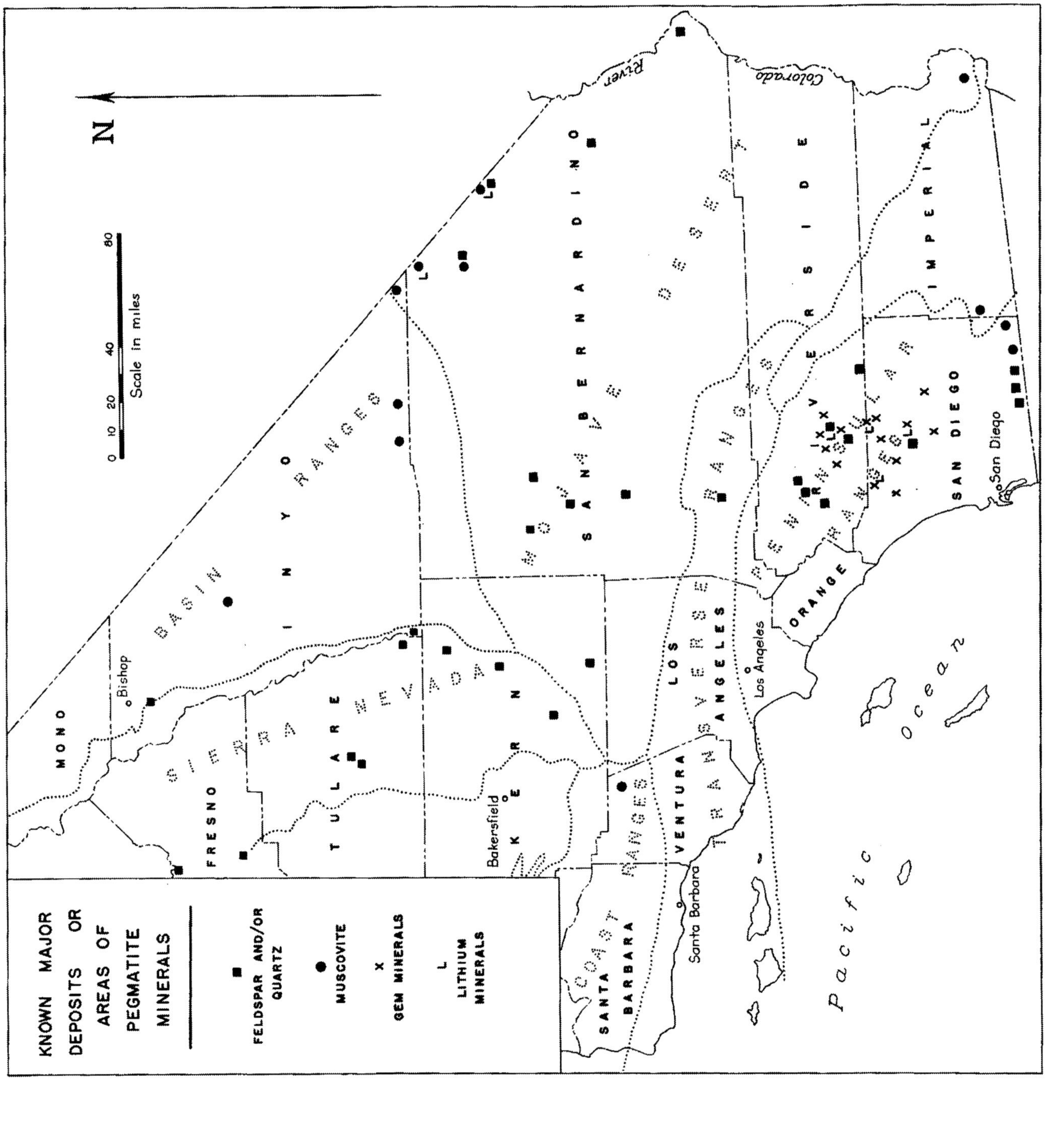

FIGURE 1. Index map of southern California, showing natural provinces and major pegmatite areas.

Table 1. Distribution and occurrence of pegmatites in southern California.

Province	General nature of pegmatite bodies	General composition of pegmatite bodies	Age of pegmatite	General abundance of pegmatite bodies
Sierra Nevada	Small to large dikes, sills, and pod-like masses	Tonalitic to granitic	Jurassic?	Sparse; abundant in only a few areas
	Very small stringers and pod-like masses in igneous rocks	Gabbroic	Jurassic?	Rare
	Numerous small, irregular masses in *lit-par-lit* gneisses and other hybrid rocks	Tonalitic to granitic	Jurassic?	Abundant
Basin Ranges	Small to large dikes, sills, and pod-like masses	Mainly granitic	Pre-Cambrian	Sparse to locally abundant
	Small to moderately large dikes and pod-like masses	Mainly monzonitic to granitic	Mesozoic	Sparse to very rare
	Numerous small, irregular masses in *lit-par-lit* gneisses and other hybrid rocks	Tonalitic to granitic	Mainly pre-Cambrian	Sparse; abundant in only a few areas
Mojave Desert	Small to large dikes, sills, and pod-like masses	Mainly granitic*	Pre-Cambrian	Sparse to locally abundant
	Small to moderately large dikes and pod-like masses	Mainly monzonitic to granitic	Mesozoic	Sparse to very rare
	Numerous small, irregular segregations in igneous rocks	Mainly granitic; some gabbroic	Pre-Cambrian and Mesozoic	Abundant in a few areas; very rare elsewhere
	Numerous small, irregular masses in *lit-par-lit* gneisses and other hybrid rocks	Tonalitic to granitic	Mainly pre-Cambrian	Sparse; abundant in only a few cases
Transverse Ranges	Small dikes, sills, and pod-like masses	Tonalitic to granitic	Cretaceous or older	Sparse to moderately abundant
	Very small stringers and pod-like masses	Dioritic to gabbroic	Cretaceous or older	Abundant in a few areas; absent elsewhere
	Numerous small, irregular masses in *lit-par-lit* gneisses and other hybrid rocks	Dioritic to granitic	Cretaceous or older	Rare to locally abundant

Table 1. Distribution and occurrence of pegmatites in southern California—Continued.

Province	General nature of pegmatite bodies	General composition of pegmatite bodies	Age of pegmatite	General abundance of pegmatite bodies
Peninsular Ranges	Relatively continuous dikes and elongate pod-like masses	Mainly granitic†	Cretaceous	Very abundant as swarms in many well-defined areas; sparse elsewhere
	Small to moderately large dikes and pod-like masses	Tonalitic to quartz-monzonitic	Cretaceous and older	Sparse to very abundant
	Very small stringers and pod-like masses	Dioritic to gabbroic	Cretaceous	Locally abundant; absent elsewhere
	Numerous small, irregular masses, swarms, and stockworks in igneous, metamorphic, and hybrid rocks	Tonalitic to granitic	Cretaceous and older	Rare to locally abundant

* Includes a few bodies of lithium-bearing pegmatite.
† Includes numerous bodies of lithium-bearing pegmatite.

muscovite and biotite. Some of the potash feldspar contains graphic intergrowths of quartz. Minor amounts of garnet, tourmaline, magnetite, apatite, beryl, sphene, chlorite, epidote, and calcite are present locally.

The major minerals generally form anhedral to subhedral crystals that are $\frac{1}{2}$ inch to 6 inches in maximum dimension. Most of the gabbroic pegmatites are uniform aggregates of such crystals, but in others the minerals show a distinct zonal distribution with respect to the walls of the containing body. This is in every way similar to the zoning described for granitic pegmatites by Cameron, et al. (1949). In general the pyroxene is near the walls, the hornblende is nearer the central part of the body, and the plagioclase is more widely distributed (fig. 3). Quartz and alkali feldspars, where present, are centrally disposed.

A similar zonal structure appears in most of the less basic segregation masses, and typical wall-to-center sequences include oligoclase—perthite—quartz, graphic granite—perthite—quartz (fig. 4), and oligoclase—graphic granite—perthite—muscovite and sodic albite—quartz.

Many of the segregation masses appear to have been formed in place, and they probably represent local concentrations of mineralizer-rich fluids that were developed during the end stages of crystallization of the enclosing igneous rock. Others evidently were injected

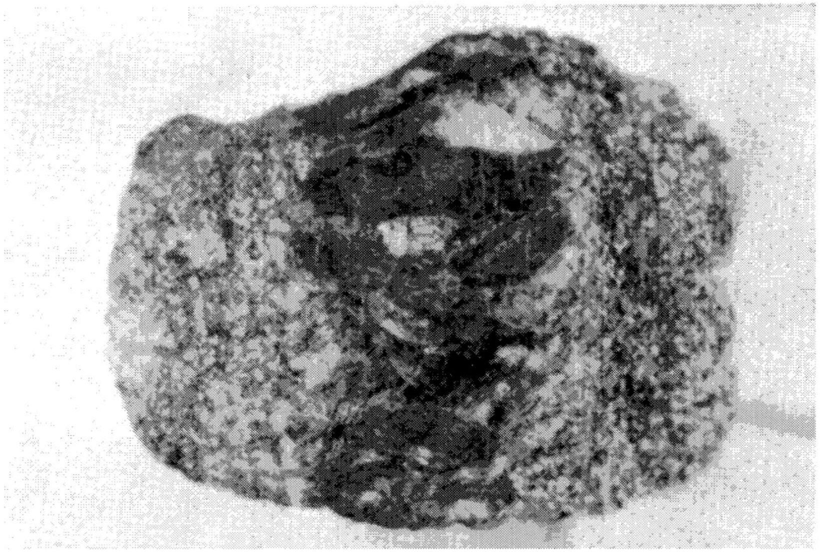

FIGURE 2. Three-inch stringer of hornblende-rich pegmatite in medium-grained gabbro, western San Gabriel Mountains.

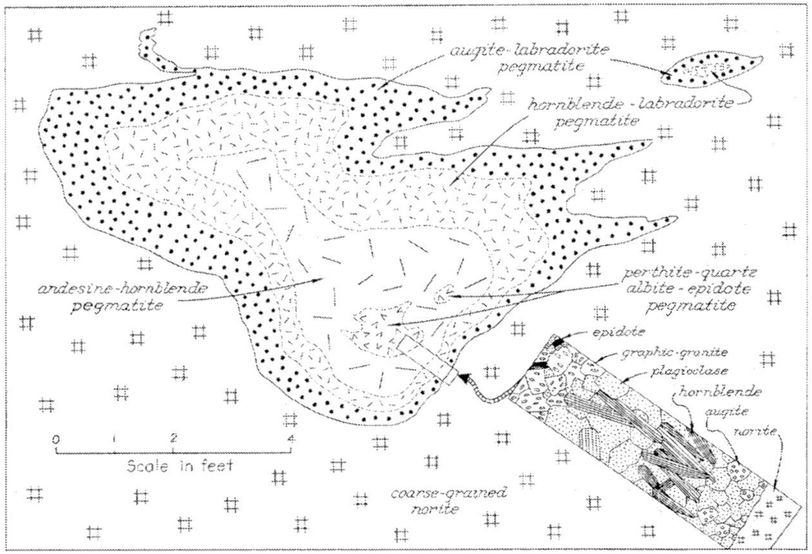

FIGURE 3. Pod-like mass of gabbroic pegmatite in coarse-grained norite, western San Gabriel Mountains.

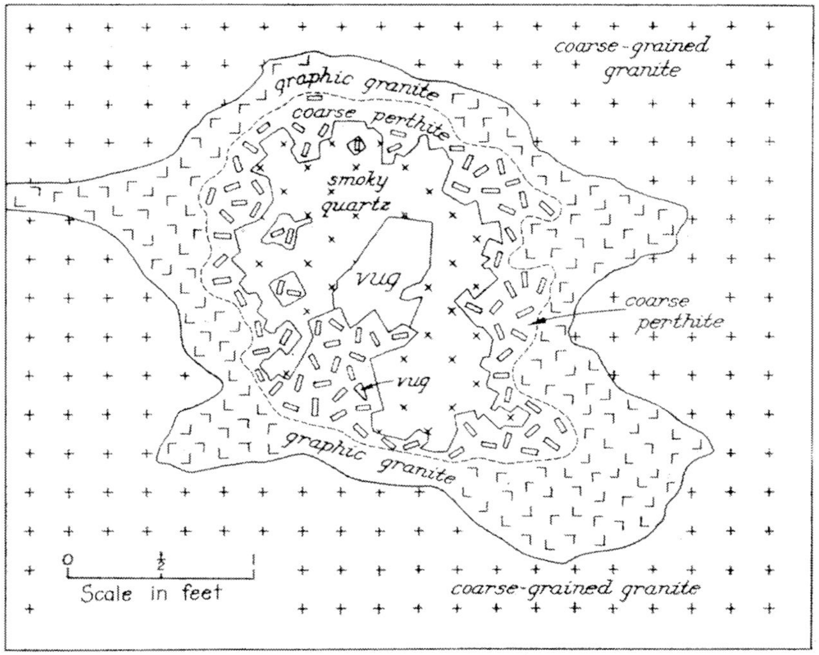

FIGURE 4. Pod-like mass of granitic pegmatite in coarse-grained granite, Whipple Mountains.

into solid or nearly solid rock, and perhaps were derived from pockets of fluid that had formed elsewhere in the crystallizing intrusive mass. A variety of these auto-injection features has been described by F. S. Miller (1938, pp. 1220-1222, 1230) from gabbroic rocks of the Peninsular Range province. In no occurrences do the pegmatitic fluids appear to have moved for great distances from their sources.

In the gabbroic complex of the western San Gabriel Mountains, both anorthosite and norite are extremely coarse grained (see Higgs, Contribution 8, this chapter), and hence might be regarded as special varieties of pegmatite. These rocks form gigantic segregation masses, but no complete gradations in size are known between these masses and the small ones described above.

Most of the segregation masses of pegmatite in southern California appear to be genetic associates of Mesozoic igneous rocks, especially in the western parts of the region. Farther east, in the Mojave Desert and Basin-Range provinces, many of the pegmatite-bearing intrusives are pre-Cambrian in age.

PEGMATITE COMPLEXES AND HYBRID ROCKS

Pegmatite complexes, consisting in general of numerous small, irregular masses (fig. 5), are significant parts of the older crystalline rocks that form the so-called "basement complex" of many areas in southern California. They occur, for example, in the pre-

FIGURE 5. Typical complex of granodiorite pegmatite in hybrid gneiss, western San Gabriel Mountains.

batholithic rocks of the Peninsular Ranges and the Sierra Nevada, in the metamorphic rocks of the San Gabriel and San Bernardino Mountains, and in the early pre-Cambrian rocks of several desert ranges in the southeastern part of the State. In addition, they appear as parts of hybrid rocks and pegmatitized zones at and near the margins of many bodies of intrusive rocks. These contact complexes are chiefly of Mesozoic age in southwestern California, but both pre-Cambrian and Mesozoic ages are represented in areas farther east and northeast.

Though mentioned repeatedly in descriptions of crystalline rocks in southern California (for example, W. J. Miller, 1946), the pegmatite complexes have not been described in detail. They generally are referred to as groups of subparallel, anastomosing, or scattered small pegmatites, and all too often are dismissed as complicating elements in rocks that already seem too complex for interpretation by ordinary methods of study. The few detailed observations thus far made suggest that different pegmatites in some of these complexes were injected into, or were formed within, the host rocks at different times, and that the age differences commonly are reflected by differences in composition (fig. 6). The pegmatites also show significant relations to episodes of deformation, recrystallization, and reconstitution of the host rock, and further detailed study of such occurrences might well provide valuable information concerning the development of the rock as a whole.

Most of the pegmatite complexes are mineralogically simple, and consist of quartz, plagioclase, and potash feldspar, with or without micas and accessory constituents. In gross composition they range from tonalitic to granitic, and most are in the general granodioritic field. Many individual pegmatites in the complexes appear to be internally homogeneous, whereas others, especially the largest and most bulbous ones, show well-developed zonal structure like that described for the discrete pegmatites farther on.

Many of the pegmatite complexes are demonstrably intrusive into the containing rocks, whereas others just as clearly were developed in place by replacement of the host rock. Most appear to have been formed by some combination of the two general processes, but the quantitative relations are not easily determined, especially where the pegmatites form parts of lit-par-lit gneisses and other hybrid rocks. Further, it is evident that various complexes differ from one another in age, in conditions of emplacement, and in source of the pegmatite material. Much more study is needed before the problems can be properly defined, let alone solved!

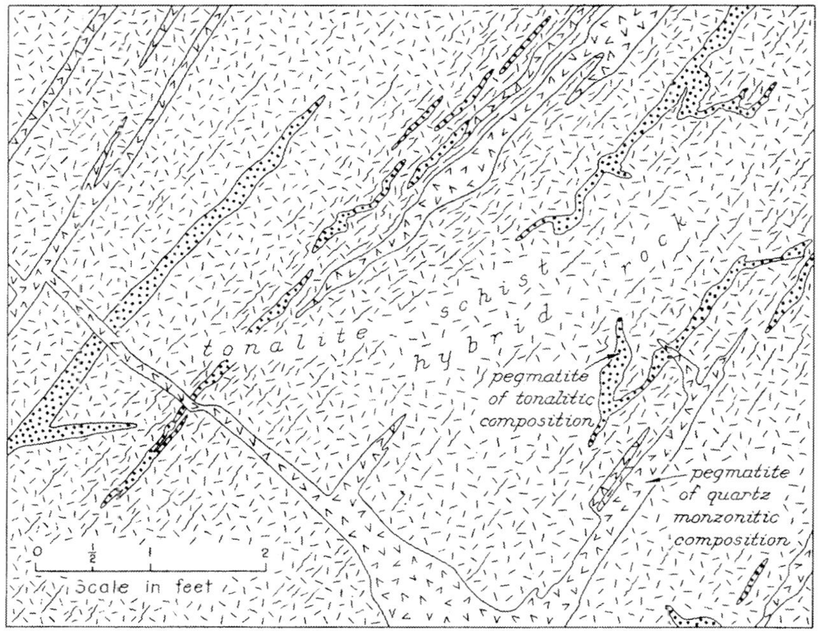

FIGURE 6. Pegmatite complex in tonalite-schist hybrid rock, near Escondido. Pegmatites of two different ages are represented.

DISCRETE PEGMATITE BODIES

Distribution and Occurrence. The discrete pegmatite bodies in southern California appear as dikes, sills, lenses, and masses of bulbous or highly irregular form. They are widely distributed in the Sierra Nevada, in the Basin-Range province to the east, in the Mojave Desert region, and in the Transverse Ranges and Peninsular Ranges in the southwestern part of the State (fig. 1).

As noted in table 1, these individual masses are particularly abundant in the Peninsular Range province, where most of them are genetically related to rocks of the great composite batholith of southern California (see Larsen, Contribution 3, this chapter). Pegmatites of tonalitic, granodioritic, and quartz-monzonitic composition are closely associated with stocks and other large plutons of similar respective compositions, and also appear in septa, screens, and larger masses of pre-batholithic metamorphic rocks. Somewhat younger pegmatites of granitic composition also are genetically related to the batholith, and occur mainly as subparallel dikes in certain of the gabbroic and tonalitic plutons. These include numerous gem- and lithium-bearing dikes. A few other pegmatites, mostly of granodioritic and quartz monzonitic composition, appear to antedate rocks of the batholith.

On the basis of their geologic relations, the pegmatites of the southern California batholith probably are middle or late Cretaceous in age (Jahns and Wright, 1951, pp. 18, 44). Isotopic analyses of lepidolite from the Stewart dike, at Pala, suggest ages ranging from about 110 million years (Ahrens, 1949, p. 250) to 147 ± 5 million years (Davis and Aldrich, 1953, p. 380), which would indicate a range from middle Cretaceous to middle Jurassic. An age of 100 million years, based on analyses of clean zircon concentrates, has been calculated for a tonalite (Larsen, et al., 1952) that represents the batholith. This tonalite is older than most of the pegmatites.

Individual masses of pegmatite also are abundant in parts of the southern Sierra Nevada, where they appear in plutons of the Sierra Nevada batholith and in older masses of schist, quartzite, and other metamorphic rocks. The batholith rocks are commonly regarded as Jurassic in age, and these pegmatites are therefore tentatively assigned to this period (see table 1).

Dikes, sills, and other bodies of pegmatite also are present in the Transverse Ranges, and are locally abundant in the San Gabriel, San Bernardino, and Little San Bernardino Mountains. Few of them are as large as many of the pegmatites in the region to the south and east, but they show the same general ranges in composition. Some of them may be as young as Cretaceous, but it is quite possible that many of them are Jurassic or older.

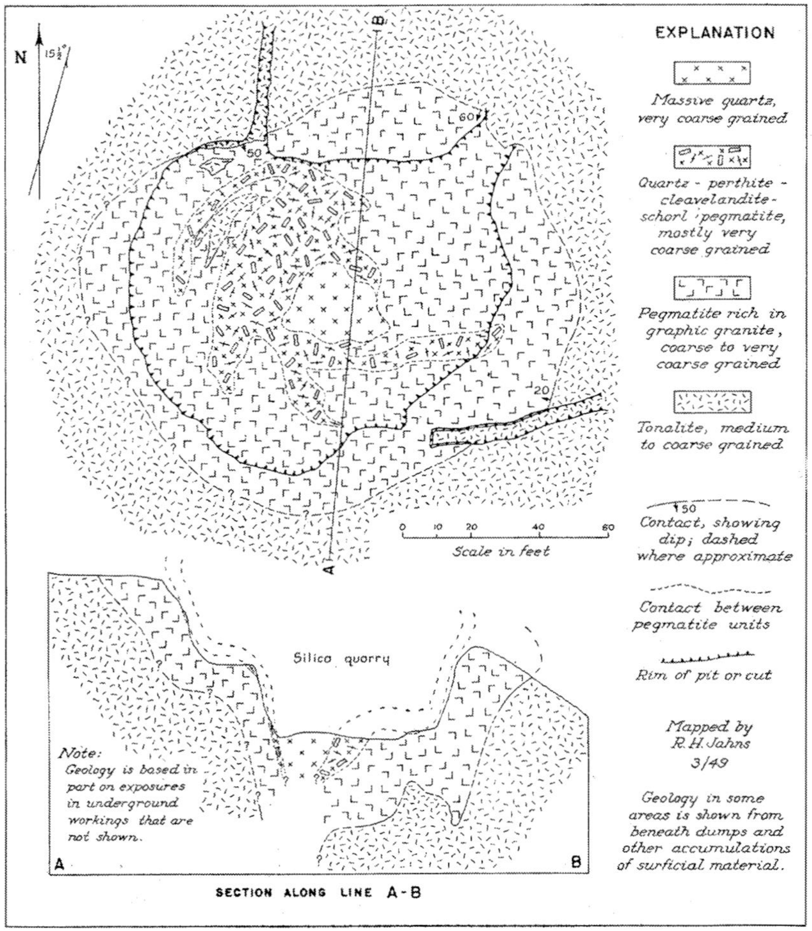

FIGURE 7. Geologic sketch map and cross section of the Southern Pacific pegmatite, near Nuevo in western Riverside County.

Although crystalline rocks are well represented in the Basin-Range and Mojave Desert provinces of southeastern California, remarkably few pegmatites are present in these large areas (fig. 1). Scattered occurrences are known from the Inyo Mountains, the Panamint Range, the Amargosa Range, the New York Mountains, the Whipple Mountains, the Granite Mountains, and many other ranges and buttes in the region, but still other broad terranes of igneous and metamorphic rocks appear to be virtually barren of pegmatites. Among the bodies of pegmatite that have been observed, many are probably of Mesozoic age, especially in the western parts of the

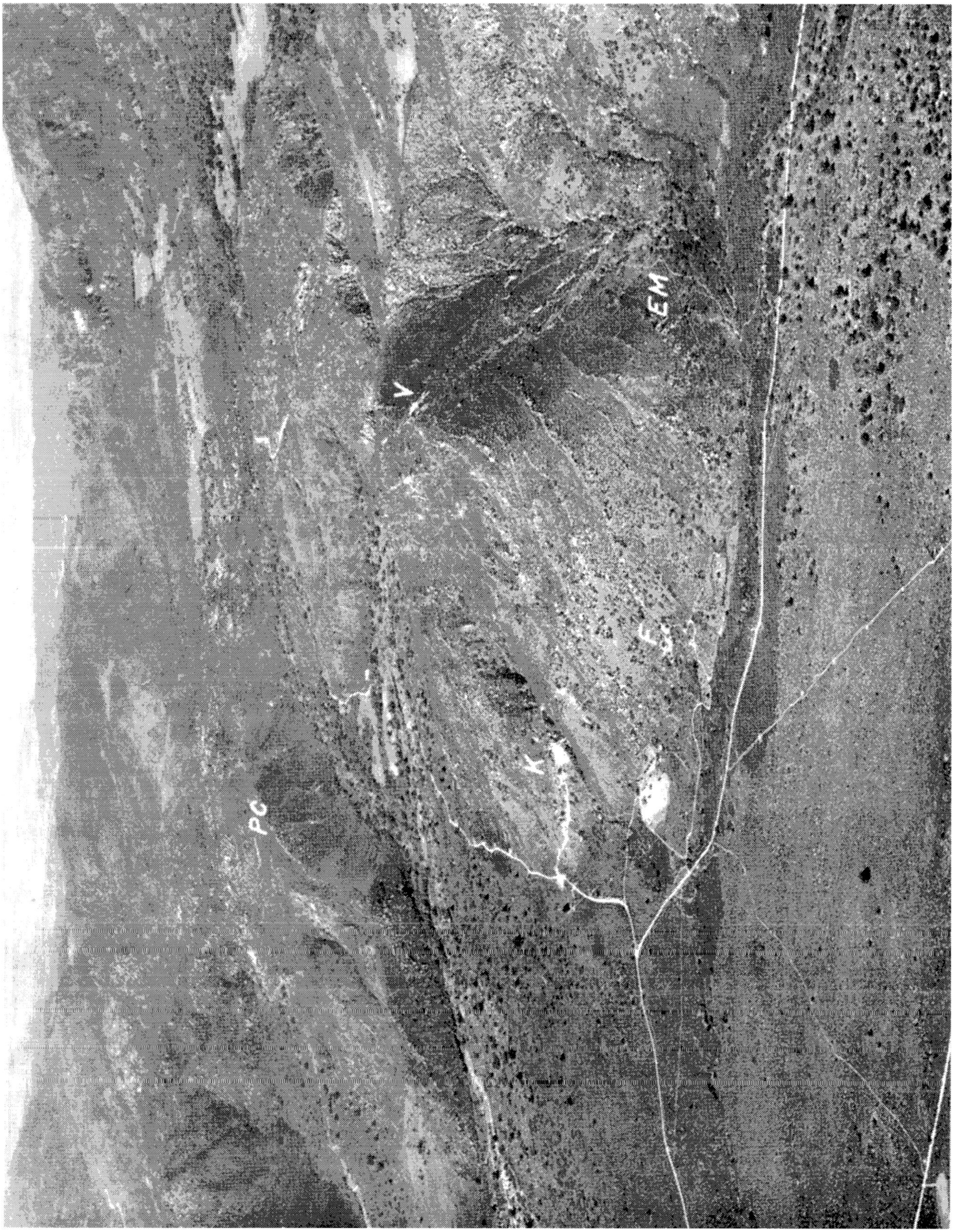

FIGURE 8. Aerial view north-northwest over a part of the Pala district, San Diego County. Subparallel dikes of gem-bearing pegmatite form rib-like outcrops on Hiriart Mountain, in foreground, and on Little Chief Mountain, immediately beyond wash in foreground at left. Gem mines in the view include the Katerina (K), Vanderburg (V), El Molino (EM), Fargo (F), and Pala Chief (PC). Elsinore fault zone at upper right. ***Pacific Air Industries photo.***

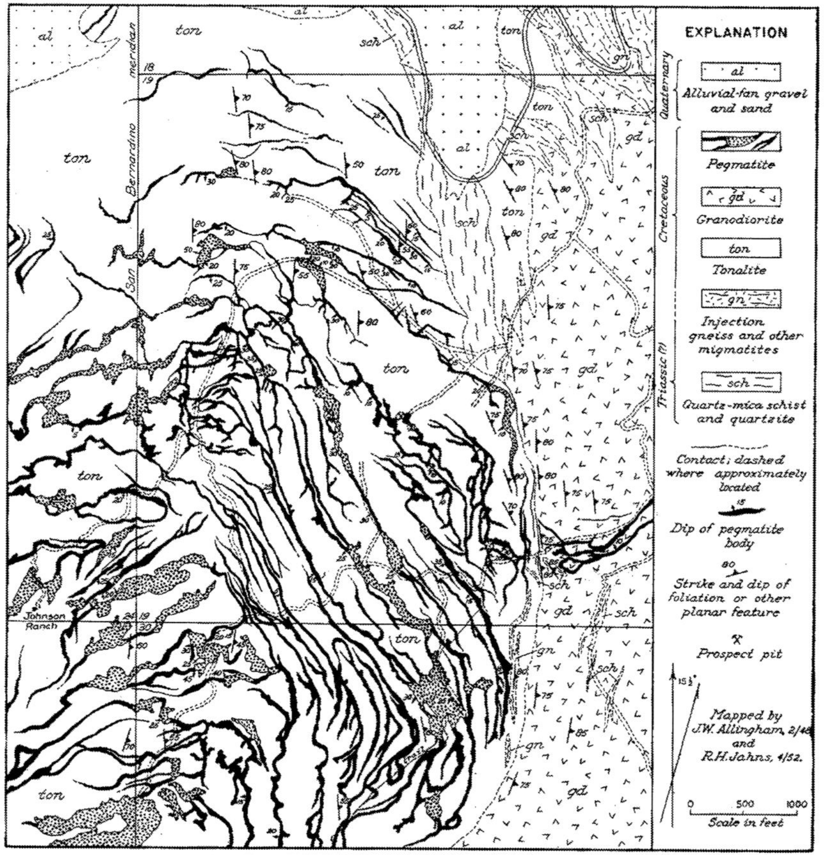

FIGURE 9. Swarm of pegmatite dikes on Rincon Mountain, San Diego County.

region. Most of the remainder are pre-Cambrian in age, and some of those in the extreme northeastern part of San Bernardino County and the southeastern part of Inyo County may represent the northwestern edge of the so-called Arizona pegmatite belt, which embraces a large number of pre-Cambrian pegmatites in west-central Arizona.

Structural Features. The discrete pegmatites in southern California are highly variable in general form, attitude, and size, and do not differ markedly in these features from pegmatite bodies in many other parts of the world. Most are a few feet to about 500 feet in maximum dimension, and are moderately thick with respect to their length and width (fig. 7). Discordant bodies are predominant, and even the few that are concordant transect the country-rock structure in detail. The pegmatite-wallrock contacts generally are sharp, but in several occurrences they are gradational over distances of a few inches.

Noteworthy exceptions to some of the above generalizations are provided by most of the granitic pegmatites of the Peninsular Range province, and by a few of those in the southern Sierra Nevada. These are dikes with unusually large along-strike and down-dip dimensions relative to their thickness. They range in length from a few feet to nearly a mile, and are slightly less than 10 feet in average thickness. In most districts they are very uniform in attitude, and have moderate to gentle dips. They commonly appear as swarms, in which the individual dikes are essentially parallel to one another but tend to branch and join in detail (figs. 8, 9).

These thin and remarkably continuous dikes evidently were emplaced along subparallel fractures that are best developed in plutons of gabbroic and tonalitic rocks. The fractures are independent of schistosity, foliation, lineation, and primary flow layering in these rocks, and are not related systematically to the walls of individual plutons. They are thought to have been formed as tensional features during the end stages of cooling in the southern California and Sierra Nevada batholiths (Jahns and Wright, 1951, p. 18).

Many of the discrete pegmatites, regardless of their general form, appear to be essentially homogeneous internally, in that they are simple aggregates of quartz and feldspars, with or without micas and accessory minerals. Many others, in contrast, show a systematic internal arrangement of their constituent minerals, which ordinarily appears as a well-defined zonal structure (Cameron, et al., 1949; Jahns and Wright, 1951; Hanley, 1951). The pattern of zoning is remarkably consistent from one pegmatite body to another, and even from one pegmatite district to another. Each zone, which can be distinguished from the immediately adjacent zone on the basis of differences in mineralogy or texture, reflects to some degree the shape and size of the containing pegmatite body. The zonal structure in some pegmatite bodies is complicated by younger fracture fillings and replacement masses of closely related pegmatite, and locally by younger pegmatite dikes and sills that commonly are also zoned (fig. 10).

Composition and Texture. The gross composition of the discrete pegmatites in southern California ranges from tonalitic to granitic. Gabbroic pegmatites are by no means rare, but most of these seem best classified as segregation masses.

Quartz, perthite (generally microcline with lenticular lamellae of albite), and sodic plagioclase are the dominant minerals, and commonly are accompanied by muscovite, biotite, apatite, beryl, garnet, sulfide minerals, and tourmaline, either singly or in some combination. Rarer accessory constituents include bismuth minerals, cas-

FIGURE 10. Composite pegmatite mass, Pala district, San Diego County. Three-foot sill of coarse-grained pegmatite is flanked by older, much finer-grained pegmatite. Note the sharp contact above hammer head. The margins of the sill are fine-grained quartz-albite-perthite pegmatite with muscovite, garnet, and schorl, and the inner parts consist mainly of coarse-grained graphic granite and quartz.

siterite, columbite-tantalite, epidote, microlite, monazite, spinel, topaz, and triplite (Murdoch and Webb, 1948, 1952; Jahns and Wright, 1951). Zeolites and clay minerals are widespread, but rarely are abundant. Lithium minerals are locally abundant in granitic pegmatites of the Peninsular Range province, and are present in a few pegmatites in the eastern Mojave Desert region.

The pegmatites that show no pronounced zonal structure generally are medium to coarse grained, and contain no marked concentrations of the less common minerals. Minor amounts of apatite, beryl, garnet, and other accessory minerals are disseminated through some of these rocks.

The outer parts of the zoned pegmatites generally are granitoid in texture, and are fine- to coarse-grained aggregates of feldspars and quartz, with or without other minerals. The inner zones, in contrast, are essentially monomineralic or consist of two or more minerals in coarse- to giant-textured aggregates (figs. 10, 11). Some of the crystals are truly enormous, and have dimensions measured in feet or even in tens of feet.

Perthite, much of it graphically intergrown with quartz, is abundant near the walls of many pegmatites, but where both plagioclase and potash feldspar are present in a given body, the plagioclase ordinarily lies nearer the walls. A noteworthy exception is the cleavelandite variety of albite, which has a wider distribution and is a common constituent of fracture fillings and replacement bodies. Quartz and potash feldspar are the most abundant minerals of the inner zones (figs. 7, 10), and the quartz is accompanied by spodumene and lepidolite in most of the lithium-rich pegmatites.

FIGURE 11. Very coarse-grained perthite and quartz, with large prisms of schorl. Perthite crystals are fringed with aggregates of muscovite and albite. Stewart mine, San Diego County.

The distribution of the less abundant minerals is highly irregular in detail, but does follow broadly consistent patterns. A given mineral, for example, commonly is restricted to a zone or other lithologic unit that is quite distinct from adjacent units in the containing pegmatite body. Where the mineral occurs in more than one unit in the body, it generally shows consistent differences in composition and physical properties from one unit to the next. These features have been discussed in detail elsewhere (for example, Cameron, et al., 1949), and need no further review here.

Masses of fine-grained rock that is essentially quartz-albite-microcline aplite occur in some of the pegmatite bodies, and are by far most abundant in the dikes of granitic composition in the Peninsular Range province. They are most prominent in the footwall parts of these dikes, but also occur elsewhere. Some varieties of the aplite are distinctly layered, and the planar structure is accentuated in many occurrences by thin, subparallel layers that are rich in tiny crystals of garnet and/or schorl. This variety of aplite is known in several districts as "line rock" (fig. 12).

Origin. The discrete pegmatites appear to have been formed by injection of liquid material along fractures or other planes of weakness in the host rocks, in some occurrences accompanied by minor digestion of the older rocks. This is in sharp contrast to many of the pegmatite complexes already described, in which much of the pegmatite clearly was developed by replacement of the older rock. The intrusive origin of the discrete pegmatites is evidenced mainly by combinations of the following features:

1. The borders of most masses are sharp, and commonly can be traced across contacts between different types of country rock without change in attitude.
2. The pegmatite bodies themselves show no changes in composition or internal structure that can be correlated with differences in country-rock lithology, nor does the internal structure of these bodies reflect the detailed structure of the country rock.
3. Schist and other thinly foliated types of country rock commonly are crumpled or otherwise disturbed immediately adjacent to many pegmatite contacts.
4. The structure, mineralogy, and age relations of the units within the zoned pegmatites are not compatible with a process of development through replacement of country rock (Cameron, et al., 1949, pp. 97-106).

Following their emplacement, the discrete pegmatites appear to have crystallized from their walls inward, with much the same mineral sequence as in any ordinary plutonic or hypabyssal rock. The structure of the zoned pegmatites seems best attributed to fractional crystallization and incomplete reaction between successive crops of crystals and rest-liquid (Cameron, et al., 1949, pp. 97-106; Jahns, 1953, pp. 580-595), chiefly because of the remarkable consistencies in age relations and textural and compositional trends from one zone to another.

FIGURE 12. Sharply layered quartz-albite-garnet-microcline aplite (line rock) in contact with very coarse-grained quartz-rich pegmatite, near Pala, San Diego County.

Complications in the development of many of the pegmatites, especially during the end stages, are evidenced by widespread replacement features. In particular, much albite, muscovite, lepidolite, and quartz appears to have been formed in part by replacement of other pegmatite minerals, and evidences of corrosion and replacement are especially striking in the inner, pocket-bearing parts of many of the gem-bearing pegmatites. The total amount of replacement material rarely is large relative to the pegmatite body as a whole, but it may be moderately large in a few of the mineralogically complex bodies. Much of the replacement can be explained in terms of deuteric action during the intermediate and end stages of crystallization with a given mass of pegmatite.

The origin of the aplitic rocks within many of the pegmatite bodies is less clear. This fine-grained material has been variously interpreted as an early product of simple crystallization from a hydrous magma (Waring 1905, p. 366), as an early, probably primary aplite, in some of which a rhythmic replacement yielded line rock and other layered varieties (Merriam, 1946, pp. 242-243), and as a much later result of replacement of pre-existing graphic granite by soda-rich solutions (Schaller, 1925, pp. 274-277). Evidence now at hand does not warrant a single definite conclusion, although it is clear that the aplitic rocks are integral parts of the pegmatite bodies in which they occur, and that some of these rocks show impressive evidence of mineral replacements.

GEOCHEMICAL FEATURES OF THE PEGMATITES

In general, the segregation pegmatites and the discrete pegmatites were formed from fluids that traveled varying distances from their sources, and can be regarded as late-stage products of magmatic differentiation. As such, they not only reflect the gross composition of larger igneous masses with which they are genetically associated, but also contain concentrations of rare elements that were present in the original magmas. Thus they provide samples of various magmas in the region—samples in which both major and minor constituents are relatively easy to recognize and study. Unfortunately, some of the pegmatite fluids seem to have moved such great distances from their sources that no genetically related rocks are exposed nearby. Certainly this appears to be the case with the granitic pegmatites of the Peninsular Range province.

Nearly all of the pegmatite bodies, including those of great mineralogic and structural complexity, are astonishingly similar to ordinary igneous rocks in gross composition. Even those with prominent masses of quartz rarely contain unusually high percentages of SiO_2, in terms of their entire bulk. Similarly, the percentages of alumina, lime, and the alkalies are in the normal ranges, and only iron is present in consistently low concentrations.

The minor constituents show some interesting trends. On the basis of mineralogic and chemical studies thus far made on the pegmatites of the Peninsular Range province, the magmas of the southern California batholith appear to have contained notable concentrations of boron, relative to magmas in other parts of the southwestern United States. Tourmaline is both widespread and abundant, and axinite is an accessory constituent of numerous pegmatites that are flanked or surrounded by limestone. Barium and strontium also are unusually abundant, and are concentrated mainly in the alkali feldspars. Lithium, phosphorous, and rare-earth elements are prominent minor constituents of many dikes, and thorium is well represented in several of them. Beryllium, bismuth, cesium, fluorine, manganese, columbium (niobium), rubidium, tantalum, titanium, and tin also are present, but not in unusual concentrations. Conspicuously rare are uranium, vanadium, and zirconium.

These observations are in good agreement with the results of trace-element studies that have been made on the various non-pegmatitic rocks of the batholith (see Larsen, Contribution 3, this chapter), except for the case of lithium, which has a low concentration in these rocks relative to the average igneous rock. Perhaps the concentrations of this element in the Peninsular Range pegmatites, relative to the average pegmatite, in part reflect the great difficulty with which it can be fixed in the earlier crystallizing minerals of the non-pegmatitic rocks, thus resulting in a high partition factor for lithium in favor of pegmatite fluids that separated from crystallizing masses of these rocks.

The trends in minor-element distribution are less well known for pegmatites in other parts of southern California, but a few generalizations can be made. Many of the Mesozoic pegmatites contain notable concentrations of boron, phosphorous, rare-earth elements, and titanium, and some of them are relatively rich in thorium-bearing minerals, as well. Although uranium-bearing minerals are known from a few of the pegmatites (for example, Hewett and Glass, 1953; Neuerburg, 1954), this element appears to be present in concentrations far below the average for pegmatites in general.

The pre-Cambrian pegmatites, and especially those in the eastern parts of the Basin-Range and Mojave Desert regions, evidently represent an entirely different geochemical province. They are relatively poor in boron, but are remarkably rich in fluorine. Many also contain moderate concentrations of rare-earth elements and tungsten.

The distribution of quantitatively minor elements within the southern California pegmatites is typical of that for pegmatites in general. Some of these elements form minerals of their own, and others appear as guests in minerals that generally were developed at a late stage during crystallization of the containing pegmatite body. Thus beryllium forms beryl, bavenite, bertrandite, gadolinite, phenakite, and helvite; bismuth occurs as a sulfide, oxide, vanadate, and in several carbonate minerals; tantalum and columbium (niobium) form tantalite-columbite, microlite, pyrochlore, hatchettolite, and several other complex minerals; and the rare-earths appear in

*Table 2. Occurrence of commercially desirable minerals in the pegmatites of southern California.**

Mineral or commodity	Province or region				
	Sierra Nevada	Basin Ranges	Mojave Desert	Transverse Ranges	Peninsular Ranges
Coarse potash feldspar	Common	Common	Common	Rare	Common
Mixed soda and potash feldspars, coarse†	Common	Common	Common	Sparse	Sparse
Sheet muscovite	Rare	Sparse	Very rare	Very rare	None known
Scrap muscovite	Rare	Common	Sparse	Rare	Sparse
Commercial quartz†	Common	Common	Common	Rare	Common
Commercial beryl	Sparse	Sparse	Rare	Very rare	Rare
Commercial spodumene	None known	None known	Very rare	None known	Sparse
Lepidolite	Very rare	Very rare	Very rare	None known	Common
Tantalum-columbium minerals	Very rare	Rare	Very rare	None known	Rare
Rare-earth minerals	Very rare	Rare	Sparse	Very rare	Sparse
Gem minerals	Very rare	Very rare	Very rare	None known	Common

* The terms used in this table apply to the individual pegmatite bodies within a given province or region; for an appraisal of the general abundance of each mineral or commodity within such a province or region, the terms should be combined with the data in the right-hand column of table 1. The terms are intended to have commercial application as follows:
Common—major or widespread constituent of several pegmatites, or lesser constituent of many; commercially significant.
Sparse—major or widespread constituent of only one or two pegmatites, or moderately abundant in a few, or a minor constituent of many; of limited commercial significance.
Rare—minor constituent of a few pegmatites; of little commercial significance.
Very rare—known constituent of one or more pegmatites, but of mineralogic interest only.
† Rarely or never marketed because of geographic occurrence and/or low commercial demand.

monazite, allanite, euxenite, gadolinite, pyrochlore, samarskite, and xenotime. Cesium forms very rare pollucite, and is present as a guest element in much of the beryl. Fluorine forms fluorite in many of the pegmatites, and also is a widespread minor constituent of apatite, topaz, amblygonite, micas, and other minerals. Rubidium is a guest constituent of lepidolite and potash feldspar, and both strontium and barium appear in the alkali feldspars. Boron is mainly an essential constituent of the tourmaline and axinite.

ECONOMIC FEATURES OF THE PEGMATITES

Commercial operations in the pegmatites of southern California have yielded substantial amounts of quartz, feldspars, mica, lithium minerals, gem minerals, and other commodities, the total value of which probably exceeds 4 million dollars. Slightly more than half of this value is represented by the output of gem materials, consisting mainly of tourmaline, spodumene, beryl, topaz, garnet, and quartz. The geographic occurrence of these and other commercially desirable minerals in southern California is indicated in table 2.

The gem-bearing pegmatites of the Peninsular Range province have been mined intermittently since about 1890, but much of the easily recoverable material appears to have been worked out. Additional masses of gem-bearing ground are present in several districts, and detailed geologic studies have disclosed numerous structural and mineralogic features that should aid further prospecting and exploration; nevertheless, new concentrations of gem minerals undoubtedly will be more difficult and expensive to find than those already known and worked.

Feldspar has been mined from at least 40 large bodies of pegmatite in the eastern and southern parts of the Peninsular Range province, and from a few scattered occurrences in the Mojave Desert and Basin-Range regions. Production has amounted to about 150,000 tons, nearly all of which has consisted of very coarse-grained potash feldspar obtained from the inner parts of zoned pegmatite bodies (figs. 7, 13). Additional reserves of high-grade material are present, but commercial operations have been seriously limited by the distances of most deposits from centers of demand, as well as by keen competition from talc and other minerals, especially in the ceramic industry.

Small tonnages of pegmatite quartz have been obtained, chiefly for special ceramic uses, from several deposits in the Peninsular Range, Sierra Nevada, Basin-Range, and Mojave Desert provinces. The quartz ordinarily mined is very coarse grained and relatively pure, and commonly is associated with coarse potash feldspar in the central parts of large pegmatite bodies (figs. 7, 13). Demands for this material rarely are large.

No deposits of sheet muscovite are known from most pegmatite areas in southern California, and only a few small lots of this material have been obtained from the deposits that are sparsely scattered through the southeastern part of the State, and through parts of the Sierra Nevada and Transverse Range provinces. Mining for scrap mica has been more widespread, especially in the southeastern part of the Peninsular Range province, but production never has been large.

The Stewart mine, in San Diego County, yielded large tonnages of lepidolite, as well as some amblygonite and spodumene, during the period 1895-1928, and at one time was the largest domestic producer of lithium minerals. The two main ore bodies have been worked out, but additional concentrations of lepidolite and spodumene may be present elsewhere in this large pegmatite body. Small quantities of lithium minerals are known to occur in several other pegmatites

FIGURE 13. Very coarse-grained pegmatite, rich in quartz and perthite. Dark-colored rock is a septum of quartz-biotite schist. Hauser Canyon feldspar quarry, San Diego County.

in the Peninsular Range province, as well as in the easternmost parts of the Mojave Desert and Basin-Range regions in the State.

The known pegmatitic occurrences of bismuth minerals, cassiterite, non-gem beryl, rare-earth minerals, tantalum-columbium minerals, and thorium minerals in southern California thus far have been of mineralogic interest only.

REFERENCES

Ahrens, Louis, 1949, Measuring geologic time by the strontium method: Geol. Soc. America Bull., vol. 60, pp. 217-266.

Cameron, E. N., Jahns, R. H., McNair, A. H., and Page, L. R., 1949, Internal structure of granitic pegmatites: Econ. Geology, Mon. 2, 115 pp.

Davis, G. L., and Aldrich, L. T., 1953, Determination of the age of lepidolites by the method of isotope dilution: Geol. Soc. America Bull., vol. 64, pp. 379-380.

Donnelly, M. G., 1936, Notes on the lithium pegmatites of Pala, California: Pacific Mineralogist, vol. 3, pp. 8-12.

Hanley, J. B., 1951, Economic geology of the Rincon pegmatites, San Diego County, California: California Div. Mines Special Rept. 7-B, 24 pp.

Hewett, D. F., and Glass, J. J., 1953, Two uranium-bearing pegmatite bodies in San Bernardino County, California: Am. Mineralogist, vol. 38, pp. 1040-1050.

Jahns, R. H., 1948, The gem deposits of southern California: Eng. and Sci. Monthly, vol. 11, no. 2, pp. 6-9.

Jahns, R. H., and Wright, L. A., 1951, Gem- and lithium-bearing pegmatites of the Pala district, San Diego County, California: California Div. Mines Special Rept. 7-A, 72 pp.

Jahns, R. H., 1953, The genesis of pegmatites. I. Occurrence and origin of giant crystals: Am. Mineralogist, vol. 38, pp. 563-598.

Kunz, G. F., 1903, On a new lilac colored spodumene, San Diego County, California: Science, new ser., vol. 18, p. 280.

Kunz, G. F., 1905, Gems, jewelers' materials, and ornamental stones of California: California Min. Bur. Bull. 37, 155 pp.

Larsen, E. S., Jr., 1948, Batholith and associated rocks of Corona, Elsinore, and San Luis Rey quadrangles, southern California: Geol. Soc. America Mem. 29, 182 pp.

Larsen, E. S., Jr., Keevil, N. B., and Harrison, H. C., 1952, Method for determining the age of igneous rocks using the accessory minerals: Geol. Soc. America Bull., vol. 63, pp. 1045-1052.

Macdonald, G. A., and Merriam, Richard, 1938, Andalusite in pegmatite from Fresno County, California: Am. Mineralogist, vol. 23, pp. 588-594.

Mackevett, E. M., 1951, Geology of the Jurupa Mountains, San Bernardino and Riverside Counties, California: California Div. Mines Special Rept. 5, 14 pp.

Merriam, Richard, 1946, Igneous and metamorphic rocks of the southwestern part of the Ramona quadrangle, San Diego County, California: Geol. Soc. America Bull., vol. 57, pp. 223-260.

Merrill, F. J. H., 1914, Mines and mineral resources of San Diego County, California: California Min. Bur. Rept. XIV, pp. 691-700, 706-708.

Miller, F. S., 1938, Hornblendes and primary structures of the San Marcos gabbro: Geol. Soc. America Bull., vol. 49, pp. 1213-1232.

Miller, W. J., 1946, Crystalline rocks of southern California: Geol. Soc. America Bull., vol. 57, pp. 457-542.

Murdoch, Joseph, 1936, Andalusite in pegmatite: Am. Mineralogist, vol. 21, pp. 68-69.

Murdoch, Joseph, and Webb, R. W., 1948, Minerals of California; 1952, Supplement: California Div. Mines Bull. 136 (and 1952 Supplement), 448 pp.

Murdoch, Joseph, 1951, Notes on California minerals: neuvite=samarskite; trona, and hanksite; gaylussite: Am. Mineralogist, vol. 36, pp. 358-361.

Neuerburg, G. J., 1954, Allanite pegmatite, San Gabriel Mountains, Los Angeles County, California: Am. Mineralogist, vol. 39, in press.

Norman, L. A., Jr., and Stewart, R. M., 1951, Mines and mineral resources of Inyo County: California Jour. Mines and Geology, vol. 47, pp. 99-100, 103, 127.

Rogers, A. F., 1910, Minerals from the pegmatite veins of Rincon, San Diego County, California: Columbia Univ., School of Mines Quart., vol. 31, pp. 208-218.

Schaller, W. T., 1903, Spodumene from San Diego County, California: Univ. California, Dept. Geol. Bull., vol. 3, pp. 265-275.

Schaller, W. T., 1904, The tourmaline localities of southern California: Science, new ser., vol. 19, pp. 266-268.

Schaller, W. T., 1912, New manganese phosphates from the gem tourmaline field of southern California: Washington Acad. Sci. Jour., vol. 2, pp. 143-145.

Schaller, W. T., 1922, The Mesa Grande, California, pegmatites, in Merrill, G. P., Handbook and descriptive catalogue of the collections of gems and precious stones in the U. S. National Museum: Smithsonian Inst., Bull. 118, pp. 138-139.

Schaller, W. T., 1925, The genesis of lithium pegmatites: Am. Jour. Sci., 5th ser., vol. 10, pp. 269-279.

Schaller, W. T., 1927, Mineral replacements in pegmatites: Am. Mineralogist, vol. 12, pp. 59-63.

Schaller, W. T., and Fairchild, J. G., 1932, Bavenite, a beryllium mineral, pseudomorphous after beryl, from California: Am. Mineralogist, vol. 17, pp. 409-422.

Schaller, W. T., 1933, Pegmatites, in Ore deposits of the Western States, pp. 144-151, Amer. Inst. Min. Met. Eng., New York.

Sperisen, F. J., 1938, Gem minerals of California: California Jour. Mines and Geology, vol. 34, pp. 34-78.

Sterrett, D. B., 1904, Tourmaline from San Diego County, California: Am. Jour. Sci., 4th ser., vol. 17, pp. 459-465.

Waring, G. A., 1905, The pegmatyte veins of Pala, San Diego County: Am. Geologist, vol. 35, pp. 356-369.

Webb, R. W., 1943, Two andalusite pegmatites from Riverside County, California: Am. Mineralogist, vol. 28, pp. 581-593.

Wright, L. A., 1950, Beryllium, feldspar, gem stones, mica, quartz crystal, in Mineral commodities of California: California Div. Mines, Bull. 156, pp. 132-133, 158-161, 164-168, 184-186, 206-208.

Wright, L. A., Stewart, R. M., Gay, T. E., Jr., and Hazenbush, G. C., 1953, Mines and mineral deposits of San Bernardino County, California: California Jour. Mines and Geology, vol. 49, pp. 163-165.

Reprinted from *Geology of Southern California*, California Division of Mines Bulletin 170, 1954, Chapter VII, "Mineralogy and Petrology."

Gem-Bearing Pegmatites of San Diego County, California

Eugene E. Foord
Branch of Central Mineral Resources, U.S. Geological Survey Denver, Colorado

David London
School of Geology and Geophysics, University of Oklahoma

Anthony R. Kampf
Mineralogy Section, Natural History Museum of Los Angeles County, California

James E. Shigley
Research Department, Gemological Institute of America, Santa Monica, California

Lawrence W. Snee
Branch of Isotope Geology, U.S. Geological Survey, Denver, Colorado

Introduction and Background Sketch

E. E. Foord U.S. Geological Survey
J. E. Shigley Gemological Institute of America

The gem-bearing granitic pegmatites of southern California have been world famous for nearly a century. Descriptions of the abundant pegmatite mineral wealth of this region have been provided by numerous authors [*Kunz*, 1905; *Merrill*, 1914; *Donnelly*, 1936; *Jahns and Wright*, 1951; *Sinkankas*, 1959, 1976; *Weber*, 1963; *Rynerson*, 1967; *Johnson*, 1969]. Magnificent mineral specimens and gem-quality crystals of tourmaline, beryl, topaz, spodumene, and garnet have come from pegmatite bodies in a number of mining districts (Figure 1). These districts are situated along a central mountainous belt, known as the Peninsular Ranges geologic province, that roughly parallels the Pacific coastline. Photographs of some of the more attractive gemstones and crystal specimens from the pegmatites of this area are often seen in popular books and articles about minerals [*Bancroft*, 1973, 1984; *Larson*, 1977]. Few major mineral collections lack specimens from these pegmatites. In addition to their mineral and gem production, these pegmatites have yielded a wealth of scientific information in the form of both mineralogical and geological studies [*Schaller*, 1916, 1925; *Jahns and Wright*, 1951; *Simpson*, 1965; *Foord*, 1976, 1977; *Foord et al.*, 1986, 1989, in press; *Shigley and Brown*, 1985; *Stern et al.*, 1986].

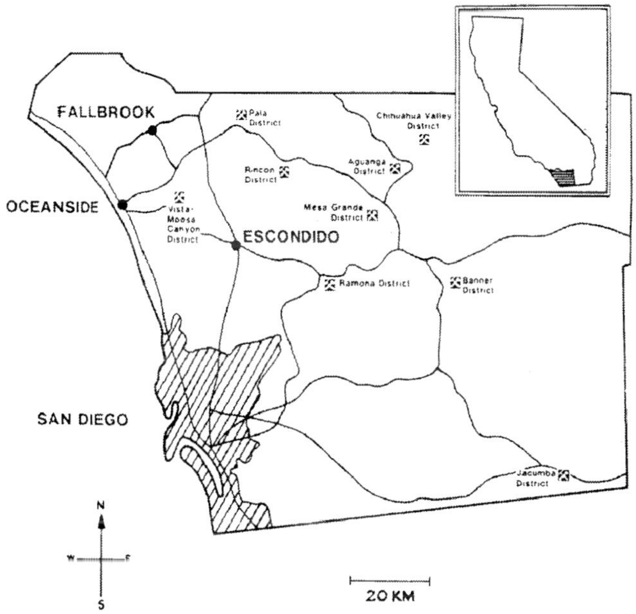

FIGURE 1. Index map of locations of principal pegmatite districts in San Diego County, California.

Local Indians undoubtedly collected crystal fragments of tourmaline and other pegmatite minerals in this area, but the first documented discovery of gem tourmaline in southern California was made in 1872 on Thomas Mountain in southwestern Riverside County [*Kunz*, 1905]. Gem mining in San Diego County can be traced to the report of pink tourmaline in lepidolite from the Stewart pegmatite at Pala in 1892 [*Kunz*, 1905; *Merrill*, 1914]. Development of the Stewart Lithia mine as a lepidolite source stimulated the search for pegmatite occurrences elsewhere in the county, and this exploration resulted in several discoveries of lepidolite and gem minerals during the next few years. Subsequent production of gems and other pegmatite minerals was greatest in the Pala, Mesa Grande and Ramona pegmatite districts.

The Pala pegmatite district is located in northwestern San Diego County approximately 40 km northeast of Oceanside

(Figure 1). Several hundred granitic pegmatites are exposed in a hilly area of about 10 square km along the north side of the San Luis Rey River. Important pegmatite occurrences are found on Tourmaline Queen, Chief and Hiriart Mountains within several kilometers of the small village of Pala. The pegmatites, chiefly subhorizontal tabular dikes enclosed in gabbroic host rocks, strike northwest and dip moderately to the southwest. This district is principally known for its production of lepidolite, tourmaline and spodumene. Both the kunzite variety of spodumene and morganite variety of beryl were first found in this district. Pegmatite mining was most vigorous during the periods 1900-1926 and from 1947 to the present. Intervening periods were characterized by only sporadic exploration activity. The Pala district provided a major portion of the gemstones that made California famous as a source of gem material during the early years of this century.

The Mesa Grande pegmatite district, in the hilly country southwest of Lake Henshaw and some 30 km southeast of Pala (Figure 1), was for a period the world's foremost source of gem and specimen tourmaline. Approximately 90 pegmatite bodies occur within an area of several square kilometers. Similar to the pegmatites in the Pala district, most of them strike northward to northwestward, dip to the west or southwest, are tabular in shape, and are hosted by gabbronorite. The most important of the pegmatite mines in this district is the Himalaya mine, which has probably produced the greatest quantity of gem tourmaline--more than 100,000 kg--of any single gem pegmatite mine in the world. Mining operations began here in 1898 and were most active up until about 1913, when the loss of important markets in the Orient, the onset of World War I, and the discovery of major Brazilian pegmatite deposits combined to cause a cessation of most pegmatite mining in southern California. Attempts to reopen the mines and to revive production faltered until the early 1950's, when the rising prices of gem and specimen materials led to a successful resumption of pegmatite mining.

The granitic pegmatites of the Ramona pegmatite district were discovered in 1903. This district lies about 8 km east-northeast of Ramona (Figure 1), in the central part of San Diego County. A single swarm of pegmatite dikes is exposed over an area of several square kilometers. These dikes, like those in the Pala and Mesa Grande districts, strike northward to northwestward and dip gently to moderately west to southwest. Host rocks, however, are tonalite or diorite rather than gabbronorite. Total gem and specimen production is much smaller in this district than in Pala or Mesa Grande, but quality is equal to those from the other districts. The most productive mines have been the Little Three and the ABC. The district is best known as a source of topaz, beryl and spessartine in contrast to Pala and Mesa Grande. Mining took place principally from 1903-1912, during the late 1950's and has been done on a regular basis during the past 15 years.

The greatest mining activity and production in all three districts occurred prior to 1913. A new period of intermittent mining activity over the past three decades has yielded thousands of pounds of gem crystals and mineral specimens. In addition to economic considerations, this period of renewed pegmatite prospecting and mining has presented opportunities for scientific study. Results of this recent period of investigation can be seen in papers by [*Foord*, 1977; *Taylor et al.*, 1979; *Jahns*, 1982; *Marcusson*, 1985; *Shigley and Brown*, 1985; *Foord et al.*, 1979, 1986, 1989, in press; *Shigley et al.*, 1986; *Stern et al.*, 1986; and *Snee and Foord* (in preparation)]. Further details on other pegmatites and pegmatite mining operations in southern California can be found in [*Jahns and Wright*, 1951; *Sinkankas*, 1959, 1976; and *Weber*, 1963]. In general, most of the pegmatite mines in southern California that are not being actively worked at the present time are in extremely poor condition and are virtually inaccessible.

Geologic Setting

E. E. Foord U.S. Geological Survey

The gem-bearing pegmatites of southern California are located within what was formerly referred to as the Southern California batholith [*Larsen*, 1948], but is now termed the Peninsular Ranges batholith (PRB) [*Todd and Shaw*, 1979; *Walawender et al.*, 1990]. This batholith forms the core of the Peninsular Ranges which extend 1600 km from the Los Angeles Basin in the north to the tip of Baja California, Mexico in the south. The batholith averages at least 80 km wide. It is a major link in the chain of late Mesozoic batholiths that rim the Pacific Basin.

The Peninsular Ranges batholith is a large igneous complex consisting of many generally well-defined plutons ranging in exposed diameter from a few hundred meters to nearly 40 km. Igneous rock types range from peridotite to granite with granodiorite and tonalite being the most common. Older metamorphic rocks, chiefly schists and quartzites, occur as screens, septa, and pendants among the plutons. The gem-bearing pegmatites are for the most part hosted by gabbros or norites, and major outcroppings of these rock types correspond closely to the locations of the important pegmatite districts.

Based on geochemical, geophysical and lithologic parameters, the PRB batholith has been divided by [*Todd and Shaw*, 1979, 1985] and others [(e.g. *Gastil*, 1975; *Silver et al.*, 1979; *Walawender et al.*, 1990)] into two zones: an older, syntectonic intrusive series to the west, and a younger, late syntectonic or post tectonic plutonic series to the east. The western portion of the batholith is characterized by smaller diameter epizonal plutons (less than 10 km), moderate grades of metamorphism, and the presence of gabbro. This contrasts with the larger (up to 40 km) mesozonal plutons, sillimanite-bearing and migmatitic prebatholithic rocks, and the absence of gabbroic plutons to the east.

Within the western gabbro sub-belt [*Walawender*, 1979], three major rock series can be delineated--gabbro, quartz diorite to tonalite, and granodiorite to adamellite. The gabbroic plutons appear to be the oldest, followed by the quartz diorite to tonalite series, and finally by the granodiorite to adamellite series. K-Ar cooling ages for biotite and hornblende [*Krummenacher et al.*, 1975] ranged from 143 Ma to 101 Ma. On the basis of recent geochronologic and isotopic studies by [*Walawender et al.*, 1990], magmatism of the style and character of the western gabbroic zone effectively ceased but overlapped in time with the onset of La Posta-type magmatism. Details of the subduction model for the origin of the PRB are given in [*Walawender et al.*, 1990]. A discussion of Ar-Ar geochronologic studies of pegmatites from the three main districts (Pala, Mesa Grande and Ramona) [*Snee and Foord*, in preparation] is given in the section on geochronology.

DESCRIPTION OF THE PEGMATITES

E. E. Foord U.S. Geological Survey
J. E. Shigley Gemological Institute of America

The granitic pegmatites of San Diego County are genetically related to the PRB, and were interpreted to represent products of the final stages of its magmatic differentiation and crystallization [*Jahns and Wright*, 1951; *Jahns*, 1979]. More recent studies suggest that the pegmatites are related to the emplacement of the younger La Posta-type plutons in the central and eastern portions of the PRB (98 Ma to 89 Ma), [*Walawender et al.*, 1990]. The pegmatites range from very thin to fairly thick tabular dikes with northerly to northwesterly trends and gentle to moderate westerly to southwesterly dips averaging about 20° in surface exposure. With few exceptions, the gem-bearing pegmatites are confined to gabbroic or tonalitic plutons, and on approaching the contacts with other rock types, they are commonly observed to either pinch out or abruptly terminate. The pegmatites exhibit a general uniformity in attitude and tabular shape. This consistency and close relationship with the relatively early-crystallized members of the batholithic assemblage is ascribed to a regional structural control [*Jahns and Wright*, 1951; *Jahns*, 1979]. The pegmatites were evidently emplaced as late-stage residual magmas which crystallized as vein fillings along well-developed sets of fractures in the older and more competent gabbro. The fractures are thought to represent regional cooling and contraction within the batholith as a whole [*Jahns*, 1979]. In addition, they appear to have been formed only in plutons that were essentially unjointed or unfractured during earlier stages in their respective emplacement and cooling histories. As such, the fractures postdate the main period of batholith emplacement. The original attitude of the majority of the fractures was sub-horizontal based on interpretation of 'snow-on-the-roof' depositional features within the pegmatites [*Foord*, 1976]. Subsequent tilting to the west of about 20-30° occurred due to continuing tectonic activity and uplift of the batholith.

Although typically exhibiting a uniform attitude, [*Jahns and Wright*, 1951] reported that some pegmatite dikes do vary substantially in strike and dip. This variation is most clearly evident in the gentle rolls or terrace-like ledges that can be seen in some pegmatite outcrops. Axes of these rolls may either parallel the strike of the pegmatite body or may be inclined to the strike and plunge at an angle to the dip.

Many of the pegmatites are well exposed. Because of their relative resistance to erosion, they tend to stand out on the hillsides as light-colored, rib-like projections extending across otherwise smooth slopes. A uniform tabular shape is most common, but pegmatites that either branch or pinch and swell along their length are frequently observed. Exposures of most pegmatite dikes indicate that they are remarkably persistent, with strike lengths extending up to several kilometers (e.g. Himalaya dike system). The pegmatites do vary in thickness from narrow stringers only a few centimeters wide to major dikes with enlarged bulges as much as 30 m across (e.g. Stewart dike). Some of the dikes are closely spaced as subparallel dike swarms in which individual dikes may branch and rejoin along strike or downdip (e.g. Himalaya dike system). In some instances, dike swarms converge to form a single composite pegmatite. Many of the pegmatites, however, appear to be isolated within large volumes of the host rock. Most dikes taper somewhat near their ends where they terminate as abrupt stubs or elongate prongs. While the outer edges of some pegmatites contain septa and inclusions of host rock, the contacts between pegmatite and host rock are generally sharp and regular in detail. Host rock alteration adjacent to pegmatite bodies has been noted in several instances ([e.g. *Foord et al.*, 1986; *Shigley et al.*, 1986]) but does not appear to be extensive. Alteration envelopes around pegmatite dikes typically range from nearly nil to a few centimeters. Details of the mineralogy and chemistry of the alteration process are given in [*Foord et al.*, 1986]. Boron metasomatism of the host gabbro is localized above and below areas where tourmaline-filled pockets have ruptured, releasing boron (see section on The Role of Boron).

The gem-bearing pegmatites of San Diego and Riverside Counties, California are granitic in composition, and have as principal mineral constituents potash feldspar (microcline and orthoclase), plagioclase (oligoclase to albite), quartz and muscovite. Most pegmatites are neither sufficiently exposed nor adequately sampled to permit an accurate determination of their bulk composition. [*Jahns*, 1953a], however, published the following data on the composition of the southern portion of the Stewart pegmatite (in weight % oxides): SiO_2 74.9, Al_2O_3 14.9, CaO 0.1, Na_2O 3.6, K_2O 5.2, Li_2O 0.7, F 0.4, H_2O 0.4, total 100.2. Presumably these data can be applied as a first approximation to other gem-bearing pegmatites in the region. The approximate 100 minerals reported from these pegmatites also reflect the abundance of B, P, Be, Bi, Cs, Nb, Ta, Sn, Rb, Sb, Th, V and to a lesser extent the REE elements in these pegmatite bodies. Table 1 compares the mineralogy of the three major pegmatite districts that are the subjects of this guidebook.

TABLE 1. Minerals of the Pala, Ramona, and Mesa Grande Districts, San Diego, California

Mineral or Mineral Group	P	R	MG
amblygonite	C	-	-
apatite (fluorapatite)	R*	R*	R*
axinite (inc. possible tinzenite)	-	R#	VR#
bavenite	VR*	-	R*
beryl (goshenite, morganite, etc.)	R	R*	R*
beyerite	VR#	VR#	VR#
bismite	R*	VR#	-
bismuth	R*	VR#	-
bismuthinite	R*	-	-
bismutite	R*	VR#	-
bismutocolumbite			

TABLE 1. Minerals of the Pala, Ramona, and Mesa Grande Districts, San Diego, California--Continued

Mineral or Mineral Group	P	R	MG
bismutotantalite	R*	-	-
calcite	-	-	R#
cassiterite	VR(?)	VR	R#
cerussite	-	ER*	-
clays:			
Li-tosudite	-	-	R#
beidellite	-	-	C#
Ca-Mg montmorillonite	C*	R#	C#
nontronite	-	-	R#
kaolinite (ordered and disordered)	C	R	C
clinobisvanite[1]	ER*	-	-
cookeite	R*	VR#	C#
cryptomelane	R*	-	-
eosphorite	VR	-	-
epidote (incl. clinozoisite)	R	R*	-
fairfieldite	ER	-	-
feldspars:			
albite (incl. cleavelandite)	A	A	A
microcline (maximum, intermediate)	A	A	A
oligoclase	C	C	C
orthoclase	VR#	R#	R#
ferrocolumbite	-	VR*	-
fersmite	-	VR#	VR#
garnets:			
almandine-spessartine	C	C	C
spessartine	R	C*	C*
gahnite	R	-	R
hafnon[1]	-	-	ER#
hambergite	-	VR#	VR#
hematite (incl. goethite)	R	R	R
heterosite	R	-	-
hopeite	ER	-	-
hureaulite	R	-	-
jahnsite	ER	-	-
lithiophilite	C	-	-
magnetite	R	R	R
malayaite[1]	-	-	ER#
micas:			
biotite (in altered wallrock inclusions)	C	R	R
boromuscovite	-	C#+	-
lepidolite	A	C*	C*
muscovite	A	A	A

Mineral or Mineral Group	P	R	MG
microlite-uranmicrolite	VR#	VR#	VR#
mitridatite	VR	-	-
monazite	R	VR#	VR#
montebrasite	VR	-	-
namibite[1]	ER	-	-
palygorskite	-	-	C#
petalite	R	-	-
phosphosiderite	R	-	-
pollucite	VR#	-	R#
pseudomalachite	ER#	-	-
pucherite	R	VR#	ER#
purpurite	R	-	-
pyrite	VR	VR#	-
quartz	A	A	A
robertsite	VR	-	-
rynersonite	-	-	VR#
sicklerite	R	-	-
spodumene (including var. kunzite)	A	-	-
stewartite	VR	-	-
stibiocolumbite-stibiotantalite	R#	R#+$	R#
stokesite	-	-	ER#
strengite	VR	-	-
thorogummite	-	-	VR#
todorokite	-	R#	R#
topaz	R#	C#	R#
tourmalines:			
schorl-dravite	-	-	R
schorl-elbaite	C	C	C
vermiculite (alteration of biotite)	-	R	R
xenotime	-	-	ER#
zeolites:			
heulandite	VR#	-	R#
laumontite	VR#	R#	C#
stilbite	VR#	R#	C#
zircon (incl. cyrtolite)	-	VR#	VR#
unidentified Ca-arsenate	-	-	ER#
unidentified Pb-Bi-Ag sulfosalt	-	VR	-
Ta-analogue of fersmite	-	ER#+	-
unidentified U-Ta-Nb oxide hydrated	-	ER#	-

A abundant and widespread
C common or locally abundant
R uncommon or rare
\+ new mineral species
VR very rare
ER extremely rare, only a few specimens
\- not known or reported
* chiefly or wholly in pocket pegmatite
\# only from pocket pegmatite
$ bismuthian and tungstenian variety
1 not previously reported from San Diego County or California.

Other minerals reported from these pegmatites in Jahns and Wright (1951) or Jahns (1979), but not observed by the authors are: allanite, andalusite, andradite, arsenopyrite, bornite, bertrandite, chalcocite, chrysocolla, eucryptite, fluorite, grossular, halloysite, helvite, herderite, loellingite, malachite, molybdenite, phenakite, siderite, triphylite, triplite, vivianite, and zinnwaldite.

Most pegmatites in this area can be considered simple pegmatites on the basis of their common mineralogy and homogeneous internal structure. However, the gem-bearing dikes are internally segregated and have a more complex and diverse mineralogy. In these complex pegmatites, the upper (hangingwall) and central portions are characteristically coarse- to very coarse-grained and relatively potassic in composition. In contrast, the lower (footwall) portions are relatively sodic, much finer grained, and commonly saccharoidal in texture. A typical hangingwall to footwall sequence of rock types (slightly modified from [*Jahns*, 1979]) is as follows:

1. Medium- to very coarse-grained pegmatite: crystals of microcline perthite with graphically intergrown quartz, along

with interstitial quartz, plagioclase, and perthite, with or without muscovite, biotite, schorl, and/or garnet.

2. Very coarse-grained pegmatite: euhedral to subhedral crystals of microcline perthite set in quartz, commonly with some large prisms of schorl and clusters of muscovite plates.

3. Pocket pegmatite (present locally): chiefly subhedral to euhedral potash feldspar (microcline and/or orthoclase), quartz, albite (cleavelandite), muscovite, lepidolite, and elbaite. Minor amounts of exotic minerals are present.

4. Coarse-grained pegmatite: microcline, albite, quartz, and muscovite, with or without schorl and garnet (almandine-spessartine).

5. Saccharoidal aplite or fine-grained pegmatite: albite and quartz with or without microcline, muscovite, almandine-spessartine, and/or schorl. Aplite may be layered or uniform.

The very fine-grained, sugary-textured footwall rocks are so widespread that many of the dikes can be described as pegmatite-aplite bodies of the type discussed by [*Jahns and Tuttle*, 1963; and *Foord*, 1976]. The aplitic rocks are essentially massive in some dikes, but in others they exhibit a vague to sharply defined layering widely recognized in the region as 'line-rock'. This is usually an expression of mineral segregation into thin, alternating layers of quartz and feldspar and garnet- and/or schorl-rich and -poor layers.

The internal distribution of lithologic units within the pegmatite bodies is remarkably consistent throughout the pegmatite districts. An orderly downward progression, observed from the hangingwall to the central pocket-bearing horizons of the dikes, typically involves increasing grain size, increasing numbers of euhedral crystals, and a change from perthite crystals with graphically intergrown quartz to those without quartz. Additional changes involve a transition from schorl to elbaite and from muscovite to lepidolite, commonly within single crystals ([e.g. *Brock*, 1973; *Foord*, 1976, 1977; *Foord et al.*, 1989]).

The interior portions of some dikes, especially most of the larger ones, consist of additional lithologic units with very coarse-grained textures. These units include large, anhedral crystals of quartz, masses of coarse, blocky perthite, and rocks composed of large, lath-like crystals of spodumene set in a mosaic of anhedral quartz. When they are present, the main pocket zone or horizon ordinarily is found along the footwall side of these coarse-grained units.

Pala District

The Stewart Lithia Mine

J. E. Shigley Gemological Institute of America

General features. The Stewart pegmatite occurs as a bold, elongate outcrop of light-colored granitic rock on the southeasterly slope of Tourmaline Queen Mountain, 2 km north of Pala. It is exposed as a narrow dike along a distance of more than a kilometer with a northerly strike and a westerly dip of 20° to 25°. Toward the north, the pegmatite splits into two subparallel thin prongs that eventually terminate where the enclosing gabbroic rocks are in contact with pre-batholithic schist and quartzite. The main portion of the pegmatite is located at its southern end where the dike forms an enlarged bulge with a maximum thickness of 25 m. It is in this thickest portion of the dike where the Stewart Lithia mine, a major gem tourmaline producer and once the most important domestic source of lithium ore, is located. The mine can easily be reached by paved and dirt road from nearby Pala.

In the area of the Stewart Lithia mine, the pegmatite is exposed high on the east side of a prominent ridge that projects southward from the mountain. At the surface, the pegmatite appears as a very gentle to moderate dip slope of hangingwall graphic granite over much of the west side of that ridge. The contact of the pegmatite and the overlying gabbro passes beneath the ground surface in the bottom of a small canyon immediately west of the ridge. The upper half of a vertical cross section of the pegmatite is visible along the steeper east face of the ridge, but is today much obscured by dump material and other loose debris. The internal structure of the pegmatite is best observed in the underground mine workings that have been greatly extended in the past 10 years.

Petrology and lithologic units. As with other complex granitic pegmatites in this area and in other pegmatite districts, the Stewart pegmatite exhibits a pronounced internal zonation of different mineral assemblages, textures and chemistry. The pegmatite is comprised of mineral assemblages arranged in subparallel, originally subhorizontal (but now tilted) layered zones above and below a discontinuous quartz-spodumene core. Zones above the core are generally coarser grained and are rich in microcline, whereas those beneath contain both coarse- and fine-grained material and have more albitite. This internal zonal arrangement corresponds to the generalized internal zonal sequence noted in many granitic pegmatites and described by [*Cameron et al.*, 1949; and *Norton*, 1983]. This zonal arrangement is most clearly visible at the southern end of the dike as the result of underground mining operations.

Within the core zone, the segments of quartz-spodumene pegmatite form lens-like masses up to 50 m long and 3 to 5 m in maximum thickness. Proceeding upward within the dike, relatively thin lenses of massive quartz overlie the segments of the quartz-spodumene core. This inner intermediate quartz zone thickens enormously along strike beyond the edges of the core segments and is at least 10 m thick in some exposures. Much more continuous is the middle intermediate zone consisting of massive quartz with large subhedral to euhedral crystals of microcline perthite. Within this zone, the ratio of quartz to perthite is variable. Large, subhedral crystals up to 10 cm across of lithiophilite, $Li(Mn,Fe)PO_4$, and associated secondary phosphates are locally abundant within this zone [*Shigley*, 1982; *Shigley and Brown*, 1985]. The phosphate assemblage was also studied by [*Jacobson*, 1973].

Above the quartz-perthite zone is a 3 to 5 m thick unit designated as the outer intermediate zone. It is composed of coarse, blocky microcline perthite with some quartz, muscovite, and albite. This in turn is overlain in most places along the dike by a coarse-grained graphic granite with minor albite and muscovite which together comprise a thick hangingwall zone.

Underneath portions of the quartz-spodumene core is a discontinuous zone of massive quartz. Beneath this is a zone of quartz-perthite followed by a zone of coarse, blocky perthite. These three zones correspond to the three intermediate zones located above the core. Much of the lower portion of the dike, however, consists of a coarse-grained assemblage of albite-quartz-muscovite and a sugary, fine-grained albite-quartz-muscovite-schorl unit. In contrast to other differentiated granitic pegmatites in the area, the rocks in the footwall aplitic zone of the Stewart pegmatite do not exhibit a planar 'line-rock' structure (see [*Jahns and Tuttle*, 1963, and *Foord et al.*, 1989]).

Mining operations in the early years of this century uncovered two large pod-like masses of lepidolite beneath the quartz-spodumene core. One of these pods measured 70 x 40 x 6 m in dimension. The two pods were the principal focus of the initial mining operations for lithium ore in the pegmatite since most of the lepidolite that they contain is quite pure. The best known mineral specimens from this pegmatite are the pieces of massive lepidolite with numerous sprays of radiating pink tourmaline crystals that have come from these two pods. Along strike and downdip, the two lepidolite masses pinch out or rapidly become discontinuous. They grade outward into veinlets of lepidolite in adjacent quartz or other coarse-grained pegmatite. [*Jahns*, 1979] regarded these pods as gigantic "pockets" that contain little or no open space. It is interesting to note that these large lepidolite masses are situated along or near a marked bench-like roll in the dike, a location within the dike where pocket formation may have been focused during pegmatite crystallization.

This internal mineral segregation, so well developed in the near-surface parts of the pegmatite penetrated by the early underground workings, appears to continue downdip within the pegmatite. However, recent mining operations that have been concentrated in the intermediate zones and along the gem pocket-bearing core of the dike have not presented an opportunity for extensive study of the remaining portions of the pegmatite.

History and mining activity. During its first year of operation in 1892, approximately 18,000 kg of lepidolite were removed from the Stewart Lithia mine. This level of output was increased somewhat during the following years. A substantial annual production of lithium ore was maintained from 1900 to 1907 when the mine was idled due to competition from other lithium sources. Mining was resumed during World War I and reached a peak in 1920. It was during these early years of this century that the pegmatite was also an important source of gem tourmaline, much of which was sent to China as carving material. Activity dwindled thereafter, and the mine was finally closed down in 1928. Minor exploratory work was carried out at the mine intermittently between 1928 and 1968. However, for all practical purposes the mine was completely closed.

A new period of mining operation began in 1968 under the direction of Pala Properties International, Inc. New underground workings were begun in a northeast direction to intersect an early exploratory drift that reportedly had been driven northwestward from the main tunnel in search of lepidolite. This new effort was successful in locating a gem tourmaline-rich pocket zone, and recent mining activity has extended workings downdip along the central portions of the pegmatite. [*Sinkankas*, 1976] provided a good summary of this renewed activity. During the past 15 years, the mine changed ownership several times, and most recently under a company called Gems of Pala, there has been continued mining of the central portion of the pegmatite for gem tourmaline and other minerals.

Mining and production. The Stewart pegmatite was initially worked along its prominent, east-facing outcrop by open-cut methods. During the early years of this century, most of the underground workings that crisscross the near-surface portion of the pegmatite were developed. The two large lepidolite masses encountered during the early stages of mining were essentially removed in their entirety by the late 1920's. Small pockets of gem tourmaline and beryl were occasionally found during the excavation of underground workings, but the production of gem material from the pegmatite was of secondary importance. Today, most of the early underground workings are inaccessible.

The recent period of mining operations has uncovered numerous gem tourmaline pockets in the area to the north of the early workings. The gem-bearing pocket zone in this new part of the mine is located in the core parts of the pegmatite that appear to lie downdip from the large lepidolite masses described earlier. The pockets occur mainly in highly fractured quartz-spodumene-albite-muscovite pegmatite and in the overlying quartz-perthite-albite-muscovite-schorl pegmatite. A few gem pockets have been found in massive quartz. According to [*Sinkankas*, 1976], the largest gemmy pink tourmaline crystal, found in 1971, measured 25.5 cm in length and 10 cm in width. This crystal was attached to a matrix of albite, perthite, quartz, and lepidolite. As a result of recent (1968-1990) mining activity, more than 150 m of new underground workings have been completed. Production figures indicate that some 500 kg of tourmaline were mined since 1968.

Mineralogy. Table 1 lists the minerals reported from the Pala district, and most of them occur in the Stewart mine. The bulk mineralogy of the Stewart pegmatite is similar to that of most complex granitic pegmatites, with the more unusual minerals found in the pegmatite coming from the pocket-bearing core zone. Within the core zone or along its margin, the gem pockets are highly irregular in form, and in general they are less than 1 m in maximum dimension. They are unsystematically distributed within the enclosing pegmatite. The typical pocket is a cavity that is tightly filled with white to reddish hypogene smectite (Ca-Mg montmorillonite) or interstratified chlorite-smectite (Li-tosudite) with minor amounts of other minerals [*Foord et al.*, 1986]. Within the smectite clay are scattered abundant flakes and small crystals of muscovite and lepidolite, many crystal fragments of quartz and albite, rare splinters and relicts of spodumene, and a few prismatic fragments of tourmaline crystals. In some instances, these crystals are attached to the surrounding walls of the pockets. Most of the mineral fragments in this clay-filled rubble evidently were derived from much larger, pocket-lining crystals that were broken and separated during the final stages of evolution of the pocket (see [*Foord*, 1976, 1977; *Foord et al.*, 1989; and *Stern et al.*, 1986] for more information).

The irregular pocket walls typically consist of firmly attached and inward-directed euhedral crystals of quartz, perthite, albite and tourmaline, accompanied in places by beryl, spodumene, and occasionally rarer minerals. Most of these crystals are deeply rooted in the solid pegmatite that surrounds the pocket. The ends of these crystals that extend into the pocket terminate in smooth faces (quartz, beryl), slightly to highly corroded faces (perthite, spodumene), or broadly irregular surfaces representing gross detachment and separation of material from the originally larger crystals (quartz, tourmaline, some perthite, and albite). Within the pocket, the crystals are commonly coated, wholly or in part, by fine-grained aggregates of albite, muscovite, lepidolite, or some combination of these minerals.

The Pala Chief Mine

A. R. Kampf Natural History Museum of Los Angeles County

General features. The Pala Chief mine is located just northwest of the summit of Pala Chief Mountain, about 2 miles northeast of the town of Pala. The pegmatite is exposed as a series of bench-like open cuts along the southwest face of a nearly flat-topped ridge near the summit.

The Pala Chief dike is part of a complex of branching and joining dikes. The main pegmatite mass ranges in thickness from 5 to more than 10 m and can be traced at the surface for about one kilometer. The dike trends north-northwest to northwest and dips 5° to 55° southwest.

Petrology and lithologic units. [*Jahns and Wright*, 1951] described the pegmatite in some detail. Because relatively little additional pegmatite has been exposed in the intervening years and no further detailed study has been undertaken, we will draw heavily from their report in the following description.

The upper portion of the dike is a 2 to 5 m thick unit consisting of coarse-grained graphic granite with smaller amounts of albite and muscovite. The lower portion of the dike is approximately equal to the upper portion in thickness and consists of a fine-grained aplitic mixture of quartz, albite, perthite, and garnet (almandine-spessartine). This line rock unit grades along strike from very distinctly layered to unlayered. The lowermost meter or so of the dike is a fine-grained homogeneous rock of similar mineralogy.

In some areas of the dike very coarse-grained zones are present between the upper and lower units. Most commonly these consist of massive quartz with euhedral to subhedral crystals of perthite. In the southeastern part of the mine a series of discoidal masses of quartz-spodumene pegmatite forms the core unit of the dike.

Pockets occur mostly within and along the edges of the quartz-spodumene core. They contain lepidolite, albite, elbaite tourmaline, quartz, spodumene, beryl, microcline, and several rare accessory minerals. Underlying most pockets is fine- to medium-grained albite-quartz-perthite with plumose muscovite. This unit with local cleavelandite and schorl also occurs intermittently elsewhere in the dike as a fracture-filling or replacement.

History and mining activity. In May 1903 the Pala Chief claim was staked by John Giddens, Pedro Feiletch, Bernardo Hiriart, and Frank A. Salmons. The next several years saw the greatest mining activity after which only intermittent work continued until 1915. The claim was patented in 1956 by Mrs. Margaret S. Moore and Mrs. Mildred S. Ware, daughters of Frank A. Salmons, although it remained essentially inactive until 1959. The last three decades have seen sporadic small scale working of the mine by several owners with the production of relatively little gem material. Since 1989 Bob Dawson has leased the mine from Oscar Nukka with an option to buy.

Mining and production. Historically the Pala Chief mine has produced fine gem crystals of elbaite tourmaline and the morganite variety of beryl. It is most noteworthy, however, for producing some of the world's finest examples of the pink variety of spodumene known as kunzite. Other mines in the district, particularly several on Hiriart Mountain, have produced large kunzite crystals; however, the Pala Chief is the only one which has yielded large gem-quality crystals of both tourmaline and kunzite. The best kunzite, tourmaline, and morganite were removed during the early years of production, the finest kunzite coming in major finds in 1912 and 1914.

The mine has been exploited principally by open cuts. [*Jahns and Wright*, 1951] described the early workings. The main open cut is 85 m in length and 6 to 20 m wide. In places the cut is nearly 8 m deep at the face, but in most cases this dimension is limited to less than 5 m. Large dumps rim the series of southwest facing cuts. Irregular, interconnected underground workings consisting mostly of gently sloping drifts and inclines extend generally northeastward from the central and southeastern portions of the cuts. A variety of less important shallow cuts, trenches, etc. are found elsewhere along the dike. Modern work has involved relatively minor selective expansion of portions of the early workings.

Mineralogy. The mineralogy of the Pala Chief pegmatite has not received serious study since the work by [*Jahns and Wright*, 1951]. They note quartz, potash feldspar/perthite (microcline), albite, muscovite, lepidolite, garnet (almandine-spessartine), beryl, schorl, spodumene, and alkali tourmaline (elbaite). They also mention "several rare accessory minerals" occurring in pockets, but they do not identify them. In light of the occurrence of triphylite/lithiophilite in nearby pegmatites (notably the Elizabeth R), it would not be surprising to find primary and secondary phosphate minerals here, although we have not observed any in the current exposures. A small mass of namibite and clinobisvanite, similar to that found in the White Queen mine, was recently found.

The Elizabeth R Mine

A. R. Kampf Natural History Museum of Los Angeles County

General features. The Elizabeth R mine is located on the north-northwest slope of Pala Chief Mountain, about 3.5 km northeast of the town of Pala. The mine is about 200 m north of the Pala Chief mine.

The Elizabeth R mine exploits a portion of the Ocean View dike. In the area of the mine the Ocean View dike strikes approximately north-south and dips 25° to 40° to the west. The thickness of the pegmatite ranges from 3 to 6 m. The mine's workings are marked by three adits, each surrounded by a dump. Two are at the north and south ends of the upper workings and the third is about 25 m lower on the hillside.

Petrology and lithologic units. The only geological study of the Elizabeth R mine is an unpublished senior thesis by Charles Houser (San Diego State Univ., 1986). He describes the petrology as follows: "Fine-grained rocks generally occur in the footwall of the dike and may or may not exhibit layers rich in garnet or tourmaline. Graphic granite occurs as masses up to several feet across in all parts of the dike, but primarily in the upper portion near or along the hanging wall. Where no core zone is present, the rest of the dike consists of tourmaline-bearing and muscovite-bearing pegmatite made up of quartz, microcline and albite. Core zones occur in the central or lower portions of the dike and contain the largest crystals in the pegmatite. Primary core zone minerals are quartz, perthite and spodumene. Pockets or cavities within the core zone contain euhedral crystals of quartz, perthitic microcline, albite, muscovite, tourmaline, beryl, columbite, apatite and a variety of other phosphate minerals."

The spodumene is found as decomposed laths consisting largely of clay in which remnant slivers of spodumene are embedded. Large quartz groups commonly exhibit distinctive spodumene casts. Subhedral crystals of lithiophilite are locally abundant along the upper core margin. These crystals typically exhibit pronounced alteration to a variety of secondary phases.

History and mining activity. In November 1973 Roland Reed purchased and patented the adjoining Hazel W. (Hazel Wand) and Ocean View claims. These claims are immediately north of the Pala Chief and Goddess claims, respectively, and all were originally part of the Pala Chief group, which dates back to 1903. In January 1974 Reed began work on a portion of the Ocean View pegmatite dike which crosses these properties. He named his new workings the Elizabeth R mine.

The mine has been essentially a part-time one-man operation since its inception, although Roland Reed has periodically enlisted additional help. Reed has been remarkably persistent in his efforts despite having only limited and inconsistent success.

Mining and production. There was no reported production from the Hazel W. claim prior to its acquisition by Roland Reed. [*Sinkankas*, 1976] reported on the first of the modern production in 1974 and 1975. He describes "some small peach-colored morganite crystals of unusual sharpness of form from a series of very small cleftlike pockets along the centerline of the pegmatite body." The largest crystal found at that time was 50 x 38 x 25 mm and was largely clear. A fair amount of good crystals and cutting rough was removed during this period.

Subsequently production was very low until 1982 when Reed made a major discovery. From a single pocket he removed about 100 kgs of specimens with morganite crystals ranging in diameter up to 13 cm. The crystals are equant to tabular with well formed lustrous faces. They are moderately included with a pale to medium pink color and many have aquamarine cores.

Shortly after this discovery Reed began working the pegmatite about 30 m downdip from his original site. Since the 1982 discovery, production from the mine has again been quite low. The upper workings include about 200 m of tunnel with an adit at either end. The lower workings include about 75 m of tunnel.

Mineralogy. Early reports note the Ocean View pegmatite to contain beryl and quartz. These are the most important products of the Elizabeth R mine. [*Houser*, 1986] also notes spodumene, orthoclase, perthitic microcline, albite, muscovite, tourmaline, columbite, apatite and a variety of other phosphate minerals. We have found pods of lithiophilite altered to a variety of secondary minerals. These include sicklerite, purpurite, hureaulite, phosphosiderite, strengite, stewartite, eosphorite, robertsite, mitridatite, jahnsite, fairfieldite, hopeite, pseudomalachite, and cryptomelane.

The White Queen Mine

A. R. Kampf Natural History Museum of Los Angeles County

General features. The White Queen mine is located high on the west slope of Hiriart Mountain, about 3.5 km east-northeast of the town of Pala. The pegmatite workings are underground creating a prominent dump on the surface. Two adits are visible, the original or upper adit and the lower adit about 30 m lower on the hillside.

The White Queen dike is part of an intricate pattern of narrow subparallel dikes which branch out and converge. The pegmatite ranges in thickness from about 3 to 5 m and can be traced at the surface for about 200 m. The dike trends north-northwest and dips about 35° southwest.

Petrology and lithologic units. Very few specific details concerning the geology of the White Queen pegmatite have been published. The following general description is the result of discussions with Bob Dawson and a cursory inspection of the mine. The lower adit intersects albitic layered aplite and pegmatitic aplite; the undulatory layering being defined by alternating schorl- and albite-rich layers. Both the footwall and hangingwall contacts of the pegmatite are not exposed in the underground workings. Approximately 3.5 m of aplite and overlying pegmatite is exposed in the adit. The aplite grades upward into sodic pegmatite that is faintly layered. The upper adit intersects massive perthite-quartz pegmatite above the aplitic footwall and centrally disposed pocket zone.

History and mining activity. Frederick Sickler has the distinction of being the first to discover the pink variety of spodumene, kunzite. He picked crystals up as float upon what several years later in 1901 was staked as the White Queen claim by him and his father. The claim was prospected by the Sicklers without significant positive results. This and several other claims on Hiriart Mountain were purchased by George Ashley in 1947. The claim was sold to Norman Dawson in 1948, but active development did not commence until 1959. The most important finds made by Norm Dawson were in the early 1960's and early 1970's. Ownership of the mine is currently in the hands of Bob and Ken Dawson (Norm Dawson's sons), Marty Misslin, and Maury Morgenstein who made an important discovery of morganite in 1989.

Mining and production. [*Sinkankas*, 1976] described the most productive period in the mine's history which began in 1959 with the initiation of active mining by Norm Dawson. A 7.5 m tunnel was driven into the pegmatite 22.5 m west of a prospect pit originally sunk by the Sicklers. This tunnel intersected a large clay-filled pocket which when excavated was large enough for a man to stand in. The pocket yielded some of the finest crystals of morganite beryl ever found in California as well as many large quartz crystals remarkable for their beautifully intricate patterns of pink and white clay (montmorillonite) inclusions. Approximately 5600 kg of quartz crystals and 80-120 kg of beryl (mostly morganite) were removed from this pocket.

Dawson subsequently followed the lower lead from this pocket encountering several similar but smaller pockets until in 1973 he found another large pocket. This pocket yielded about 6000 kg of quartz crystals and 30 kg of beryl. The morganite was generally of even better quality than that produced by the first pocket. By the end of 1973 the White Queen workings extended 60 m beneath the surface.

The mine saw only sporadic and minor production until recently. A new tunnel was initiated about 30 m downdip from the original workings. About 100 m of new tunnel were excavated connecting with the upper workings. On Labor Day 1990 a pocket measuring 2.5 m in diameter and 1 m thick was discovered. This pocket produced 150 to 200 specimens of peach-salmon orange morganite. The better crystals are well formed and lustrous and measure 5 to 8 cm in diameter and 4 cm in thickness. Some have typical basal terminations, while others are elongate perpendicular to the c-axis. The best crystals from this find are regarded as the finest ever produced by the mine.

Mineralogy. The mineralogy of the White Queen dike has never received detailed scientific study. [*Sinkankas*, 1976] notes the following minerals from the pockets found between 1959 and 1973: beryl (morganite and aquamarine), quartz, lepidolite, albite (cleavelandite), dark blue tourmaline (presumably elbaite), apatite, columbite-tantalite, amblygonite, lithiophilite, stilbite, and native bismuth. Namibite, clinobisvanite, and minor bismutite were identified by E. E. Foord in 1987. Considering the presence of lithiophilite, it is safe to assume that at least a small variety secondary phosphates are also present.

RAMONA DISTRICT

The Little Three Mine

E. E. Foord U.S. Geological Survey

General features. The Little Three main dike is one of five principal pegmatite dikes (with or without aplite) on the Little Three property. These dikes include the Little Three main dike, the Hercules-Spessartine dike, the Spaulding dike, and the Hatfield Creek-Sinkankas dike. All apparently formed as locally anastomosing but essentially separate intrusives.

All of the dikes strike northwest and dip southwest at 10° to 65°. Most crop out along hillside exposures where they are folded on dip-slope orientation and run perpendicular to the hillside, presenting irregularly rolling surfaces in cross-sectional view.

Petrology and lithologic units. The Little Three main dike displays marked variation in texture, composition, zoning patterns and mineralogy. In these aspects, it is similar to layered pegmatite-aplite intrusives exposed in neighboring districts (Pala, Rincon, Mesa Grande, and others). The 0.5-1 m-thick aplitic footwall comprises one-third to locally, one-half of the zoned intrusive. The lowermost massive, sometimes indistinctly layered, albite-rich aplite contains major quartz and albite, with less abundant potash-feldspar and schorl. Almandine-spessartine garnet is the only other minor, sparsely distributed phase.

The lower massive section compositionally and texturally grades upwards to banded aplite or 'line-rock,' contains somewhat more schorl, and garnet; the section ranges from 10 to 36 cm in maximum composite thickness The banded aplite consists of 0.1- to 4-cm-thick layers of varying shades of white, gray and nearly black (schorl-rich layers).

The layered aplite grades upwards to a zone of increasingly coarse-grained quartz-perthite-albite pegmatite. Coarse-grained albite and intermediate-microcline hosted perthite make up the bulk of this zone, along with subordinate and widespread schorl and lesser amounts of muscovite and garnet. Potash-feldspar megacrysts, as much as 10 cm long, flare outward toward the pocket zone. A light to dark color transition is coincident with a symmetry inversion to the monoclinic state (orthoclase) in the potash-feldspar. Schorl, muscovite, and spessartine garnet increase in size and concentration as the pocket zone is approached.

Graphic quartz-perthite pegmatite is the predominant unit in the hanging wall of the dike. The perthite is composed of a maximum microcline host containing exsolved low albite lamellae. [*Simpson*, 1962] has documented the three-dimensional nature of the quartz-feldspar perthite. Distribution of schorl, spessartine and muscovite is similar to that in the underlying pegmatite; they increase in grain size and abundance towards the approximately centrally located pocket line.

The pocket zone is characterized by the presence of large, euhedral crystals of quartz, feldspar, tourmaline, topaz and other minerals which either occur in open miarolitic cavities (pockets) or mutually enclose one another. Larger pockets commonly extend up into the graphic granite pegmatite hanging wall, and the floors of such pockets may extend down into the layered aplite.

History and mining activity. The Little Three mining property was discovered in May 1903 by H. W. Robb of Escondido who took two partners, Dan McIntosh of Ramona and Charles F. Schnak of Escondido. The partnership purchased the 40-acre plot from John Ferguson, who owned a ranch of several hundred acres. The name 'Little Three' refers to the three original owners. As was true for the other pegmatite gem mines in San Diego County, the collapse of the Chinese tourmaline market in 1912 halted mining activity. Only sporadic high-grading occurred until the acquisition of the property by Louis B. Spaulding and his son, Louis Jr., in 1951. No intensive mining was undertaken until 1955. Several dikes on the property were mined by Mr. Spaulding until his death in 1973. [*Simpson*, 1962, 1965] examined the structure and petrology of the dikes, particularly of the graphic granite, but no detailed mineralogical work was done. One cash lease was obtained by J. F. Scripps and W. F. Larson in 1972 to work a portion of the Little Three main dike, and John Sinkankas was granted permission in 1957 to work the Sinkankas-Hatfield Creek dike. Since resumption of nearly full-time mining activity in March 1975 by Louis B. Spaulding,

Jr., an extensive amount of work has been done, and approximately 70 m of underground tunnel have been driven. In the last 10 years, principally surface mining has been done. Efforts have been primarily devoted to the main Little Three dike and to the Hercules-Spessartine dike. Perhaps the most scientifically and economically significant discoveries (June 1976) and perhaps for the entire mining history of the dike, have been made within the upper parts of the Main Little Three dike, and in particular within one pocket called the 'New Spaulding' pocket [*Stern et al.*, 1986; *Foord et al.*, 1989]. There has been significant production of gem-quality spessartine and mineral specimens from the Hercules-Spessartine dike as well.

Mining and production. Most of the mining on the Little Three property has been on the surface. Only one 18 m drift on the Hercules dike near the northern property line was underground work. A total of approximately 200 m of underground and surface work has been done in the Little Three main dike and the Hercules-Spessartine workings. The dip-slope configuration of large parts of many of the dikes has made surface stripping practicable.

Production of gem and specimen-quality material from the Little Three property has been sporadic, with boom-or-bust cycles predominating. Production records for the Little Three property are extremely scanty, and adequate records have been kept for only about the past 30 years. Production from the large 'New Spaulding pocket' consisted of approximately 90 kg of elbaite, more than 28 kg of topaz, in excess of 200 kg of large, euhedral 'books' of lepidolite, and still larger amounts of quartz and feldspar. Approximately 1 kg of tungsten-bearing stibio-bismuto-columbite-tantalite was also recovered. Total mine production of elbaite is estimated to be 1000 kg and topaz production is estimated to be 110 kg. A minimum of 550 kg of lepidolite has been produced to date (L. B. Spaulding, Jr., pers. comm., 1990). The total production of faceting-grade spessartine garnet from the Hercules-Spessartine workings and the Spaulding dike is not certain but is estimated to be about 70,000 carats.

A very fine topaz-elbaite matrix specimen weighing about 70 kg was mined in 1905 from the Little Three main dike. It was on display for many years at the San Diego Chamber of Commerce until moved to the then newly opened Natural History Museum. During World War II, the Navy took over the museum and the displays were crated and stored for the duration. The specimen was never seen again, and it is probable that it was hauled out to sea with a barge of junk and dumped! (L. B. Spaulding, Jr., written commun., 1986).

Mineralogy. Aspects of the mineralogy of the Little Three property have been treated by [*Foord*, 1982a,b; *Foord et al.*, 1986; *Stern et al.*, 1986], and most recently by [*Foord et al.*, 1989, in press]. Despite similar bulk compositions and asymmetrical zoning features among the dikes on the property, each dike has its own characteristic pocket mineralogy, suggesting at least some degree of independent formation. The Hercules-Spessartine dike yields muscovite, microcline, albite, spessartine, schorl, and quartz plus minor manganese-rich axinite as principal primary pocket minerals. In contrast, pockets from the Little Three main dike have produced elbaite (principally dark green), muscovite or fluorine-rich lepidolite, quartz, microcline, albite, and topaz (blue to colorless), plus minor apatite, microlite-uranmicrolite, tungsten-bearing stibio-bismuto-columbite-tantalite, beryl (goshenite-morganite), hambergite, columbite, native bismuth, pucherite, bismite and bismutite. The Little Three main dike has also produced perhaps the best quality lepidolite crystals of all of the pegmatite dikes in San Diego County. Sharp, well-formed crystals as much as 8 cm or more in diameter and 2.5 cm thick were recovered. The Sinkankas-Hatfield Creek dike has produced abundant hambergite and dark red-brown crystals of axinite. Gem-quality spessartine garnet was recovered from the Spaulding dike. Table 1 includes a list of minerals identified from the several dikes on the Little Three property.

MESA GRANDE DISTRICT

The Himalaya Dike System

E. E. Foord U.S. Geological Survey

General features. Although the individual pegmatite dikes comprising the Himalaya dike system are exceptionally thin (averaging about 0.5 m in thickness), they are the most continuous, both vertically and laterally, of all of the composite aplite-pegmatite dikes in San Diego County that have been and(or) are currently being worked for specimen and gem-quality minerals. A series of as many as five or more closely spaced, subparallel dikes or dikelets, averaging about 50 cm in thickness, is exposed on the upper part of the easterly facing slope of Gem Hill, above Angell Field. The outcrop belt of dikes, most of which are now well exposed only in mine workings, is nearly 2 km long.

The dikes strike slightly west of north in the southerly part of the system and swing gradually westward to the north. Their outcrop surface traces are seen to rise northward on the hillside to the crest of the northward-projecting main ridge of Gem Hill, then they appear along the flank of a low, northwesterly extending ridge. All of the dikes dip westward at an average angle of 35°, but the extreme range is 15° to 55°.

In most places, the system consists of two subparallel dikes that are 3 to 10 m apart. They converge southward and are in direct contact, without an intervening screen of gabbronorite country rock, in underground and surface workings on the San Diego mine property. The upper dike, which is the thinner of the two, has been the principal source of gem and specimen materials. This dike has been referred to as the 'Himalaya Dike' [*Jahns*, 1979]. The upper dike consists almost wholly of massive to miarolitic pegmatite, while the lower dike consists of layered and unlayered aplite-pegmatitic aplite as well as pegmatite.

The remarkable continuity of the dike system is illustrated by a continuous 1 to 3 mm thick band or layer of almandine-spessartine garnet, minor gahnite, and schorl that can be traced all along the strike at the top of a sequence of unlayered albite-schorl-quartz-microcline pegmatitic aplite.

The dikes are simple and sharply bounded tabular masses in most places but locally are complex in detail. Branching, pinching, and swelling of the dikes and of locally included masses of country rock are exposed at a number of places in the underground mine workings.

Petrology and lithologic units. The dikes are zoned texturally, compositionally, and mineralogically, and the general pattern of their internal structure changes systematically along strike and downdip. At the northwest end of the dike system, the main dike is almost bilaterally symmetrical with respect to the core. However, southward toward the San Diego mine area, the dike becomes increasingly asymmetrical because of the appearance of fine-grained pegmatitic aplite and aplite in its lower parts. A vertical section through the combined upper and lower veins as exposed in the southern end of the San Diego mine workings and at the surface, shows graphic quartz-feldspar-muscovite-schorl pegmatite in the upper part, with an expanded zone of quartz-rich pegmatite occupying the pocket zone, and distinctly layered quartz-feldspar-schorl-garnet-muscovite aplite in the basal portion. In places, the bottom one-third to one-half of the dike is composed of layered and non-layered aplite, markedly enriched in sodium, which locally is physically transected by another zone of graphic and pocket pegmatite. The layering is defined by bands of garnet, albite-quartz, and schorl. A second type of aplite, known as "pressure-quench" aplite and that shows autoinjection features, is present locally within the pocket zone and as transecting dikelets emanating from the pocket zone and cutting across the pegmatite host rocks. In some places, boron-metasomatism and(or) intense hydrothermal alteration of the host norite have occurred adjacent to the pegmatite dikes [*Foord et al.*, 1986].

History and mining activity. The Himalaya dike system is world famous for its production of gem- and specimen-quality tourmaline and other rare mineral species. It ranks as North America's largest producer of tourmaline.

Tourmaline from the Himalaya dike system was first collected by school children and local residents in the 1880's. Active mining operations began on the Himalaya claim in 1898, and involved the collecting of surface material. Shortly thereafter, opencut work was begun. In 1899, tourmaline was discovered on the neighboring San Diego mine property. In 1902, the Himalaya property was taken over by the Himalaya Mining Company; by 1904, surface work was essentially completed and subsequent mining was entirely underground.

Initial development of the San Diego property was by excavation of several opencuts and trenches. In 1902, underground work was begun and the San Diego Tourmaline Mining Company was organized in 1904 to work the deposits. Work proceeded at an extremely rapid pace and by 1910 nearly all of the old (pre-1952) underground workings on the property had been developed. The collapse of the Oriental tourmaline market in 1912 resulted in bankruptcy and cessation of work on both properties. Between 1912 and 1960, mining and excavation at the San Diego mine were limited in large part to short inclines, generally less than 30 m long. During the period between 1925 and 1928, the General Electric Company excavated two inclines and the General Electric tunnel in the Himalaya mine.

Some additional work was done at the San Diego mine between 1933 and 1950 by E. B. Rynerson and B. F. Rynerson, and in 1934 by Arthur Montgomery and Edwin Over. In 1960, Eugene B. Rynerson, initially with the assistance of his nephew, Alan Rynerson, began intermittent to full-time operation of the San Diego mine by driving a new tunnel some 10 m below the level of the old (1902-10) Naylor workings. To date, a total of approximately 600 m of workings has been developed. Some surface excavation has also been done in the last ten years by E. B. Rynerson. Currently the property is inactive.

In 1952, the Himalaya mine was acquired by Ralph R. Potter and operated by him; several old inclines and tunnels were reopened and additional new work was done until 1963. At this time, a new organization, Himalaya Gem Mines, Inc., was formed to develop the property. After the main portal of the Himalaya mine collapsed during the wet season of 1968-69, some surface work was done by the corporation. During the period 1971-74, a new portal was excavated on the east side of Gem Hill to reopen the old "hard rock" tunnel, and some mining was done from 1975-77. In 1977, a portion of the Himalaya property was leased to Pala Properties, International, and full-time mining activity commenced under the direction of John MacLean. A new portal and tunnel were constructed north of and above the hard-rock tunnel. A series of five dikes or stringers was crossed by the 236 m tunnel. Drifting was done both to the north and south on the main level and several inclines were sunk to below the water table. Currently, Pala International is continuing mining activity with a crew of three to five men. Most of the mining is being done in raises and inclines above and to the north of the main adit, on the upper or Himalaya dike. An extensive system of galleries aggregating about 800 m of workings has been developed to date.

Mining and production. Initial mining by pick-and-shovel methods along the strike of the dikes went downdip as far as was practical to strip off the overlying soil and decomposed norite. Following this mining, underground work was extended downdip in favorable areas, and thence along strike. Depth of mining was limited to about 30 m on the incline, because of the necessity of hand winching and the decrease in degree of surficial alteration of the norite host rock.

During the 15 years from 1898 to 1913, the heyday of operations on the Himalaya dike system, approximately 100,000 kg of tourmaline, then valued at more than $750,000, was produced. One pocket alone, at the north end of the dike system, produced in excess of 700 kg of gem tourmaline. Cut gemstones produced from the period 1902 to 1913 had an aggregate value of about $2,500,000.

Production since about 1960 has been aimed towards the increasing collectors' and investors' market for specimen and gem material. Production from the San Diego mine property since 1960 has amounted to several tons of tourmaline, but not all of the material was useable. From 1977 to the present, production from the Himalaya mine has been intermittent, but excellent specimen and gem quality material has been recovered [*Marcusson*, 1985]. More than 5000 kg of tourmaline of all grades have been produced, along with some morganite, Nb-Ta minerals, hambergite, and other minerals.

Mineralogy. The mineralogy of the Himalaya dike system has been extensively studied since about 1903. Some of the more important and complete references are [*Penfield and Ford*, 1906; *Schaller*, 1916; *Foord*, 1976, 1977; and *Foord et al.*,

1986]. Table 1 is a compilation of minerals observed and positively identified by us.

Phosphate minerals are nearly absent in the Himalaya dike system except for pink and blue fluorapatite and very rare monazite and xenotime. This paucity is in distinct contrast to many dikes in the Pala district, which have produced abundant primary and secondary phosphate minerals. The Himalaya dike system is enriched in B relative to many other dikes in San Diego County. This enrichment is reflected in the abundance of tourmaline produced. In contrast, the Little Three mine in the Ramona district is famous for its production of topaz and spessartine garnet, with some elbaite. Primary Li-Al silicates and phosphates such as petalite, spodumene and amblygonite-montebrasite have not been observed in the Himalaya dike system by the authors nor have they been found during the recent (post-1977) extensive mining operations at the Himalaya mine. However, some spodumene (including kunzite) and amblygonite were listed by [*Jahns*, 1979].

$^{40}Ar/^{39}Ar$ Geochronology and Thermochronology of the Granitic Pegmatites of San Diego County, California

L. W. Snee and E. E. Foord U.S. Geological Survey

The geochronology of the Peninsular Ranges batholith in southern California and northern Baja California was presented by [*Krummenacher et al.*, 1975] using more than 200 K-Ar apparent ages. Available Rb-Sr and U-Pb ages were also discussed by [*Krummenacher et al.*, 1975]. More recently, geochronologic data have been presented for the La Posta-type plutons in the eastern part of the Peninsular Ranges batholith [*Walawender, et al.*, 1990].

Within the western portion of the Peninsular Ranges batholith, complex granitic pegmatites such as those in the Pala, Ramona and Mesa Grande districts, are known to be the youngest rocks of the batholith. No direct source for the pegmatites has been located and it is well known that the distribution of pegmatites (including aplite) is strongly controlled by host lithology. The gabbro and norite of the San Marcos Gabbro is by far the favored host for the pegmatite dikes and sheets. The pegmatites and monzogranites of the La Posta-type plutons are believed to be derived from anatexis of the prebatholithic metasedimentary assemblage at moderately shallow levels (>4 kbar or so) (M. J. Walawender, pers. comm., 1991). Emplacement depths (P_T) are estimated to be between 2.5 and 3.0 kbar [*Clinkenbeard and Walawender*, 1989; M. J. Walawender, pers. comm., 1989].

We have dated approximately 40 samples of K-feldspar, muscovite, and lepidolite from pegmatites in each of the three main gem- and specimen-producing districts: Pala, Mesa Grande and Ramona, as well as biotite, and hornblende from immediately adjacent or nearby host rocks to the pegmatites and one sample of Woodson Mountain Granodiorite well removed (15 km) from any pegmatite dikes.

To summarize, it appears that the Woodson Mountain Granodiorite was emplaced before 113 Ma (hornblende age), it cooled below 500-550°C (hornblende argon-retention temperature) by 113 Ma, it cooled below 300°C (biotite argon-retention temperature) by 107 Ma (biotite apparent age), and had cooled to below 150°C (microcline argon-retention temperature) by 104 Ma (microcline apparent age). The Hercules-Spessartine dike on the Little Three property, Ramona district, was emplaced >98.7 Ma and <100 Ma. The age of the main Little Three pegmatite dike may be 96 Ma (lepidolite age) or earlier. The 96 Ma age may be a result of the different retention (setting temperature) age for lepidolite vs. muscovite. Muscovite from the Pala Chief mine, Pala district, gave an age of 99.4 Ma in agreement with the age for the Hercules-Spessartine dike at the Little Three. The host rocks (San Marcos Gabbro and mafic and felsic phases of the Green Valley Tonalite are equivalent to Bonsall Tonalite) adjacent to the pegmatite dikes on the Little Three property have had their ages (both biotite and hornblende) reset to 100 Ma by the emplacement of the dikes. The same is true for the host rocks (San Marcos Gabbro) in the Mesa Grande [*Foord*, 1976] district and the Pala district.

After emplacement of the pegmatite dikes, cooling to an ambient temperature of about 100°C was achieved by about 91 Ma (based on K-feldspar apparent ages).

Additional details and discussion regarding the Ar-Ar geochronology and thermochronology of the pegmatites are given in Snee and Foord (in preparation).

Genesis of the Pegmatites of San Diego County

David London University of Oklahoma

General Considerations

There are few aspects of pegmatite crystallization that can be taken for granted as known. It is reasonably certain that the pegmatite dikes of San Diego County, like most others, crystallized from intrusive silicate melts that were qualitatively granitic in composition. Although these pegmatites share many features in common with pegmatite bodies worldwide, they differ from most in their unusual abundance of miarolitic cavities, which indicates that these magmas became vapor saturated prior to complete crystallization.

Pressure-Temperature Constraints

The determination of emplacement pressures and liquidus-solidus intervals is not straightforward for these or other similarly evolved plutonic granitic bodies. The compositions of the pegmatites are not strictly eutectic in the haplogranite minimum: they are strongly peraluminous and contain abundant boron, fluorine, some phosphorus, and elevated concentrations of rare alkalis and beryllium. The experimental liquidus-solidus relations of rare-element-enriched peraluminous rhyolite, an obsidian from Macusani, Peru, provide the best available constraints for liquidus relations in comparably evolved and peraluminous systems [*London et al.*, 1988, 1989]. At 200 MPa, the solidus of hydrous macusanite melt is near 450°C, which places an approximate lower limit on the equilibrium solidus of the San Diego County pegmatite magmas (this does not preclude the possibility that the magmas experienced significant undercooling below the stable solidus). The H_2O-saturated liquidus of macusanite lies at 645°C, but the liquidus temperatures increase rapidly with decreasing H_2O content below saturation. Using equilibrium relations calculated from mineral chemistries, [*Stern et al.*, 1986] proposed a liquidus near 700°C for the Little Three pegmatite. [*Taylor et al.*, 1979] utilized stable isotope systematics to propose a liquidus temperature of 700°-730°C for the Himalaya dike. Both proposed liquidi are higher than reasonable values for H_2O-saturated magmas of the compositions of these dikes (which are particularly boron- and fluorine-rich) at pressures ≥ 200 MPa. Hence if these liquidi are valid, and if pressures were sufficient to prevent devolitalization through decompression, then the pegmatite magmas were initially undersaturated in H_2O.

Estimates of pressures attending crystallization can be determined from the lithium aluminosilicate stability relations [*London*, 1984]. A few San Diego County pegmatites (Stewart, Pala Chief, Vanderburg, and Katerina-Pala district; Victor-Rincon district; Esmeralda-Mesa Grande district; Bebe Hole-Jacumba district) contain primary spodumene + quartz assemblages (though petalite also occurs at the Stewart mine; both primary spodumene and petalite from the Stewart dike have been confirmed by XRD). Minor contents of Fe and Mn in spodumene do not substantially change the stability relations of this assemblage, so that the minimum pressures of pegmatite consolidation are constrained to values above the univariant reaction spodumene + 2 quartz = petalite at temperatures above the solidus [*London*, 1984, 1986a]. Taking 450°C as the minimum feasible solidus, pressures of crystallization in spodumene-bearing pegmatites (and by implication, associated pegmatites that lack lithium aluminosilicates) were >250 MPa [*London*, 1986a]. Both [*Stern et al.*, 1986, and Taylor et al., 1979] have proposed pegmatite solidi near 550°C, which would correspondingly raise the minimum pressure estimates to >340 MPa. At pressures in the range of 250-350 MPa, and with the evolved compositions of these pegmatite magmas, the liquidus temperatures cited above (700°-730°C) would correspond to initially H_2O-undersaturated melts.

Internal Evolution of the Pegmatites

The relationships of textural, chemical, and zonal evolution of these and other pegmatite dikes are intimately related. [*Jahns*, 1953b] recognized this fact in his attempts to identify the underlying processes of pegmatite development. His conclusions [*Jahns*, 1953b] were that the zonal features and progressive coarsening of pegmatite minerals could be "attributed to fractional crystallization and incomplete reaction between successive crops of crystals and rest-liquid...All known features of pegmatites seem reasonably explainable on the basis of crystallization from a silicate melt of low viscosity, with or without end-stage deuteric or hydrothermal activity. Many of these features also suggest that masses of pegmatite crystallized inward from the walls of an original chamber, under restricted-system conditions, rather than in some sort of channelway or thoroughfare under more open-system conditions...nearly all the giant crystals were formed during what has been designated by most investigators the primary, or magmatic, stage of pegmatite development, and hence by crystallization from liquid under conditions that permitted the remarkable growth of a relatively few individuals...The [giant] crystals are thought to have formed rapidly under restricted-system conditions involving rather delicate temperature and chemical balance. Temperatures were almost certainly below 600°C, and the confining pressures were sufficiently great to prevent major escape of volatile constituents during the period of giant-crystal development." These interpretations by [*Jahns*, 1953b], which were subsequently abandoned by [*Jahns and Burnham*, 1969] for a radically different concept, are virtually identical to the recently proposed model of [*London et al.*, 1989; *London*, 1990, 1991]. Jahns' belief that pegmatite dikes form by restricted-system crystallization is supported by the fact that most pegmatites of a given field lack a high degree of interconnection that makes open-system flow viable. After emplacement, groups of pegmatite bodies become isolated, and internal differentiation of melt occurs in increasingly restricted volumes of magma. Early Rb-Sr isotopic studies by [*Methot and Brookins*, 1973] demonstrated that typical granitic pegmatites of Middletown, Connecticut, are not in isotopic equilibrium with their host rocks, and hence could not have been derived from them by in situ anatexis or other processes; moreover, Methot and Brookins noted that the zone of isotopic mixing between pegmatites and host rocks is very narrowly confined to pegmatite margins. The lack of a pervasive isotopic reequilibration of host rocks in regions of dense pegmatite swarms provides partial evidence that the conduits for pegmatite magma emplacement are not pervasively interconnected through fracture systems. Subsequent studies of stable isotope systematics ([e.g., *Taylor et al.*, 1979]) and other geochemical features ([e.g., *London*, 1990]) now reveal that pegmatite magmas are open to influx of material from their host rocks during the earliest and latest stages of their crystallization (discussed further below under The Role of Boron), but that the chemical differentiation that typifies pegmatite fields is inherited from the source during dike emplacement.

Jahn's hypothesis that pegmatites crystallize rapidly is consistent with a recent model for the consolidation of the large Harding pegmatite, New Mexico [*Chakoumakos and Lumpkin*, 1990], which probably crystallized in <100 years. The generally smaller, thinner dikes of San Diego County could have crystallized in a shorter time frame, if the temperature difference between pegmatite liquidus and host rock was >200°-300°C. Although this may seem like a surprisingly short time interval in which to grow giant crystals, note that normal silicate crystal growth rates from feldspathic or haplogranitic melts (10^{-7} to 10^{-9} cm/s) are sufficient to grow crystals 1 m in dimension in 3-4 years, if the nucleation density of crystals in melt is virtually nil [*London*, 1991].

In the model of [*London et al.*, 1989; *London*, 1990, 1991], the transition from granite to pegmatite with increasing development of zonation reflects a state of increasing disequilibrium or metastable persistence of melt. In turn, metastable persistence of melt is caused by increasing melt undercooling with magma emplacement farther from source, and especially by the effects of quartz-feldspar incompatible components (H_2O, B, P, F, or any combination of these) on decreasing the probability of forming crystal nuclei. All of these components interact strongly with the major components of melt, especially the feldspar-forming components of alkalis and aluminum. Experimentally, each of these components has been shown to suppress crystal nucleation (decreased number of nuclei and increased lag times before the onset of crystallization). Because of speciation between these components and the major constituents of melt, there is a decreased probability of forming critical crystal nuclei that lack these excluded components. Once the initial crystal nuclei form, the local increase in concentration of excluded components diminishes the probability of forming new nuclei in their immediate vicinity. As the bulk concentration of excluded components in melt increases with fractional crystallization, the probability of nucleating new phases is further decreased, which allows a few individual crystals to grow to huge dimensions at normal rates of crystallization. As temperature decreases, the diffusion rates of components through melt are probably decreased as well. This allows for the possibility of chemical zonation of melt on a comparatively large scale, as seen in experiments with Macusani glass [*London et al.*, 1989], which contributes to the formation of mineralogically distinct zones. The components B, P, and F further augment pegmatite zonation by expanding the liquidus field of quartz, and hence promoting increasingly abundant quartz deposition and the formation of essentially monomineralic cores. Implicit in this model is the fact that the concentrations of these excluded components, including H_2O, increase without buffering in melt through phase saturation. Through the application of lithium aluminosilicate phase relations, we now know that the pressures are sufficiently high (as suggested by [*Jahns*, 1953b]) to prevent the wholesale loss of volatile components from melt at early stages of crystallization. Similarly, recent experimental studies have demonstrated that the components of B, F, and P are not truly volatile ([e.g., *London*, 1987; *London et al.*, 1990]). The latter two components could actually be regarded as "magmaphilic".

The textural zonation of pegmatites, and the San Diego County pegmatites in particular, is also indicative of and derived from strongly disequilibrium crystallization. With increasing disequilibrium (i.e., metastable supersaturation caused by increasing undercooling or other suppression of crystal nucleation), all liquid systems (silicate, metal, aqueous) display a transition from internal (homogeneous) crystal nucleation to sidewall (heterogeneous) crystallization, with an increasing development of anisotropic textures. The most prevalent of these textures is comb structure, a form of unidirectional solidification texture, which in pegmatites is commonly manifest by primary microcline. Additional features of such disequilibrium growth are the development of graphic and skeletal/branching crystals, which are also prevalent in these and many other pegmatites. Both [*Fenn*, 1986] and [*London et al.*, 1989] have experimentally generated pegmatitic fabrics of comb structure, sidewall crystallization, and graphic quartz-microcline intergrowths at reasonable states of melt undercooling, but only at H_2O-undersaturated conditions of melt. At high H_2O contents of melt, diffusion of components in the melt boundary layer in front of growing microcline crystals is sufficiently rapid (again, at reasonable undercoolings) to inhibit the local saturation of quartz at microcline growth surfaces.

As noted by [*Jahns*, 1953b], this process involves a delicate balance of temperature and chemistry. A critical aspect of pegmatite crystallization, which is largely unknown, is the rate of magma cooling after emplacement. Even if it could be assumed that heat loss from magma through host rocks is purely conductive (which simplifies calculations), an important unknown parameter remains: the temperature of host rocks at the time of magma emplacement. The temperatures recorded by mineral assemblages in metamorphic (or igneous) hosts merely provide an upper limit to conditions at magma intrusion. Metamorphic alteration of host rocks by pegmatite magmas is invariably retrograde, and mostly indicative of conditions in the greenschist facies (approximately bounded by 300°-550°C) ([e.g., *London*, 1986c; *Morgan and London*, 1987b]). In the model of [*Chakoumakos and Lumpkin*, 1990], host rock temperatures 300°C below the inferred pegmatite liquidus promote cooling of the pegmatite magma to temperatures <550°C (the approximate equilibrium solidus) in <2-3 years. This suggests that the magma experienced significant undercooling upon emplacement relative to crystal nucleation and growth rates. If this model is even qualitatively correct, then it is likely that the Harding body, like most other differentiated pegmatite magmas emplaced as dikes into cooler host rocks, experienced significant and sharp undercooling. The lag time for crystal nucleation to commence would surely be long enough that such rapid cooling would metastably supersaturate the melt in normal crystalline phases (quartz, feldspars). Eventual crystallization would have occurred far from equilibrium, and would have promoted crystal growth features (sidewall nucleation, comb structure, and graphic/skeletal crystal habits) indicative of disequilibrium systems. For the San Diego County pegmatites, host rock temperatures at the time of emplacement were probably around 100°C, based on the sequence of mineral argon retention temperatures determined for the Woodson Mountain Granodiorite. Cooling from approximately 550°C to <150°C occurred in 9 m.y. The pegmatites themselves also appear to have taken about 9 m.y. to cool to about 100°C. The San Diego County pegmatites, however, are much thinner than the large Harding body, and hence cooling of the magma likely has occurred at comparable or higher rates. Disequilibrium conditions, promoted by rapid undercooling and suppression of crystal nucleation caused by incompatible components, represent a likely cause of the strong textural development of comb structure, graphic/skeletal crystals, and sharp zonation within these pegmatites.

The Role of Boron

An important feature of the San Diego County pegmatites is their high concentration of boron, manifested as an abundance of tourmaline. The effects of boron on the properties of granitic melt, partly summarized from [*London*, 1986a, 1986b, 1987, 1990], include:

(1) reduced melt viscosities,

(2) reduced liquidus-solidus temperatures,

(3) reduced rates of crystal nucleation,

(4) increased solubility of H_2O in melt, and

(4) expansion of the liquidus field of quartz and K-silicates at the expense of albite.

[*London*, 1986a, 1986b, 1987, 1990; *London et al.*, 1987; *Morgan and London*, 1987a, 1987b, 1989] emphasized the importance of boron and of tourmaline stability in understanding the properties of pegmatite melts. In the San

Diego County pegmatites, for example, it is evident that the pegmatite magmas represented major sources of boron. In some pegmatites, inwardly flairing, comb-structured tourmaline constitutes 10-20 vol. % of the pegmatite. At localities such as the Himalaya and San Diego mines, the host norite shows only slight to moderate evidence of hydrothermal alteration along long stretches of exposed strike length. However, there is no evidence of tourmalinization. Tourmalinization of host norite, which would have been very reactive with boron-rich pegmatite fluids, is localized only where tourmaline-filled miarolitic cavities have ruptured and released unconserved boron to the host rocks. These and similar relationships elsewhere indicate that boron-enriched pegmatite magmas are initially open to infiltration (by fluid components and dissolved solutes including Fe) which is not initially bimetasomatic [*London*, 1990]. The total absence of tourmalinization (except in the vicinity of pockets) along pegmatite contacts with reactive mafic host rocks indicates that boron-rich fluids were not expelled from pegmatites, and [*London*, 1990] has suggested that this constitutes one more piece of evidence that differentiated S-type pegmatite magmas are not vapor saturated at early stages of crystallization. In the San Diego County pegmatites, as elsewhere, the sporadic distribution of tourmalinization in host rocks reflects very late-stage fluid expulsion from B-enriched residual pegmatite melts. These late-stage melts become enriched in boron only because the initial crystallization of pegmatite borders effectively seals the remaining pegmatite melt from communication with host rocks, so that the supply of host rock-derived Fe needed to stabilize tourmaline is exhausted.

The Role of Aqueous Vapor

The recently proposed model of [*London*, 1990; *London et al.*, 1989], which is fundamentally similar to that of [*Jahns*, 1953b], questions the necessity of an aqueous vapor phase in pegmatite formation. Two different considerations must be addressed: what is the evidence that pegmatites become vapor saturated early in their evolution, and what are the effects of vapor saturation?

Evidence from pegmatites. Manifestations of vapor saturation include metasomatic replacement of primary minerals in pegmatite and host rocks, and the formation of primary miarolitic cavities. Metasomatic alteration of primary pegmatite and host rock assemblages is demonstrably late-stage, and occurs at near-solidus to subsolidus conditions within and outside of pegmatite bodies. There is no evidence for metasomatic alteration through exsolved vapor at the early stages of pegmatite emplacement and consolidation, and in fact there is evidence against an early vapor phase (e.g., the example of tourmaline and wallrock alteration discussed above). Miarolitic cavities, which are unusually abundant in the San Diego County pegmatites, are also formed during the end stages of pegmatite crystallization, and cannot be linked to the initial stages of emplacement. It is important to remember that miarolitic and nonmiarolitic pegmatites share many features in common, yet the vast majority of pegmatites do not contain cavities. This fact suggests a commonalty of process that does not hinge on the presence of an aqueous vapor phase. [*London*, 1990] provides a more complete review of the evidence for or against vapor saturation in pegmatites.

Experimental evidence. [*Jahns*, 1982] cited an experiment in which the formation of a pegmatitic run product was attributed to the attainment of vapor saturation during the course of crystallization in the experiment. [*London et al.*, 1989] showed that there was no unique causal relation between vapor saturation and pegmatite development in the experiment of Jahns (cited in 1982), and that pegmatitic run products are the normal result of disequilibrium crystallization of vapor-undersaturated melt that contains modest concentrations of the quartz-feldspar incompatible components cited above (H_2O, B, F, or P).

Consequent effects. The pegmatite model of [*Jahns and Burnham*, 1969] differed significantly from the original views of [*Jahns*, 1953b] in that pegmatite formation was portrayed as an equilibrium process involving eutectic crystallization of granitic melt. Because eutectic equilibrium crystallization produces results that are sharply dissimilar to pegmatites, the Jahns-Burnham model was forced to call on another agent, an aqueous vapor phase, to promote the necessary mass transport for mineralogical zonation and variations in texture. In the Jahns-Burnham model, upward buoyant ascent of chloride-rich aqueous vapor stripped K and rare elements from footwall melt to be concentrated in the upper hangingwall portions of pegmatite. Although [*Jahns*, 1982] held to the notion of eutectic crystallization of quartz cores from melt, [*Burnham and Nekvasil*, 1986] concluded that quartz cores could not be generated by purely magmatic equilibrium processes, and hence must be largely hydrothermal in origin. A full assessment of these concepts is beyond the intended scope of this paper, but a few points are noteworthy.

Ignoring the fact that pegmatite-wallrock relations provide no evidence for but ample evidence against early vapor saturation, how have some of the fundamental precepts of the Jahns-Burnham model stood the test of subsequent research? An important observation cited by [*Burnham and Nekvasil*, 1986], and emphasized by [*London et al.*, 1988], is that under most conditions, the elemental ratios of solutes (major elemental components) in vapor are the same as in melt. This is especially true for evolved peraluminous magmatic compositions such as the San Diego County pegmatites represent. Therefore, an aqueous vapor phase (chloride-bearing or not) is incapable of promoting the heterogeneous fractionation under most known or inferred conditions that was the precept of the Jahns-Burnham model. If aqueous vapor saturation early in the history of pegmatite consolidation actually did what [*Jahns and Burnham*, 1969] proposed, what would be the consequences? One is that melt incompatible trace elements would be "scoured" from melt [*Jahns*, 1982] to be concentrated in the early stages of pegmatite consolidation, and residual melt would become progressively trace-element depleted. This would produce reverse fractionation trends (progressive depletion rather than enrichment) that are contradicted by natural evidence. The vertical thermal profile that would result from upward ascent of aqueous vapor (see [*Burnham and Nekvasil*, 1986]) would promote alkali exchange between coexisting feldspars (connected through the vapor phase) such that sodic plagioclase would be concentrated in the hotter, upper portions and K-feldspars in the cooler, lower portions of pegmatite bodies [*Orville*, 1963]. This is also contradicted by virtually all known natural examples. [*Jahns and Burnham*, 1969] also explained the common association of potassic pegmatite overlying sodic aplite as the consequence of K removal from footwall magma, and rapid crystallization of sodic residual magma (as footwall aplite) due to the compositional shift away from the granite eutectic (toward the albite-quartz join). They (Jahns and Burnham) ignored that fact that the addition of K to the hangingwall pegmatite magma would produce a complementary shift (equal in magnitude and opposite in direction) away from the granite eutectic composition, yet this does not produce compositional quenching (development of aplite) in the hangingwall portions. As both upper and lower portions of pegmatite magma remain vapor saturated through this process, the proposed accumulation of an aqueous vapor phase in the hangingwall regions is irrelevant. Finally, [*Jahns*, 1979] noted that silicate minerals deposited in the presence of a vapor phase (in miarolitic cavities) are invariably finer grained than their counterparts in massive (nonmiarolitic) pegmatite. Though largely ignored at the time, both [*Fenn*, 1977] and [*Swanson*, 1977] demonstrated experimentally that quartz and feldspars deposited at vapor saturation of melt are actually finer grained and grow more slowly than minerals crystallized at low H_2O content of melt. [*London et al.*, 1988] further showed that the deposition of silicate minerals from aqueous vapor (as an interface between crystals and melt) yielded products that were uniform in grain size, texture, distribution, and composition, with open space between mineral grains. These features are clearly unlike those of normal pegmatites (except perhaps for the miarolitic stage).

Crystallization in Miarolitic Cavities

The conditions under which miarolitic cavities form are correctly viewed as the end stages of consolidation in some pegmatites, notably those of San Diego County. Estimated temperatures of pocket formation are <550°C, and must include thermal regimes in the stability fields of zeolites and clay minerals. Although solute-depleted aqueous vapor may represent that last fluid in contact with pocket minerals, [*London*, 1986a,b]) has shown that most silicate deposition is from a hydrous silicate fluid that is transitional between normal concepts of melt or vapor. In most miarolitic pegmatites, therefore, the transition from magmatic to hydrothermal stages is essentially continuous unless pocket rupture leads to quenching. [*Jahns*, 1982] proposed that late-stage quench effects (e.g., aplites) form by a sudden drop in pressure upon rupture of pockets, and indeed there is good evidence, especially in the pegmatites of San Diego County, that miarolitic pockets were overpressured with regard to surrounding lithostatic pressures. Perhaps more important is the fact that decreasing pressure leads to volatile loss from residual melt (e.g., of important components such as H_2O and B), and that fracturing opens a potential bimetasomatic exchange between pegmatite pockets and host rocks. As explained by [*London*, 1986a,b; 1990], contact between residual pegmatite and wallrocks through fluid creates a sink for important components in the host rocks, especially for boron in the form of metasomatic tourmaline, which begins an irreversible compositional quenching process. Fracturing also permits the influx of wallrock-derived fluids into pegmatite, which may effect cooling, addition of chlorine, and dilution of components that, unlike Cl, act as fluxes for late-stage silicate liquids. Evidence for the late influx of wallrock-derived fluids includes fluid inclusion data [*London*, 1985a, 1986b] and mineralogical assemblages [*Foord et al.*, 1986]. Some new fluid inclusion data for the pegmatites of San Diego County bears this out (Table 2).

As in previous studies of miarolitic pegmatites from Afghanistan, and some additional data on the San Diego pegmatites [*London*, 1986a], the new data show that fluids that can be reasonably construed as indigenous to pegmatites possess very low salinities with very low total chloride

TABLE 2. Preliminary microthermometric properties of fluid inclusions, San Diego County pegmatites

Phase/Pegmatite	Te	Tm(ice)	Tm(xl)	Th(LV)
Quartz, Pala Chief	-29.3(5.7)[144]	-1.2(0.9)[133]		179.6(34.2)[206]
Spodumene, Pala Chief	-37.4(4.1)[14]	-1.8(0.5)[13]		
Quartz, Eliz. R	-22.3(3.9)[68]	-2.7(1.7)[57]		235.0(41.3)[178]
Quartz, Little 3	-24.7(5.3)[70]	-2.0(0.5)[84]	69.4(15.8)[14]	204.8(50.0)[46]
Topaz, Little 3	-24.0(4.4)[54]	-2.6(0.7)[66]	> 350 (TD)	228.4(54.6)[58]

Te = temperature of eutectic; Tm(ice) = melting temperature of ice; Tm(xl) = melting temperature of daughter mineral (low relief, moderate-high birefringence); Th(LV) = temperature of homogenization of liquid and vapor (to liquid)
Values are: mean, (1 standard deviation) [number of measurements]

concentrations. Most freezing point depressions are <2°-3°C, and aqueous borate, or fluoride components account for most of this. Inclusions that are demonstrably secondary in host phases possess higher bulk salinities, lower homogenization temperatures, and contain elevated concentrations of Mg, Ca, Cl, and S, which are indicative of fluids equilibrated with metamorphic host rocks [*London*, 1985a, 1986b]. Some interesting new data (EDS-WDS analyses) from the Little Three pegmatite reveal that daughter minerals in secondary inclusions from pocket quartz (New Spaulding 1976 pocket) include Mg-Ca-borates, a borate that shows only B and O by electron beam methods (e.g., boric acid, or a borate of Li or Be), and cryolite. The presence of Mg-Ca-borates may represent the entrapped products of a mixed fluid, containing components indigenous to the pegmatite (boron) and from the wallrocks (Mg and Ca). The general abundance of borate daughter minerals reflects the incredible abundance of boron in the Little Three system, even at the very latest stages of consolidation after abundant tourmaline (and locally axinite) have crystallized from melt. Cryolite as a daughter mineral reflects not only a high concentration of fluorine in the Little Three pegmatite, but also a strongly alkaline character of late-stage fluids evolved from melt [*London*, 1986a, 1990]. The presence of boromuscovite with 7 wt. % B_2O_3 [*Foord et al.*, in press], which formed as a precipitate from post pocket-rupture solutions, also indicates the high level of boron.

Fluid inclusions from several other pegmatites (White Queen, Pala Chief, Elizabeth R, and Stewart) are qualitatively similar to each other, but different from those of the Little Three. The inclusions examined to date include secondary-pseudosecondary inclusions in quartz, and modified primary-pseudosecondary inclusions in spodumene (Pala Chief, Stewart). Inclusions in quartz are dominantly very low salinity aqueous liquid-vapor, but some contain small quantities of CO_2. Daughter minerals are absent. As in virtually all other localities, transparent spodumene from miarolitic pockets contains a very low-salinity aqueous fluid and an abundance of insoluble daughter minerals, presumably mostly silicates but not yet analyzed. Quartz, feldspar, and micas can be identified optically. The inclusions also contain an abundance and diversity of other anisotropic phases with high relief and birefringence. Some appear to be very sparingly soluble up to 200°C (cf. [*London*, 1986a,b], but soluble borates, fluorides, or phosphates do not appear to be present (cf. Little Three). Collectively, these pegmatites contain minor to major quantities of tourmaline, pink-lavender micas qualitatively identified as lepidolite, and minor phosphate minerals. Amblygonite-montebrasite from the Stewart and Pala Chief pegmatites is low- to moderate-fluorine montebrasite (XRD analysis of the 131 peak position), as at most other localities (e.g., [*London and Burt*, 1982]). This fact implies a comparatively low fluorine activity of melt (cf. [*Morgan and London*, 1987b]).

Unresolved Problems

Pegmatite-aplite pairs. One of the last major problems concerning the internal differentiation of individual pegmatite bodies is the generation of sodic aplite and potassic pegmatite pairs. The pegmatites of San Diego County afford abundant, classical exposures of pegmatite-aplite relations. Potassic pegmatite usually overlies sodic aplite, but exceptions are known (e.g., [*London*, 1985b]). Together, sodic aplites and potassic pegmatitic zones approach the composition of the granite minimum, which has led virtually all petrologists to assume some mode of contemporaneous differentiation of aplite-pegmatite from a typical batch of granitic magma.

Although the volatile-quench model of [*Jahns and Burnham*, 1969] is intuitively satisfying, subsequent research (some discussed above) has negated most of the essential hypotheses of this concept. Rather than considering all of the problems and possibilities in this presentation, an evaluation of pegmatite-aplite relations is better left to discussions in the field (i.e., in the presence of aplite-pegmatite systems that will be viewed on this field trip). In the end, the model of [*Jahns and Burnham*, 1969] may prove correct, reflecting that our current knowledge of melt-vapor systems is incomplete or erroneous. For now, however, alternative hypotheses and processes are needed.

The magmatic-hydrothermal transition. Throughout the history of pegmatite study (now comprising voluminous work spanning more than a century), a major stumbling block has been the resolution and common agreement on basic processes and products: the recognition and meaning of primary versus secondary crystallization, crystallization from melt versus vapor, replacive versus non-replacive assemblages. It is the textural and mineralogical complexities of pegmatites such as those in San Diego County that pique the curiosity of petrologists and mineralogists, and that afford so many different views of pegmatite origin. An underlying theme in the recent research of [*London*, 1986a, and subsequent papers] is that the transition from dominantly magmatic to hydrothermal processes in evolved, peraluminous pegmatites may be essentially continuous (i.e., supercritical with respect to melt and vapor phases), unless this evolution is interrupted by any of several processes (e.g., sudden decrease of pressure, abrupt loss of volatile or fluxing components through phase saturation within pegmatites or via communication between pegmatites and host rocks). Like most modern concepts in igneous petrology, the hypothesis of a continuous magmatic-hydrothermal transition is not new. The very term "pegmatitic stage" has been widely construed as some transitional state between magmatic and hydrothermal (e.g., [*Lindgren*, 1933]). Archaic and arcane terms such as "ichor" [*Sederholm*, 1926] also embody the longstanding recognition that some states of terrestrial matter may not fit neatly into existing concepts of "magmatic" and "hydrothermal". It is generally agreed that the compositions of most common pegmatites can be expressed fairly well within the hydrous haplogranite system, in which there is very limited miscibility between aqueous fluid and melt at crustal pressures corresponding to pegmatites. This is not the case for peraluminous magmas that are comparatively enriched in boron, fluorine, and phosphorus in addition to H_2O. Evidence from fluid inclusions (e.g., [*London*, 1986a,b]) and experiments (e.g., [*London et al.*, 1989]) has demonstrated that assemblages of albite + Li-mica ± quartz, tourmaline, phosphates, topaz, etc. can be precipitated directly by fractional crystallization of melt in the absence of a coexisting vapor phase. In pegmatites such as those of San Diego County, assemblages of this sort are common, and usually lack evidence of replacement. (This does not preclude resorption of preexisting minerals during deposition of albite-Li mica assemblages. In systems that manifest such strong indications of disequilibrium fractional crystallization, reaction of early-formed assemblages with derivative melts along localized fracture fillings should be the rule rather than the exception.) The significance of this distinction in pegmatite studies is to avoid the trap of using simple systems to explain complex ones where they do not apply, and to prevent preconceived notions from obscuring the distinction between fact and interpretation.

References

Bancroft, Peter, 1973, The world's finest minerals and crystals. Viking Press, New York, 176 pp.

Bancroft, Peter, 1984, Gem and Crystal Treasures. Western Enterprises and the Mineralogical Record, 488 pp.

Beus, A. A., 1960, Geochemistry of beryllium and genetic types of beryllium deposits (in Russian). Akademii Nauk SSSR, Moscow [transl.] Freeman, San Francisco, 1967.

Brock, K. J., 1974, Zoned lithium-aluminum mica crystals from the Pala pegmatite district. American Mineralogist, 59, 1242-1248.

Burnham, C. W., 1979, Magmas and hydrothermal fluids. In Barnes, H. L., ed., Geochemistry of hydrothermal ore deposits, 2nd edition. John Wiley and Sons, New York, 71-136.

Burnham, C.W. and Nekvasil, H., 1986, Equilibrium properties of granite pegmatite magmas. American Mineralogist, 71, 239-263.

Cameron, E. N., Jahns, R. H., McNair, A. H., and Page, L. R., 1949, Internal structure of granitic pegmatites. Economic Geology Monograph, No. 2, 115 pp.

Chakoumakos, B. C. and Lumpkin, G. R., 1990, Pressure-temperature constraints on the crystallization of the Harding pegmatite, Taos County, New Mexico. Canadian Mineralogist, 28, 287-298.

Chorlton, L. B. and Martin, R. F., 1978, The effect of boron on the granite solidus. Canadian Mineralogist, 16, 239-244.

Clinkenbeard, J. P., and Walawender, M. J., 1989, Mineralogy of the La Posta pluton: Implications for the origin of zoned plutons in the eastern Peninsular Ranges batholith, southern and Baja California. American Mineralogist, 74, 1258-1269.

Donnelly, Maurice, 1936, Notes on the lithium pegmatites of Pala, California. Pacific Mineralogist, 3, 8-12.

Fenn, P. M., 1977, The nucleation and growth of alkali feldspars from hydrous melts. Canadian Mineralogist, 15, 135-161.

Fenn, P. M., 1986, On the origin of graphic granite. American Mineralogist, 71, 325-330.

Foord, E. E., 1976, Mineralogy and petrogenesis of layered pegmatite-aplite dikes in the Mesa Grande district, San Diego County, California: Ph. D. thesis, Stanford University, Stanford, Calif., 326 pp.

Foord, E. E., 1977, The Himalaya dike system, Mesa Grande district, San Diego County, California. The Mineralogical Record, 8, 461-474.

Foord, E. E., Martin, R. F., and Long P. E., 1979, Potassic feldspars from pocket pegmatites, San Diego County, California [abs.]. Geological Society of America Annual Meeting, Abstracts with Programs, v. 11, no. 7, p. 427.

Foord, E. E., 1982a, Bismuthian stibiocolumbite-tantalite and other minerals from the Little Three mine, Ramona, California. Fourth annual MSA-FM Symposium, February 1981, Tucson, Arizona. In Brown, G. E., Jr., ed., The mineralogy of pegmatites. American Mineralogist, 67, 181-182.

Foord, E. E., 1982b, Minerals of tin, titanium, niobium and tantalum in granitic pegmatites. In Černý, P., ed., Mineralogical Association of Canada Short Course Handbook, 8, Granitic Pegmatites in Science and Industry, 187-238.

Foord, E. E., Starkey, H. C., and Taggart, J. E., Jr., 1986, Mineralogy and paragenesis of 'pocket clays' and associated minerals in complex granitic pegmatites, San Diego County, California. American Mineralogist, 71, 428-439.

Foord, E. E., Spaulding, L. B., Jr., Mason, R. A., and Martin, R. F., 1989, Mineralogy and Paragenesis of the Little Three mine pegmatites, Ramona District, San Diego County, California. Mineralogical Record, 20, 101-127.

Foord, E. E., Martin, R. F., Fitzpatrick, J. J., Taggart, J. E., and Crock, J. G., (in press), Boromuscovite, a new member of the mica group, from the Little Three mine pegmatite, Ramona district, San Diego County, California. American Mineralogist.

Gastil, R. G., 1975, Plutonic zones in the Peninsular Ranges of southern California and northern Baja California. Geology, 3, 361-363.

Houser, Charles, 1986, Structural controls and mineralogical indicators for the formation and distribution of pockets in the Elizabeth R mine, Pala, San Diego County, California. Senior thesis, San Diego State University, 43 pp.

Jacobson, M. I., 1973, The paragnesis of the lithium-iron-manganese phosphates of the Stewart Lithia Mine, Pala, California. Unpublished manuscript on file at Pennsylvania State University Library, 26 pp.

Jahns, R. H., 1953a, The genesis of pegmatites. II. Quantitative analysis of lithium-bearing pegmatite, Mora County, New Mexico. American Mineralogist, 38, 1078-1112.

Jahns, R. H., 1953b, The genesis of pegmatites. I. Occurrence and origin of giant crystals. American Mineralogist, 38, 563-598.

Jahns, R. H., 1979, Gem-bearing pegmatites in San Diego County, California: The Stewart Mine, Pala district and the Himalaya Mine, Mesa Grande district. In Abbott, P. L., and Todd, V. R., eds., Mesozoic Crystalline Rocks: San Diego, California, San Diego State University, Department of Geological Sciences, p. 3-38.

Jahns, R. H., 1982, Internal evolution of pegmatite bodies. In Černý, P., ed., Granitic Pegmatites in Science and Industry. Mineralogical Association of Canada Short Course Handbook, 8, 293-327.

Jahns, R. H. and Burnham, C. W., 1969, Experimental studies of pegmatite genesis. I. A model for the derivation and crystallization of granitic pegmatites. Economic Geology, 64, 843-864.

Jahns, R. H., and Tuttle, O. F., 1963, Layered pegmatite-aplite intrusives. Mineralogical Society of America Special Paper, no. 1, 78-92.

Jahns, R. H., and Wright, L. A., 1951, Gem- and lithium-bearing pegmatites of the Pala district, San Diego County, California. California Division of Mines Special Report 7A, 72 pp.

Johnson, P. W., 1969, Common gems of San Diego County, California. Gems and Gemology, 11, (7), 274-296.

Krummenacher, Daniel, Gastil, R. G., Bushee, J. and Doupont, J., 1975, K-Ar apparent ages, Peninsular Ranges Batholith, Southern California and Baja California. Geological Society of America Bulletin, 86, 760-768.

Kunz, G. F., 1905, Gems, jeweler's materials, and ornamental stones of California. California State Mining Bureau Bulletin 37, 171 pp.

Larsen, E. S., 1948, Batholith and associated rocks of Corona, Elsinore, and San Luis Rey quadrangles, southern California. Geological Society of America Memoir 29, 182 pp.

Larsen, E. S., 1951, Crystalline rocks of the Corona, Elsinore, and San Luis Rey quadrangles, southern California and Baja California. California Division of Mines Bulletin 159, 7-50.

Larson, W. F., 1977, The best of San Diego County. The Mineralogical Record, 8, 507-515.

Lindgren, W., 1933, Mineral Deposits, 4th Edition. McGraw Hill, New York, 930 p.

London, D., 1984, Experimental phase equilibria in the system $LiAlSiO_4$-SiO_2-H_2O: a petrogenetic grid for lithium-rich pegmatites. American Mineralogist, 69, 995-1004.

London, D., 1985a, Origin and significance of inclusions in quartz: a cautionary example from the Tanco pegmatite, Manitoba. Economic Geology, 80, 1988-1995.

London, D., 1985b, Pegmatites of the Middletown district, Connecticut. 77th Annual Meeting, New England Intercollegiate Geological Conference, Yale University: Connecticut Geological and Natural History Survey Guidebook No. 6, 509-533.

London, D., 1986a, Formation of tourmaline-rich gem pockets in miarolitic pegmatites. American Mineralogist, 71, 396-405.

London, D., 1986b, The magmatic-hydrothermal transition in the Tanco rare-element pegmatite: evidence from fluid inclusions and phase equilibrium experiments. American Mineralogist, 71, 376-395.

London, D., 1986c, Holmquistite as a guide to pegmatitic rare metal deposits. Economic Geology, 81, 704-712.

London, D., 1987, Internal differentiation of rare-element pegmatites: effects of boron, phosphorus, and fluorine. Geochimica et Cosmochimica Acta, 51, 403-420.

London, D., 1990, Internal differentiation of rare-element pegmatites: A synthesis of recent research. Invited contribution, "Ore-bearing granite systems; Petrogenesis and mineralizing processes" (H. J. Stein and J. L. Hannah, eds.), Geological Society of America Special Paper 246, 35-50.

London, D., 1991, A dynamic chemical model for pegmatite crystallization [abs.]. Geological Society of America Abstracts with Programs, in press.

London, D. and Burt, D. M., 1982, Alteration of spodumene, montebrasite, and lithiophilite in pegmatites of the White Picacho district, Arizona. American Mineralogist, 67, 97-113.

London, D., Hervig, R. L., and Morgan, G. B., VI, 1988, Melt-vapor solubilities and element partitioning in peraluminous granite-pegmatite systems: Experimental results with Macusani glass at 200 MPa. Contributions to Mineralogy and Petrology, 99, 360-373.

London, D., Loomis, J. L., Huang, W., and Morgan, G. B., VI, 1990, Behavior and effects of phosphorus in the Ab-Or-Qz-H_2O at 200 MPa(H_2O) [abs.]. Geological Society of America Abstracts with Programs, 22, A302.

London, D., Morgan, G. B., VI, and Hervig, R. L., 1989, Vapor-undersaturated experiments in the system macusanite-H_2O at 200 MPa, and the internal differentiation of granitic pegmatites. Contributions to Mineralogy and Petrology, 102, 1-17.

London, D., Zolensky, M. E., and Roedder, E., 1987, Diomignite: natural $Li_2B_4O_7$ from the Tanco pegmatite, Bernic Lake, Manitoba. Canadian Mineralogist, 25, 173-180.

Marcusson, C. R., 1985, Recent work at the Himalaya Mine. The Mineralogical Record, 16, 419-424.

Merrill, J. F. H., 1914, Mines and mineral resources of San Diego County, California. California State Mining Bureau Report, 14, 691-700 and 706-708.

Methot, R. L. and Brookins, D. G., 1973, Rb-Sr systematics of two large granitic pegmatites, Connecticut [abs.]. Geological Society of America Abstracts with Programs, 5, 737-738.

Morgan, G. B., VI and London, D., 1987a, Behavior of boron and tourmaline stability in granitic systems [abs.]. Geological Society of America Abstracts with Programs, 19, 777-778.

Morgan, G. B., VI and London, D., 1987b, Alteration of amphibolitic wallrocks around the Tanco rare-element pegmatite, Bernic Lake, Manitoba. American Mineralogist, 72, 1097-1121.

Morgan, G. B., VI and London, D., 1989, Experimental reactions of amphibolite with boron-bearing aqueous fluids at 200 MPa: Implications for tourmaline stability and partial melting in mafic rocks. Contributions to Mineralogy and Petrology, 102, 281-297.

Norton, J. J., 1983, Sequences of mineral assemblages in differentiated granitic pegmatites. Economic Geology, 78, 854-874.

Orville, P. M., 1963, Alkali ion exchange between vapor and feldspars. American Journal of Science, 261, 201-237.

Penfield, S. L., and Ford, W. E., 1906, On stibiotantalite. American Journal of Science, 4th series, 22, 61-77.

Rynerson, F. J., 1967, Exploring and mining for gems and gold in the West. Naturegraph, Healdsburg, California, 204 pp.

Schaller, W. T., 1916, The gem tourmaline field of southern California: U.S. Geological Survey unpublished manuscript.

Schaller, W. T., 1925, The genesis of lithium pegmatites. American Journal of Science, 5th series, 10, 269-279.

Sederholm, J. J., 1926, On migmatites and associated Pre-Cambrian rocks of southwestern Finland. Part II. The region around the Barosundsfjard west of Helsingfors and neighbouring areas. Bulletin de la Commission geologique de Finlande. In Selected Works, Granites and Migmatites, J. J. Sederholm: Oliver and Boyd, Ltd., London (1967), 370-371.

Shigley, J. E., 1982, Phosphate minerals in granitic pegmatites: a study of primary and secondary phosphates from the Stewart pegmatite, Pala, California. Ph.D. dissertation, Stanford University, Stanford, California, 535 pp.

Shigley, J. E. and Brown, G. E., Jr., 1985, Occurrence and alteration of phosphate minerals at the Stewart pegmatite, Pala district, San Diego County, California. American Mineralogist, 70, 395-408.

Shigley, J. E., Kampf, A. R., Foord, E. E., and London, D., 1986, Fieldtrip guidebook to gem pegmatites of southern California, 14th General Meeting of the International Mineralogical Association, July, 1986, Stanford University, Stanford, California, 39 pp.

Silver, L. T., Taylor, H. P., Jr., and Chappell, B., 1979, Some petrological, geochemical, and geochronological observations of the Peninsular Ranges batholith near the international border of the U.S.A. and Mexico. In Abbott, P. L., and Todd, V. R., eds., Mesozoic crystalline rocks: San Diego, California, San Diego State University, Department of Geological Sciences, p. 83-110.

Simpson, D. R., 1962, Graphic granite from the Ramona pegmatite district, California. American Mineralogist, 47, 1123-1138.

Simpson, D. R., 1965, Geology of the central part of the Ramona pegmatite district, San Diego County, California. California Division of Mines and Geology, Special Report 86, 3-23.

Sinkankas, John, 1959, Gemstones of North America, vol. 1. Van Nostrand Reinhold, New York, 675 pp.

Sinkankas, John, 1976, Gemstones of North America, vol. 2. Van Nostrand Reinhold, New York, 494 pp.

Snee, L. W. and Foord, E. E., (in preparation), Geochronology and thermochronology of the complex granitic pegmatites of San Diego Co., California. Geological Society of America Bulletin, v. pp.

Stern, L. A., Brown, G. E., Jr., Bird, D. K., Jahns, R. H., Foord, E. E., Shigley, J. E., and Spaulding, L. B., Jr., 1986, Mineralogy and geochemical evolution of the Little Three pegmatite-aplite layered intrusive; Ramona, CA. American Mineralogist, 71, 406-427.

Stewart, D. B., 1978, Petrogenesis of lithium-rich pegmatites. American Mineralogist, 63, 970-980.

Swanson, S. E., 1977, Relation of nucleation and crystal-growth rate to the development of granitic textures. American Mineralogist, 62, 966-978.

Taylor, B. E., Foord, E. E., and Friedrichsen, H., 1979, Stable isotope and fluid inclusion studies of gem-bearing granitic pegmatite-aplite dikes, San Diego Co., California. Contributions to Mineralogy and Petrology, 68, 187-205.

Todd, V. R. and Shaw, S. E., 1979, Structural, metamorphic, and intrusive framework of the Peninsular Ranges batholith in southern San Diego County, California. In Abbott, P. L. and Todd, V. R., eds., Mesozoic Crystalline Rocks: Peninsular Ranges Batholith and Pegmatites, Point Sal Ophiolite. Department of Geological Sciences, San Diego State University, San Diego, p. 177-231.

Todd, V. R., and Shaw, S. E., 1985, S-type granitoids and an I-S line in the Peninsular Ranges batholith, southern California. Geology, 13, 231-233.

Walawender, M. J., 1979, Basic plutons of the Peninsular Ranges batholith, southern California. In Abbott, P. L. and Todd, V. R., eds., Mesozoic crystalline rocks: San Diego, California, San Diego State University, Department of Geological Sciences, p. 151-162.

Walawender, M. J., Gastil, R. G., Clinkenbeard, J. P., McCormick, W. V., Eastman, B. G., Wernicke, R. S., Wardlaw, M. S., Gunn, S. H., and Smith, B. M., 1990, Origin and evolution of the zoned La Posta-type plutons, eastern Peninsular Ranges batholith, southern and Baja California. In Anderson, J. Lawford, ed., Geological Society of America Memoir 174, Chapter 1, pp. 1-18.

Weber, F. H., Jr., 1963, Geology and mineral resources of San Diego County, California. California Division of Mines and Geology County Report, no. 3, 309 pp.

Reprinted from *Geological Excursions in Southern California and Mexico*, Guidebook for the 1991 Annual Meeting, Geological Society of America, San Diego, California.

A SYNTHESIS OF RECENT WORK IN THE PENINSULAR RANGES BATHOLITH

Michael J. Walawender, Gary H. Girty, Marc R. Lombardi, David Kimbrough, Melissa S. Girty and Camille Anderson

Department of Geological Sciences,
San Diego State University,
San Diego, CA 92182

INTRODUCTION

The Peninsular Ranges batholith (PRB) of southern and Baja California is an elongate body of igneous rock that extends from its junction with the Transverse Ranges near Riverside, California to about the 28th parallel in Baja California. Throughout its strike length of nearly 1000 km, the batholith averages about 100 km in width. To the west lies the continental borderland and younger, mostly Eocene, sedimentary rocks. The eastern margin slips beneath the sedimentary cover of the Salton Trough and the Gulf of California.

The batholith, unlike its Mesozoic counterparts around the Pacific Basin, is calcic, as opposed to calc-alkalic, in character. It has been subdivided into two zones (Gastil, 1975, Silver, 1979) that parallel the long axis of the batholith. The westernmost zone consists of smaller plutons that are generally less than 100 km^2 in exposed area. Compositions range from peridotite to granite with rocks of tonalitic composition being most abundant. The eastern zone consists of a more limited range of compositions (tonalite to monzogranite) and the plutons are typically several hundreds of km^2 in size. The boundary between the eastern and western zones is closely approximated by several discontinuities. Gastil and others (1990) delineated a magnetite-ilmenite line which effectively separates a western zone of magnetite- and ilmenite-bearing plutons from an eastern zone in which the only opaque phase is ilmenite. Todd and Shaw (1985) defined an I-S line which separates a western zone of I-type granitoids from an eastern zone of both I-type and S-type granitoids. This discontinuity lies somewhat west of the magnetite-ilmenite line and the boundary between synkinematically deformed plutons of the western PRB and the generally undeformed plutons of the eastern zone (Todd and Shaw, 1979).

Silver and others (1979), Silver and Chappell (1988), and De Paolo (1980, 1981) summarized the regional isotopic data for the PRB. Their data show systematic regional patterns for Pb, Nd, and Sr isotopes that become increasingly more radiogenic eastward with isopleths that are roughly parallel to the long axis of the batholith. These isopleths, which are independent of lithology, do not appear to show any pronounced jump at or near the discontinuities described above but, rather, continue to change smoothly across the batholith. Whole-rock values for $\delta^{18}O$ also increase from approximately 6 per mil in the west to more than 12 per mil in the east (Taylor and Silver, 1978) but a series of closely spaced isopleths from 8.0 to 9.5 per mil (isotope "step" of Silver and others, 1979) approximate the boundary between the western and eastern zones of the batholith.

Silver and others (1979) and Silver and Chappell (1988) drew attention to an age discontinuity in the medial zone of the batholith that defined a transition from a western "static" arc (140 to 105 Ma) in which the emplacement ages show no discernible geographic pattern to an eastern "migrating" arc (105 to 89 Ma) in which the emplacement ages become progressively younger eastward. Walawender and others (1990), however, presented additional age and isotopic data on the eastern zone of the batholith that do not support an eastward migration of the locus of magmatism and argued for two distinct periods of static-arc magmatism within the batholith with the younger magmatic event spanning a tighter range of 98 to 89 Ma.

The purpose of this field trip is to lead participants over a cross-sectional traverse of the PRB. In the following sections, the salient aspects of both the prebatholithic and batholithic components are described. In an attempt to highlight problematical or controversial issues concerning the development of the PRB, liberal use will be made of preliminary results from our ongoing, long-term projects. A road log is provided at the end of this guidebook article.

PREBATHOLITHIC ROCKS

Santiago Peak Volcanics

The Santiago Peak Volcanics (SPV) form a discontinuous belt (Figure 1) approximately 20 km in width that consists of weakly metamorphosed volcanic, volcaniclastic, and sedimentary rocks. They are exposed along the western edge of the PRB and extend from the type locality (Larsen, 1948) in the Santa Ana Mountains southward to the Agua Blanca fault in Baja California. A lithologically similar belt, the Alisitos Group (Allison, 1955; Santillan and Barrera, 1930), occurs further to the south.

The SPV range in composition from basalt to rhyolite but are predominantly andesite and quartz latite (Larsen, 1948; Kennedy and Peterson, 1975). They overlie the Triassic to Jurassic Bedford Canyon Formation north of San Diego County. The contact is thought to be unconformable and there is some evidence to indicate that it may be depositional (Larsen, 1948; Criscione and others, 1978; C. Herzig, 1991, pers. commun. to CLA). In most areas, the SPV are intruded by Cretaceous plutons of the PRB. The intrusive contact is steep and irregular.

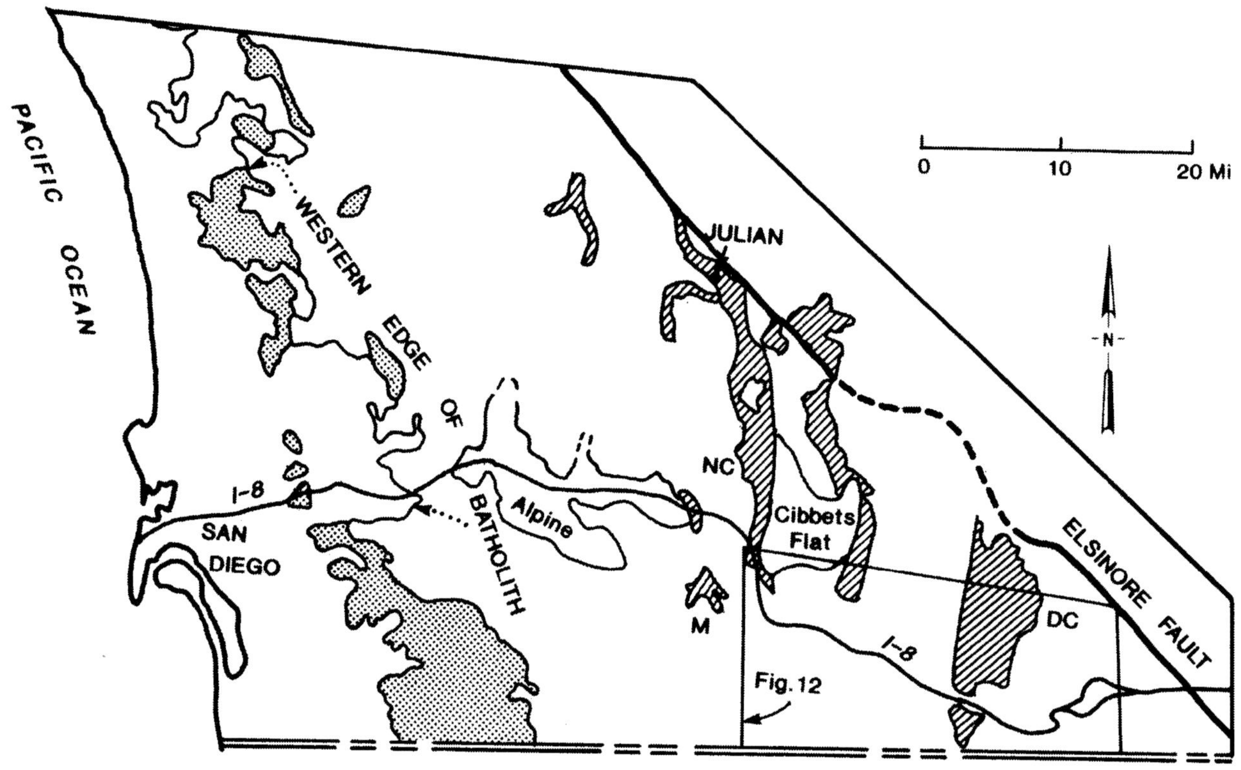

Figure 1. Location map for the Peninsular Ranges batholith and selected components. Stippled pattern represents the Santiago Peak Volcanics, ruled pattern represents the metasedimentary sequence (Julian Schist). The Alpine and Cibbets Flat plutons are outlined. See text for discussion. Locations discussed in the text are Noble Canyon (NC), Morena Reservoir (M), and Dos Cabezas (DC).

Exposures in the Santa Margarita Mountains approximately 75 km north of San Diego appear to show a somewhat greater volume of flow units relative to fragmental rocks (Gorzolla, 1988). In the Black Mountain area 30 kms north of San Diego, Adams (1979) described 12 members having a total stratigraphic thickness of about 3200 m with flow units making up about 20% of the total. Reworked fragmental units are found in both areas. In and around San Diego, the SPV is mainly fragmental with sparse flow units (Tanaka and others, 1984).

Porphyritic textures, with phenocrysts of zoned plagioclase (andesine to labradorite) and sparse orthoclase, quartz, and clinopyroxene, are common in the flow units but much less common in the breccia fragments. Phenocryst and groundmass minerals are partially converted to mixtures of quartz, albite, epidote, chlorite, carbonate, biotite, muscovite, stilpnomelane, pyrite, pyrophyllite, and tourmaline. This assemblage has been interpreted in terms of chlorite-grade metamorphism but is equally consistent with a hydrothermal overprint. Adams (1979) and Adams and Walawender (1982) argued that the SPV volcanic pile represented an emerging volcanic arc adjacent to a shallow ocean basin such that eruptions were alternately subaerial and submarine.

Tanaka and others (1984) divided the SPV near San Diego into a basalt-andesite group with tholeiitic affinities and a dacite-rhyolite group with subalkaline (calc-alkaline?) affinities and argued that these two groups represented, respectively, a younger, more mafic, volcanic arc and a more mature oceanic or continental-margin arc. Gorzolla (1988) pointed out that the SPV near San Diego are dominantly tholeiitic whereas areas to the north are mostly calc-alkaline in character. The boundary between these two chemical types, in part constrained by the structural trends in the PRB, suggested a northwesterly trend to the ancient SPV arc system.

A Late Jurassic (Tithonian) age has been assigned to the SPV on the basis of fossils, including Buchia piochii, from several isolated sections of the volcaniclastic sequence (Fife and others, 1967; Jones and Miller, 1982). These sections, however, are not representative of the massive flows and breccias which dominate the SPV. Lead-alpha ages of 150 and 155 $\pm$ 15 Ma were reported from rhyolite and dacite, respectively, from the type section in the Santa Ana Mountains (Bushee and others, 1963). Krummenacher and others (1975) obtained a K-Ar cooling age of 121 $\pm$ 8 Ma from an andesite porphyry in the SPV.

Zircon U-Pb ages have been determined on a number of samples from the SPV as part of an ongoing study being carried out at San Diego State University. Four samples have yielded slightly discordant Early Cretaceous ages (138 to 119 Ma). Silver and Chappell (1988) present a similar age range for the Allisitos Formation in Baja California. The zircon data indicate that the SPV are not restricted to a Late Jurassic (Tithonian) age as indicated by the fossil evidence but extend well into the Lower Cretaceous. The data also suggest that eruption of the SPV overlapped in time with the emplacement of at least some of the plutonic rocks of the batholith and that locally the SPV may represent a coeval volcanic cover to the batholith. In addition, Tanaka and others (1984) argued that the trace-element chemistry of the basalt and andesite portion of the SPV corresponds to that estimated by Smith and others (1983) for melts parental to the early gabbroic plutons in the western zone of the PRB.

Metasedimentary Units

The metasedimentary rocks east of the SPV are part of the sandstone-shale belt of Gastil (1975) that extends southward through the state of Baja California. In southern California, this flysch-type sequence was named the Julian Schist (Hudson, 1922) for occurences near Julian California. Todd and Shaw (1985) suggested that the SPV and Julian Schist interfinger in a 5 to 10 km wide zone east of San Diego that represents remnants of a marginal basin that received detritus from the emerging SPV arc system to the west and a continental source to the east.

The Julian Schist consists of a series of pelitic and quartzofeldspathic schists and gneisses with minor amphibolite, metaconglomerate, and calc-silicate units that are typically folded into isoclinal structures whose axial surfaces are parallel to the NNW structural grain of the batholith. The Morena Reservoir roof pendant (Figure 1) consists of a series of cordierite-biotite-andalusite- and sillimanite-bearing gneisses interbedded with quartz- and feldspar-rich diopside-actinolite gneiss. These parageneses suggest peak metamorphic conditions between 2.0 to 2.5 kb and about 600°C (Berggreen and Walawender, 1977). Germanario (1982) and Siem (1988) reported similar lithologies and conditions in Noble Canyon, about 20 kms to the north (Figure 1), and near Julian, California, respectively. Further east, amphibolite and pelitic schists become somewhat more abundant. Andalusite and cordierite are no longer present such that an andalusite- (and cordierite?-)out line can be drawn that approximates the boundary between the western and eastern zones of the PRB. East of the boundary, metamorphic conditions appear to be 3.3 to 4.3 kb and 630 to 660°C (Detterman, 1984; Dierecks, 1981).

At the eastern edge of the batholith near Dos Cabezas (Figure 1), intricately folded carbonate-quartzite units and spatially associated migmatitic gneiss and amphibolite form the wall rocks to the batholith. Sillimanite is sparse but present in more pelitic sections of the gneiss. The assemblage grossularite-diopside-plagioclase- wollastonite-calcite brackets P-T conditions to between 2.8 to 3.5 kb and 620 to 660°C.

The age of the Julian Schist is poorly constrained. An ammonite imprint on a piece of quartzite float was identified as Triassic (Hudson, 1922) but the sample was subsequently lost. Upper Triassic molluscs are reported from part of the sandstone-shale belt in the northern part of the PRB (Todd and others, 1988). The carbonate-quartzite sequence at Dos Cabezas is lithologically similar to a roof pendant found about 70 km to the south near San Marcos in Baja California and to metasedimentary rocks in the Coyote Mountains about 20 km to the east, both of which contain Ordovician conodonts (Gastil and Miller, 1984). The carbonate rocks at Dos Cabezas, however, did not yield any fossil data.

WESTERN ZONE

Introduction

Three main plutonic rock systems can be delineated within the western zone of the PRB: gabbro, quartz diorite to tonalite, and granodiorite to monzogranite. The gabbroic system has been interpreted as the oldest plutonic unit in the batholith based on field relationships but direct evidence of age relative to adjacent plutons is not always available (Walawender and Smith, 1980). The quartz diorite-tonalite and granodiorite-monzogranite series are, respectively, younger (Larsen, 1948; Todd and Shaw, 1979). Each series has its own distinct morphological character. The gabbros typically underlie high-relief areas with dark colored outcrops and reddish soil. The quartz diorites and tonalites have weathered to a lower, rolling terrane with grayish outcrops and soils whereas the granodiorite-monzogranite plutons form high-relief areas with light-colored, bouldery outcrops.

Gabbroic System

Radiometric ages in the gabbroic rocks of the PRB are limited. Krummenacher and others (1975) report K-Ar cooling ages on amphibole from 143 Ma to 101 Ma but interpreted the younger ages as reset. Silver and others (1979) argued that the older ages are due to trapping of excess Ar from an unknown source.

All of the plutons are multiple intrusive complexes with a wide range in rock types. For several (Los Pinos, Walawender, 1976; Lawson Peak, Hoffman, 1975), field evidence indicates a mutiple intrusive history with similar batches of gabbroic melt undergoing the same fractionation mechanisms at relatively shallow levels. Others (Target Range pluton, Walawender and others, 1979; Corte Madera pluton, Lillis, 1978) seem to have less complex intrusive histories and some gradational contacts.

The gabbroic rocks of the PRB are characterized by an abundance of calcic plagioclase (An_{98-52}) and pargasitic to edenitic amphibole (Smith and others, 1983). Cumulate textures dominate the early series with the mafic minerals occupying the interstices between plagioclase laths. Early formed olivine (Fo_{79-70}) and augite display resorption textures; sparse orthopyroxene (bronzite to inverted pigeonite) occurs as thin rims on augite or as uncommon interstitial patches. Amphibole is late and typically forms oikocrysts up to 5 cm in size. The crystallization sequence is indicative of emplacement levels between 2 and 5 kb (Walawender, 1976).

Crystal accumulation processes resulted in residual melts that crystallized to amphibole gabbros which show abundant textural evidence that these late-stage melts were water oversaturated. Pegmatitic pods and dikes are common in this rock type and comb layers occur locally (Walawender, 1976). Smith and others (1983) estimated that the parental gabbroic melt had at least 2-3 wt. % H_2O.

Rock types vary from peridotite and anorthosite through olivine gabbronorite to amphibole gabbro. Although the multiple intrusive history of most, if not all, plutons prevented the formation of large-scale layering, outcrop-scale layered sections do occur locally. In the Corte Madera pluton, a poorly exposed 250 m thick section of layered amphibole-bearing olivine gabbronorite contains features such as crossbedding, rip-up clasts, graded bedding, and attenuated layers that are characteristic of stratiform gabbroic complexes (Walawender and Abbott, 1982).

The gabbroic rocks have distinct geochemical signatures (Walawender and Smith, 1980). K_2O is typically less than 0.1 % whereas Rb and Zr are less than 5 ppm. Al_2O_3 and CaO are greater than 19% and 12%, respectively, and have led previous workers to characterize the parental melts as high-Al basalt (Smith and others, 1983; Nishimori, 1976; Walawender, 1976). REE patterns (Figure 2) are concave downwards with modest relative depletion in both the lightest and heaviest REE's. Silver and others (1979) report initial $^{87}Sr/^{86}Sr$ between 0.7027 and 0.7045, and $\delta\ ^{18}O$ between 6 and 8 per mil. These data are consistent with derivation of the parental gabbroic liquid by "wet" melting of a mantle wedge overlying a dehydrating subducting slab with rise and fractionation at relatively shallow levels in the crust.

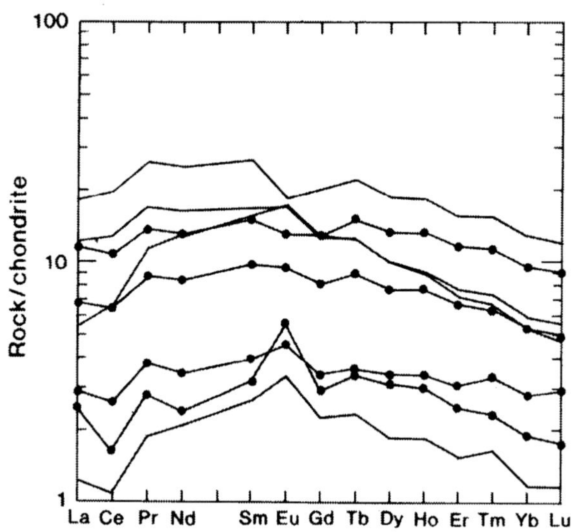

Figure 2. Gabbro REE patterns from both the Los Pinos and Corte Madera plutons. Large dots represent the olivine- and pyroxene-rich gabbros and small dots represent the hornblende-rich gabbros.

Quartz-Diorite to Tonalite Series

Tonalitic rocks in the western PRB have been mapped under a variety of formational names that include Green Valley Tonalite (Miller, 1935), Lakeview Mountain Tonalite (Dudley, 1936), Bonsall Tonalite (Hurlbut, 1935), and Alpine Quartz Diorite (Miller, 1935). Larsen (1948) and Everhart (1951) extended these units into other areas of the batholith. More recently, Todd (1980) described three spatially associated tonalitic units, the tonalites of Alpine, Las Bancas, and Japatul Valley, and placed them into a common genetic framework.

The Alpine pluton (Figure 1) is comprised of the three units identified by Todd and coworkers. Modal and geochemical variations within those three units, however, are essentially identical and suggest that the pluton is made up of a series of discrete intrusive pulses of compositionally similar melts (Lombardi, 1991). The rocks have color indices ranging from 20 to 40 and only trace amounts of alkali feldspar. Biotite crystallized late and is abundant in most samples. The main variation is due to variable proportions of the remaining three mafic silicates, orthopyroxene, clinopyroxene, and hornblende. The rocks range from hornblende-biotite tonalite to two pyroxene-biotite tonalite. Orthopyroxene and clinopyroxene formed as small discrete grains and have resorption textures. Hornblende, when present, occurs as thin rims around pyroxene and as interstitial masses within a network of plagioclase crystals. In the Cibbets Flat pluton (Figure 1), the hornblende-biotite tonalite is only found as a thin, discontinuous rim around a main core of pyroxene-biotite tonalite (Todd and Shaw, 1979). The rocks in the main body of the pluton are similar to the pyroxene-biotite tonalite in the Alpine pluton except for the presence of a small amount (trace to 4%) of interstitial alkali feldspar. Textural evidence indicates that the alkali feldspar nucleated after quartz. When combined with the experimental work of Naney and Swanson (1980), this, along with the paucity of hornblende, suggests relatively low water contents (2-4 wt.%) in the parental melt. The production of hornblende- and biotite-bearing tonalite along with pyroxene- and biotite-bearing (hornblende-free) tonalite is difficult to explain in term of a crystal fractionation mechanism using a single melt composition. We suggest that these rocks represent separate but similar melts whose water content hovered near that required to nucleate hornblende.

The Cibbets Flat pluton has a zircon U-Pb age of 105 Ma (L. T. Silver as cited in Todd and Shaw, 1979). Three fractions of zircon extracted from the Alpine pluton (Stop 2 in the following roadlog) yielded concordant ages at 108 Ma. Both plutons are in the western zone of the PRB and, as such, have Sr_i less than 0.705 and $\delta\ ^{18}O$ between 6 and 9 per mil.

Geochemical variations within the two plutons are depicted in Figure 3. The main differences involve several incompatible elements such as K_2O, Ba, and Rb which are slightly higher in the alkali feldspar-bearing Cibbets Flat pluton. REE patterns for both plutons (Figure 4,5) show similar LREE enrichment and flat HREE trends. The Cibbets Flat pluton, however, has a very tight pattern with modest negative europium anomalies whereas the Alpine pluton has a mixture of weak europium anomalies that average out to a smooth trend.

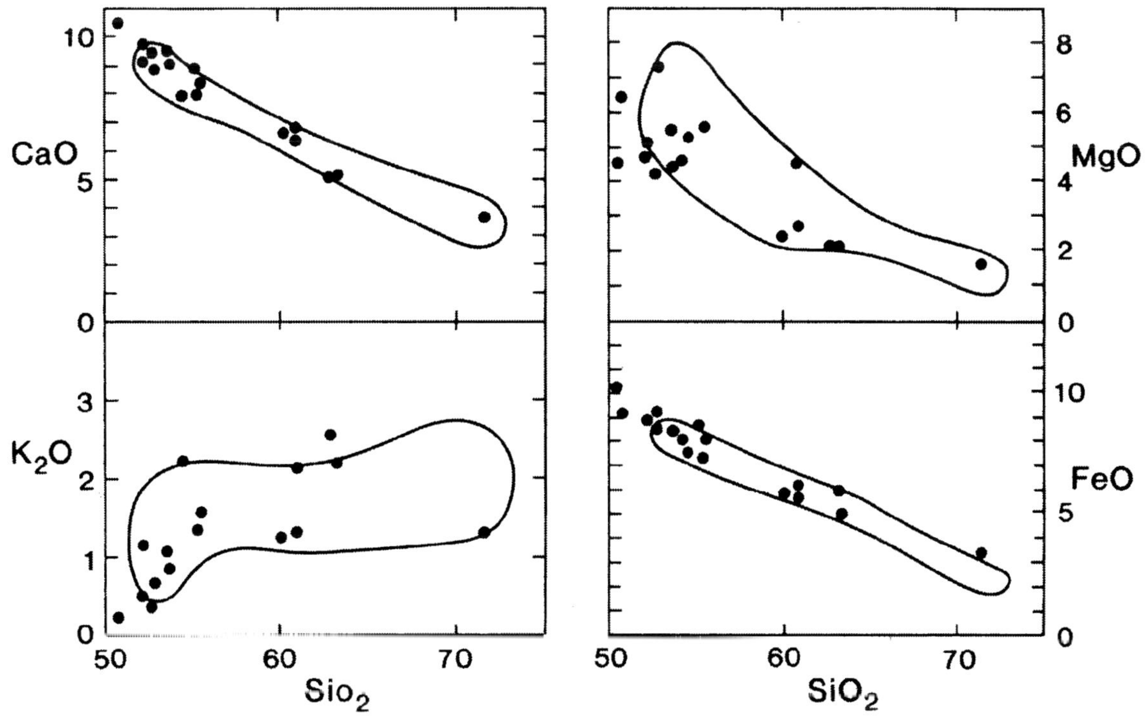

Figure 3. Major element variation diagrams for the Alpine pluton. Outlined field represents 74 host-rock analyses. Single points represents mafic inclusions.

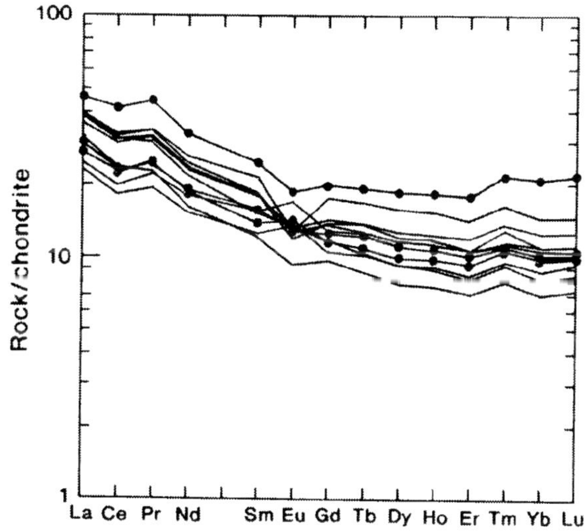

Figure 4. REE patterns for the Alpine pluton. Dots represent mafic inclusions.

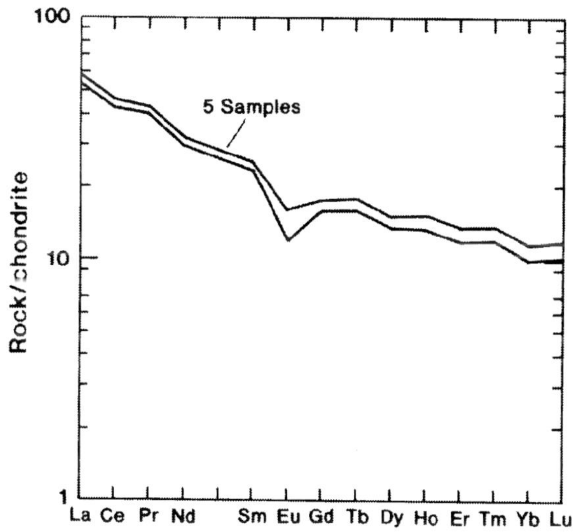

Figure 5. Five REE patterns for the Cibbets Flat pluton.

We consider that these data are consistent with the generation of tonalite by partial melting of an amphibole-plagioclase source that had not been in contact with a reservoir enriched in ^{18}O or radiogenic Sr. The most likely source region would appear to be the base of a thickened gabbroic crust formed by the accumulation of hydrous, high-Al basaltic melts. Transfer of heat via the ponding of such melts at the base of this gabbroic pile is seen as the mechanism for triggering the melting event.

Mafic inclusions in the Alpine Tonalite

Mafic inclusions are common in portions of the Alpine pluton. Their mineralogy is identical to that in the host rocks but they are finer grained, contain somewhat greater proportions of mafic minerals with color indices ranging between 26 and 42, and have a lower quartz content. Their major and trace element geochemistry (Figure 3) overlaps that of the host rocks but forms distinct groupings. This is interpreted in terms of a cognate origin with the inclusions representing fragments of older portions of the Alpine pluton through which successively younger magmatic pulses have passed.

Granodiorite to Monzogranite Series

Mapping of Todd and coworkers (1979, 1980, 1983) has led to the recognition of several lithologic and textural variants within the granodiorite-monzogranite series. These include the granites of Chiquto Peak, Pine Valley, and Corte Madera. Chiquito Peak-type is medium grained and strongly foliated, has color indicies between 5 and 12, and contains An_{29} plagioclase with both hornblende and biotite. Pine Valley-type is coarse grained and porphyritic with 2 cm phenocrysts of alkali feldspar and 2 to 3 cm lensoidal quartz aggregrates. Biotite and trace amounts of hornblende are the main mafic silicates with color indicies between 5 and 10. This unit, however, is distinct from the others in that its magnetic susceptibility is less than 25, i.e., it does not contain magnetite. Plagioclase compositions are An_{24-25}. Corte Madera-type has the same mafic silicates but with color indices less than 5. It is coarse grained with An_{11-12} plagioclase and contains 2-3 cm lensoidal quartz aggregates similar to those found in the Pine Valley type.

Based on field relationships, these three granitic types are reported (Todd and Shaw, 1979) to be younger than the spatially associated tonalitic units in the PRB. Zircon U-PB ages on two of the above, however, are older than the 108 and 105 Ma ages of the tonalitic plutons. The Pine Valley granite yielded a concordant 118 Ma age that is in good agreement with an age of 119 Ma for a similar pluton approximately 20 km to the north (L.T. Silver, as reported in Todd and Shaw, 1979). A Corte Madera-type pluton on the extreme western side of the batholith (El Cajon Mountain complex, Todd and Shaw, 1979) also gave a concordant age at 111 Ma (D.L. Kimbrough, unpublished data, 1991).

All samples reported by Todd and Shaw (1979) have greater than 70% SiO_2 and plot in similar overlapping fields on Harker diagrams. Whole-rock $\delta\ ^{18}O$ apparently ranges between 6 and 9 per mil with initial Sr ratios less than 0.705 (Silver and others, 1979, figures 6 and 7). REE patterns show moderate LREE enrichment to 60X chondritic values with pronounced negative Eu anomalies (Gromet and Silver, 1987).

CUYAMACA-LAGUNA MOUNTAIN SHEAR ZONE

Introduction

The Cuyamaca-Laguna Mountain shear zone (CLMSZ), recognized and defined by Todd and others (1988), lies approximately along the boundary between the western and eastern zones of the PRB (Figure 6). It characteristically contains highly strained, locally mylonitic plutons with a well developed, northwest-striking solid-state foliation that is coplanar with the wall-rock foliation. The shear zone varies from 5 kms to about 20 kms in outcrop width. Todd and others (1988) show it extending along strike for approximately 50 kms, although its northernmost extent is uncertain. To the east and south, the CLMSZ is intruded by younger undeformed plutons of the eastern zone of the PRB (Todd and others, 1988).

A series of petrologic, geochemical, and geophysical boundaries are approximated by the CLMSZ (Figure 6). These include the I-S line of Todd and Shaw (1988), the gabbro line (Gastil, 1975), the δ 180 step (Silver and others, 1979), and the gravity-magnetic boundary (Oliver, 1980; Silver and others, 1979; Jachens and others, 1986). Todd and others (1988) suggest that they reflect a change in source rock composition for the plutonic rocks "from primitive, oceanic crust on the west to continental crust and (or) continentally derived sediments on the east". Walawender and others (1990) argued for a change in subduction angle that shifted the source region from mantle ± oceanic crust to oceanic and continental crust.

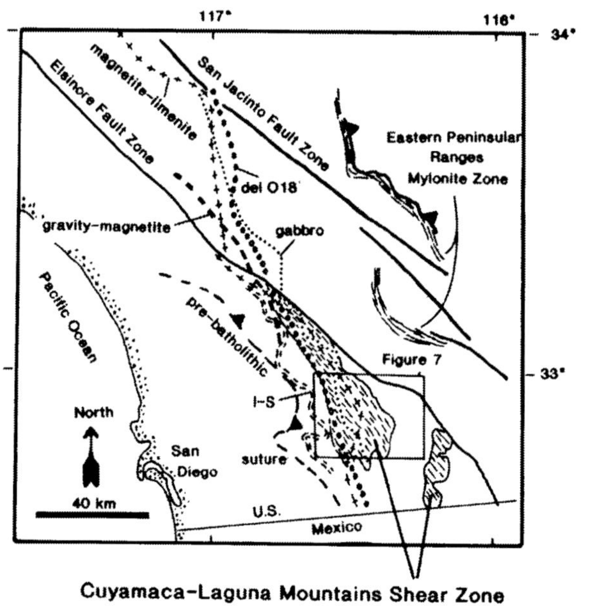

Figure 6. Location map for the CLMSZ (after Todd and others, 1988).

The CLMSZ and the various petrologic, geochemical, and geophysical boundaries are parallel to the contact zone between the Late Jurassic to Early Cretaceous SPV and the continentally derived Triassic or older metasedimentary rocks. Although the contact is nowhere well exposed, Todd and others (1988) interpret it to be an eastward-dipping, westward-verging prebatholithic suture, as did R.G. Gastil in a number of earlier publications (e.g., Gastil and others, 1978). However, Todd and others (1988) infer that following the collision between the SPV and the metasedimentary assemblage, continued displacements in the suture zone moved progressively eastward and were accompanied by magmatism to form the deformed plutonic rocks now defining the CLMSZ. This east-over-west, reverse-fault-sense movement implies that the CLMSZ brought rocks that formed deeper in the crust of the eastern PRB (9-12 kms; Grove, 1986; Clinkenbeard, 1987) up and over the shallower crustal rocks of the western PRB (6-9 kms; Grove, 1986 and pers. commun. to GHG, 1990).

Harper Creek Gneiss

The peraluminous Harper Creek Gneiss is one of the major components of the CLMSZ (Todd and Shaw, 1985; Todd and others, 1988) and occurs as four large, sheet-like units within the Cuyamaca Peak, Descanso, and Mount Laguna 7.5' quadrangle maps (Todd, 1977, 1979; Hoggat and Todd, 1977). The largest unit is approximately 18 kms long and 2.5 kms wide (Figure 7). Leeson (1989) and Leeson and others (1989) studied the mesoscopic and microscopic fabric of a small part of the Harper Creek Gneiss in the northwest quarter of the Mount Laguna 7.5'quadrangle. Here, the Harper Creek Gneiss is an S-C mylonite. Detailed study of 60 oriented thin sections cut perpendicular to the foliation and parallel to the prominent mineral elongation revealed that the foliation most commonly observed is a C-surface along which grain-size reduction has occurred. S-surfaces are common in most thin sections but are difficult to detect in outcrop because of the close spacing and pervasive character of the C-surfaces (0.50-0.25 mm). S-surfaces are defined by coarse-grained, alternating bands or lenses of quartz, plagioclase, alkali feldspar, biotite, and white mica. Twin lamellae in plagioclase and (001) faces of biotite and white mica tend to be aligned parallel to the S-surfaces which commonly intersect C-surfaces at an angle of about 10 to 20^0.

In 80% of the oriented samples, S-surfaces are deflected in a dextral sense as they enter the C-surfaces when viewed in a northwestward direction (Figure 8). Asymmetric porphyroclasts of white mica and biotite consistently indicate a dextral sense of rotation along C-surfaces as do brittly deformed and displaced feldspar fragments. In some samples, the angle between the C- and S-surfaces was as low as 5^0. These samples commonly contained secondary shear or C'-surfaces (shear bands or extensional crenulation cleavage). Both C- and S-surfaces were consistently deflected in a dextral sense as they entered the C'domains.

Orientation data for C-surfaces and for the mineral stretching lineation show that the C-surfaces on average strike N15W and dip 69NE whereas the mineral stretching lineation pitches steeply in the plane of C and clusters about an orientation of 64° at 053°. S-surfaces on oriented polished slabs and in thin sections strike parallel to C but dip more gently to the northeast. C'-surfaces also strike parallel to C but dip more steeply.

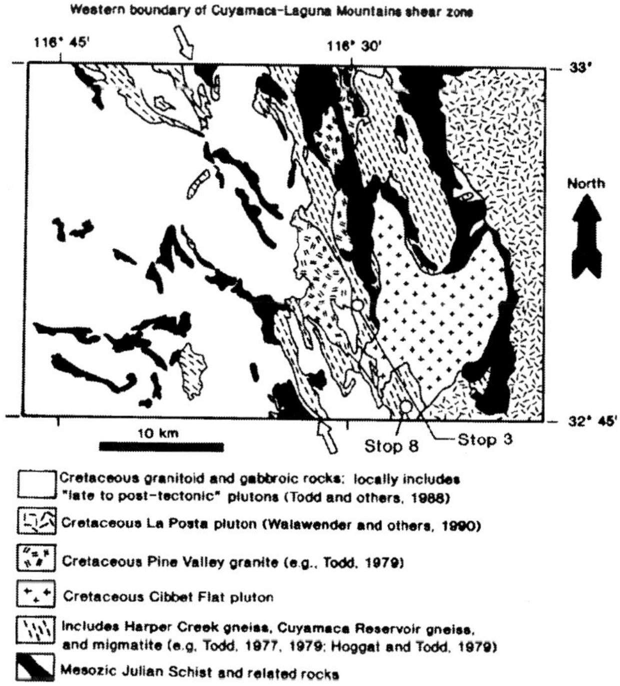

Figure 7. Geologic map of a portion of the CLMSZ (after Todd and others, 1988).

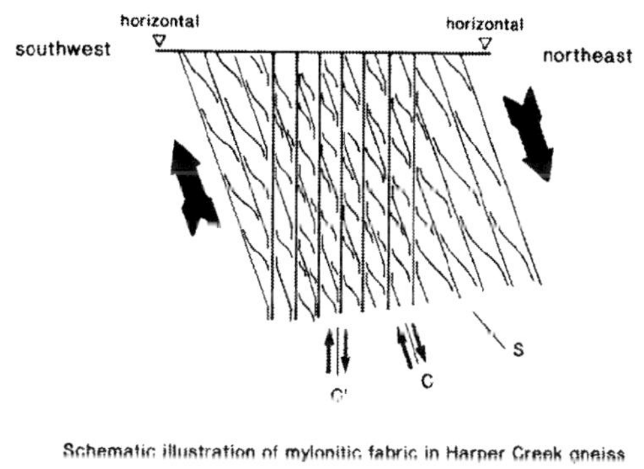

Figure 8. Schematic northwest-facing outcrop cross section of the Harper Creek gneiss.

Age Relationships

Zircon from the Harper Creek gneiss commonly contain rounded cores with clear prismatic and pyramidal overgrowths. Eleven fractions of zircon extracted from a sample of Harper Creek Gneiss (Stop 3) have been analyzed for their U-Pb compositions utilizing both air abrasion and HF-leaching techniques. These data, plotted on a concordia diagram (Figure 10) are discordant and form a linear array with a lower intercept at 156 Ma ± 12 Ma and an upper intercept at 1427 ± 49 Ma. Air-abraded or HF-leached fractions plot further toward the upper intercept along the regression line and indicate that the clear overgrowths are younger than the rounded cores. If the zircon population was not disturbed after crystallization, then the lower intercept gives the approximate age of crystallization whereas the upper intercept age is the average age of the inherited component.

Summary

The structural data clearly indicate a down-to-the-east sense of movement rather than the east-over-west movement suggested by Todd and others (1988). The geochronological data suggest that movement occurred no later than 156 Ma. The undeformed 118 Ma Pine Valley pluton appears to intrude the CLMSZ (Todd, 1979) but is reported to have a deformational fabric near its eastern margin (Todd, 1979; Leeson, 1989). Thus, deformation appears to be 118 Ma or younger. We speculate that the CLMSZ represents an Early Cretaceous synbatholithic extensional zone that accomodated, in part, the emplacement of the voluminous La Posta-type plutons in the eastern zone of the batholith. In addition, we note that the Harper Creek (ortho)gneiss is the oldest granitic rock yet recognized in the PRB and that additional zircon U-Pb studies may reveal a heretofore undocumented early magmatic history to the PRB.

EASTERN ZONE

Introduction

The large, undeformed plutons in the eastern zone of the PRB have distinct structural and petrographic characteristics that set them apart from western zone plutons with similar composition. These include:

1. Concentric zones ranging from tonalite margins to granodiorite/monzogranite cores;
2. Internal contacts that are gradational over several tens of meters;
3. Sharply euhedral crystals of biotite, sphene and hornblende;
4. Primary muscovite coexisting with biotite in the inner parts of the pluton;
5. Local development of megacrystic quartz and oikocrystic alkali feldspar;
6. Absence of foliation.

La Posta pluton

The La Posta pluton straddles the international border and has an estimated exposure area of approximately 1400 km^2. It is a single intrusive body that varies inward across gradational contacts from a metaluminous sphene-hornblende-biotite tonalite rim through two distinct biotite-bearing granodiorite facies into a weakly peraluminous core of muscovite-biotite granodiorite (Figure 9). U-Pb ages on multiple zircon fractions from three different samples yield an age of 94 ± 2 Ma. A Rb-Sr mineral isochron from one of those samples gave a regression age of 92 ± 3 Ma and an initial Sr ratio of 0.70437 ± 0.00057. Oxygen isotopic analyses on both whole rocks and quartz separates (Figure 9) show a lithologic control with the interior facies having higher values than the marginal facies. The $\delta^{18}O$ "step" of Silver and others (1979) occurs at the western margin of the La Posta pluton with quartz separates showing as much as a 2 per mil increase across the contact (Figure 9).

At least two geochemical parameters can be used to characterize the La Posta pluton. At a given SiO_2 content, Sr is several times that found in the plutons of the western zone (Walawender and others, 1990). In addition, the REE patterns of the La Posta rocks show strong LREE enrichment, depletion in HREE, and no europium anomaly. These features led Gromet and Silver (1987) to suggest that the source rock was eclogite-facies oceanic crust.

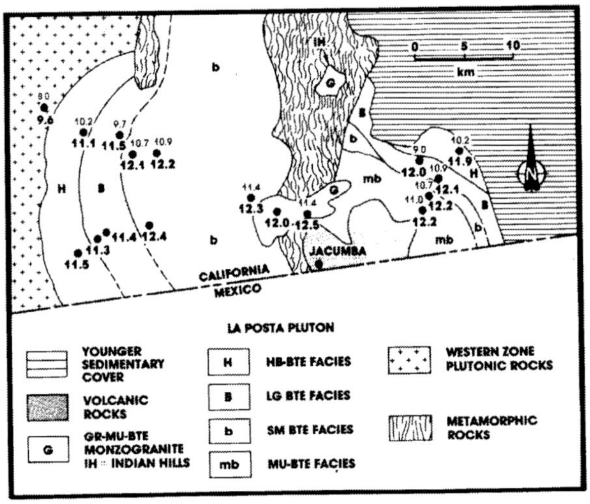

Figure 9. Geologic map of the La Posta pluton (Walawender and others, 1990). Whole-rock $\delta^{18}O$ shown in small and quartz separates in large characters, respectively.

Walawender and others (1990) argued that an amphibolitic source terrane was also consistent with the data as long as all the plagioclase had been removed from the source region during the melting event. Summarizing the data from other La Posta-type plutons in the eastern zone of the batholith, they also suggested that the innermost muscovite-bearing facies represented the tail of a diapir that had interacted with the

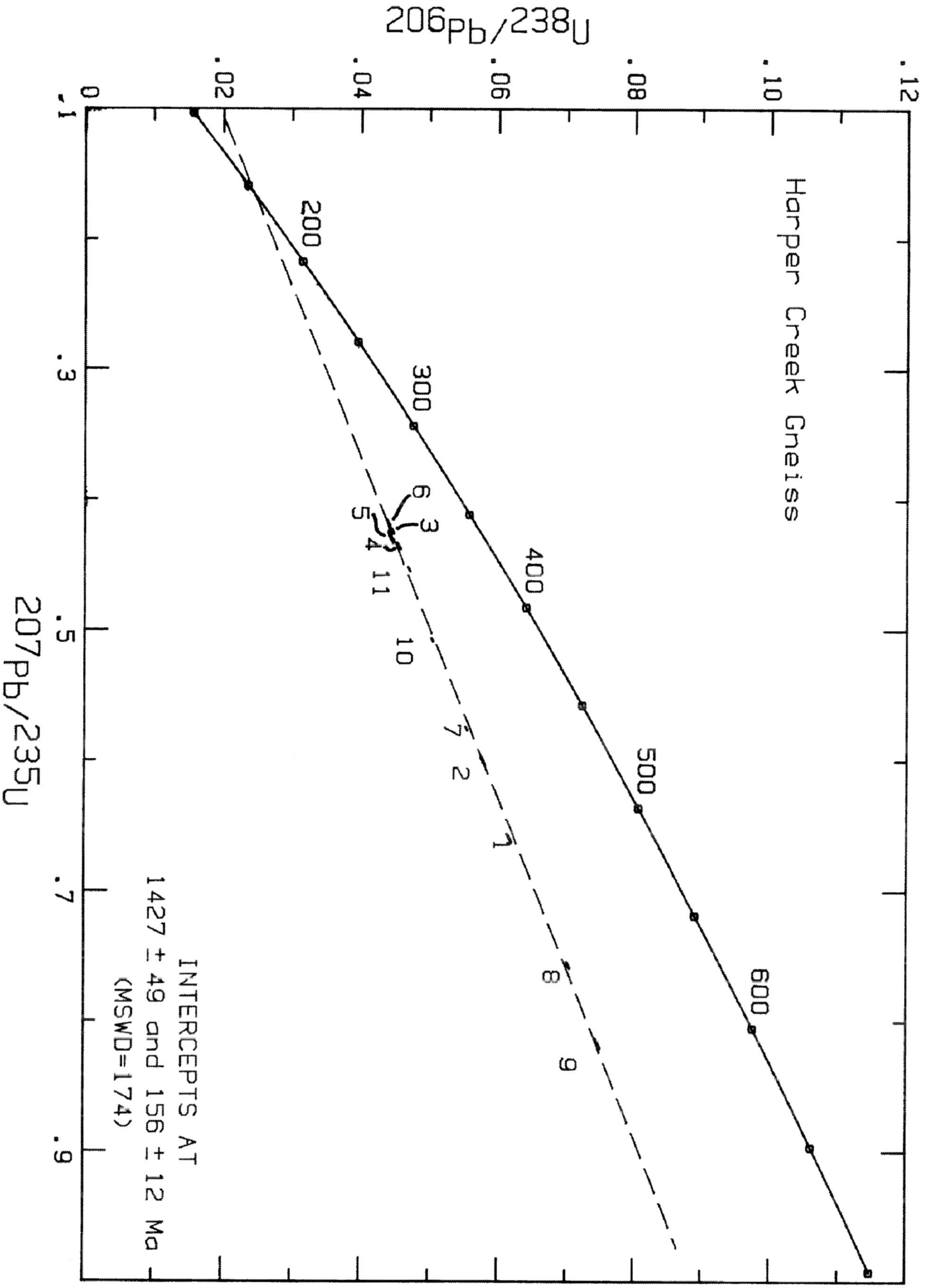

Figure 10. Concordia plot of U-Pb zircon data for the Harper Creek gneiss. Numbers refer to separate zircon fractions.

leading edge of the North American continent prior to intruding into the head of the ballooning(?) La Posta diapir. Viscosity differences between the parental La Posta melt and the contaminated tail would inhibit homogenization so that inward crystallization would still produce the observed gradational contacts between the higher temperature outer units and the lower temperature assemblages in the core. Mineral chemistries Clinkenbeard and Walawender (1989) show marked changes in plagioclase composition, and in Fe/Mg and Fe^{+2}/Fe^{+3} in biotite between the core and outer zones that are consistent with the above model.

Garnetiferous 2-mica monzogranites

Intrusive into many of the La Posta-type plutons or their associated sillimanite-grade metasedimentary screens are a series of small garnet-muscovite-biotite monzogranite plutons. Two of these intrude the large metasedimentary screen that runs through the center of the La Posta pluton (Figure 9). An unnamed body just north of the international border cuts both the metasedimentary screen and the La Posta pluton proper. Three zircon fractions yield a concordant age of 93 ± 1 Ma that is consistent with the observed field relationships (Clinkenbeard, 1987). The Indian Hill pluton (Parrish, 1990) is intrusive into the same metasedimentary screen further to the north (Figure 10). A Rb-Sr mineral isochron gave a crystallization age of 89.6 ± 2.6 Ma with an initial Sr ratio of 0.70628 ± 0.00025. Zircon U-PB systematics for the same sample were discordant with a lower intercept age of 84.4 ± 6.1 Ma and an upper intercept age of 1161 ± 430 Ma. The presence of inherited Pb seems to be a characteristic feature of many of the garnet-bearing leucogranodioritic rocks along the eastern side of the PRB.

The Indian Hill pluton consists of two facies, a medium-grained, garnet-muscovite-biotite granodiorite and a fine-grained muscovite-biotite granodiorite. Emplacement of the pluton was preceeded by injection of coarse-grained to pegmatitic leucocratic granodiorite dikes into the surrounding migmatitic metasedimentary complex. The dikes are garnetiferous and similar in composition to the garnetiferous facies of the Indian Hill pluton.

Major element analyses indicate a limited SiO_2 range (74-77%) and A/CNK values between 1.14 and 1.20. Zr appears to be a fractionation indicator in that it covaries with with other geochemical parameters (Figure 11). Zircon is readily apparent as tiny inclusions with pleochroic haloes within biotite and suggests that the Zr variation is due to the presence of varying amounts of residual biotite (restite). Thermobarometric calculations based in coexisting garnet-plagioclase-biotite-muscovite yield emplacement conditions of approximately 700°C and 2.5 ± 0.5 kb for the Indian Hill pluton, and 673 to 755°C at 2.7 to 3.0 kb for the unnamed body to the south (Clinkenbeard, 1987). Whole-rock $\delta^{18}O$ in the Indian Hill pluton varies from 11.7 to 12.1 per mil with quartz separates ranging from 12.6 to 13.2 per mil.

These data are consistent with an origin by partial melting of a metasedimentary package containing detrital zircon with an average age of approximately 1200 Ma. The role of Zr as a discriminator between the early dike system and the two facies of the Indian Hill pluton suggests that these three units represent successive melt fractions rising from the zone of melting to the present erosional level that were emplaced with little or no in situ fractionation.

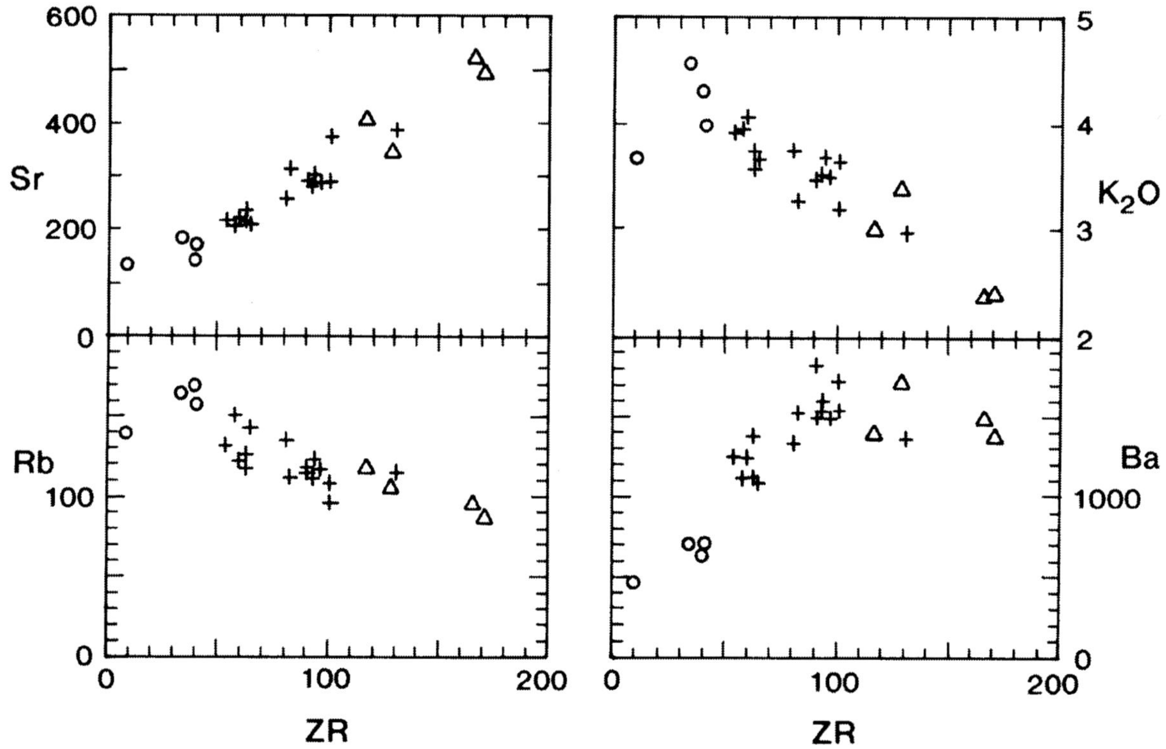

Figure 11. Trace-element variation diagrams for the Indian Hill pluton. Open circles represent the early dike system, crosses represent the garnet-bearing monzogranite, and open triangles represent the garnet-free monzogranite.

K-AR PATTERNS

Krummenacher and others (1975) showed that K-Ar ages for biotite and hornblende decreased progressively eastward across the batholith. This was interpreted in terms of a west to east emplacement sequence for the batholith as a whole and greater uplift in the eastern zone. Silver and others (1979) plotted these data, along with zircon U-Pb ages, against geographic position and drew attention to the fact that the differences between emplacement (zircon) ages and cooling (biotite) ages increased progressively to the east. This eastward-directed increase was explained in terms of a thermal model drawing upon larger magmatic bodies (i.e., heat sources) and younger emplacement ages in the eastern zone. It is interesting to note, however, that nearly all of the youngest biotite ages (<80 Ma) occur east of the large N-S trending screen that runs through the middle of the La Posta pluton and then continues southward for another 40 kms.

The 105 Ma Cibbets Flat pluton lies within a portion of the western zone that was interpreted by Silver and others (1979) to have cooling ages reset by the intrusion of the La Posta pluton. K-Ar ages on biotite within the Cibbetts Flat pluton (Heaton, 1981) vary from 85 to 102 Ma and reflect very complex, possibly multiple cooling histories. A rough N-S trending belt of >100 Ma ages occurs in the east-central part of the pluton with younger ages along the western and southeastern margins. This pattern can be interpreted in terms of resetting along the CLMSZ if its latest stage of thermal activity was younger than 105 Ma or partial and irregular resetting by the intrusion of the La Posta pluton. Interpretation of the distribution of K-Ar ages within the PRB remains problematic.

SUMMARY AND PETROGENETIC MODEL

The data cited and presented in this report permit the formulation of a working model that, like its predecessors, will be tested and refined. However, three basic points appear to be well established. First, the two magmatic zones within the PRB must reflect multiple source regions, even within any single zone. Second, at least in a broad sense, the locus of magmatism shifted eastward with time. Finally, the rate of magma generation in the eastern zone must have been significantly greater than in the western zone.

Although unsupported by reliable radiogenic data, the gabbroic rocks of the western zone are considered by most, if not all, investigators, to be the oldest of the three plutonic systems. Whether or not they continued to be emplaced during the entire history of the western zone, however, remains to be seen. We consider the gabbroic system to be essentially that described by Walawender and Smith (1980) and specifically exclude the quartz gabbros of Silver and others (1979) which have greater than 53% SiO_2 and the biotite-bearing mafic quartz diorites described by Todd and others (1979).

The low abundances of the LILE's in the gabbroic system suggest that either the source region had been depleted in these elements during a prior magmatic event or that it underwent large degrees of partial melting under elevated PH2O with the initial LILE-enriched melts forming early portions of a proto-crust. Volatiles released from subducted oceanic crust began to generate water-rich, high-alumina basaltic melts in an overlying mantle wedge (Figure 12a) at approximately 150(?) Ma. Continued generation of these melts thickened the magmatic arc until a steepened geothermal gradient led to partial melting at its base (Figure 12b). This latter event, likely

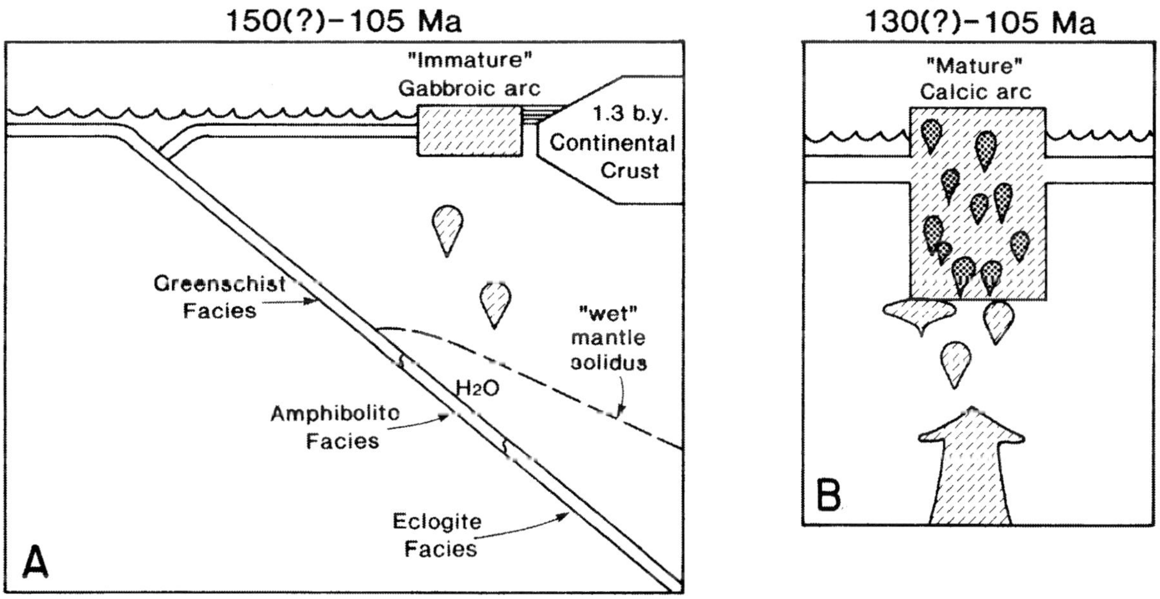

Figure 12. Schematic cross sections showing A. Development of the early gabbroic arc; B. Generation of the granitic components of the western PRB from a thickened gabbroic arc.

beginning near 130 Ma and continuing until about 105 Ma, produced the granitic rocks of the western PRB. The REE patterns and the low ^{18}O in the quartz diorite-tonalite system can be constrained in terms of partial melting of an unaltered amphibole-plagioclase source with partial melting continuing until the disappearance of plagioclase from the source region. The origin of the granodioritic system is more problematic in that (at least) two sources are possible. Anatexis of an LILE-enriched gabbroic proto-crust (for early granodiorite magmatism?) or an amphibole-bearing tonalitic source (for younger granodioritc plutons?) with plagioclase remaining in either residue could produce the REE patterns reported by Gromet and Silver (1985). Modelling of the REE patterns as well as critical zircon U-Pb systematics in well documented granodiorite plutons are presently underway and may shed further light on this problem.

Inherited lead in the interior portions of the large La Posta-type plutons and in their spatially associated garnet-bearing monzogranites as well as $\delta^{18}O$ up to 12 per mil suggest that the eastward step in the locus of magmatism after 105 Ma coincides with a change in source region from mantle and/or young mantle derivatives to one with significant crustal components, both oceanic and continental. A convenient, if not probable, mechanism involves the shallowing of the subduction plane (Figure 13) at or around 105 Ma as suggested by Walawender and others (1990). This event brings the dewatering oceanic crust above the water-saturated solidus for the overlying mantle wedge and effectively cuts off magmatism within what is now the western zone. Continued subduction and dewatering of the downgoing slab induces partial melting of the oceanic crust under either upper amphibolite- or eclogite-grade conditions to produce proto-La Posta-type melts that rise to form a younger magmatic arc stepped eastward towards the North American continent.

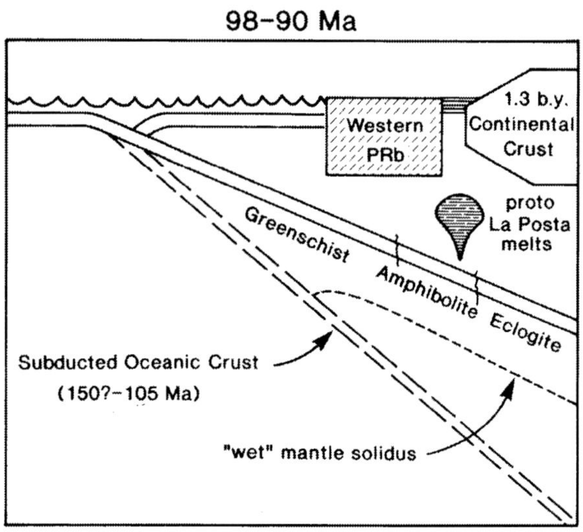

Figure 13. Schematic cross section showing initial development of the eastern zone of the PRB.

Contributions from the continental crust are apparent in the interior zones of several La Posta-type plutons and in the temporally and spatially associated garnet-bearing monzogranites. The latter are commonly associated with migmatitic sillimanite-grade metasedimentary screens and can be modelled in terms of partial melting within those systems (e.g. The Indian Hill pluton, Parrish, 1990). The greater continental signatures in the interior zones as opposed to the margins of the La Posta-type plutons, however, require a deeper source. These melts appear to have risen through the juncture of the North American craton and oceanic lithosphere (Gastil, 1975) and interacted with a terrane/source region relatively enriched in ^{18}O, radiogenic isotopes, and ca. 1200-1400 Ma zircons (Figure 14a). Some of this interaction was likely distributed through the diapir via diffusion and/or convective mixing but the main effect was to create an isotopically enriched tail that intruded the head of the diapir before solidification (Figure 14b). This, combined with inward crystallization of a now zoned magma chamber produced the concentric lithologic pattern and the gradational contacts within the La Posta-type plutons (Figure 14c).

A PARTING SHOT

Several of the investigators cited in this report have argued that the PRB is part of a larger, subduction-related magmatic arc that includes the Sierra Nevada and possibly even the infamous Salinian block of coastal California. The record of magmatism now extends at least as far back as 156 Ma. Neogene extension that produced the Salton Trough and the Gulf of California separates the PRB from magmatic rocks in the general ranges of 80 to 50 Ma in mainland Mexico and Arizona. If these redistributed younger rocks were produced by the same plate interactions as the PRB, then the dynamics of that interaction must have changed in a consistent fashion over a record of approximately 100 million years. If, as we have argued, the western and eastern zones of the PRB represent two distinct static arcs, each related to a particular angle (and rate) of subduction, then a natural extension of this model would involve a further shallowing of the subduction plane to initiate the even younger magmatism that clearly cuts through the Precambrian rocks of the North American continent.

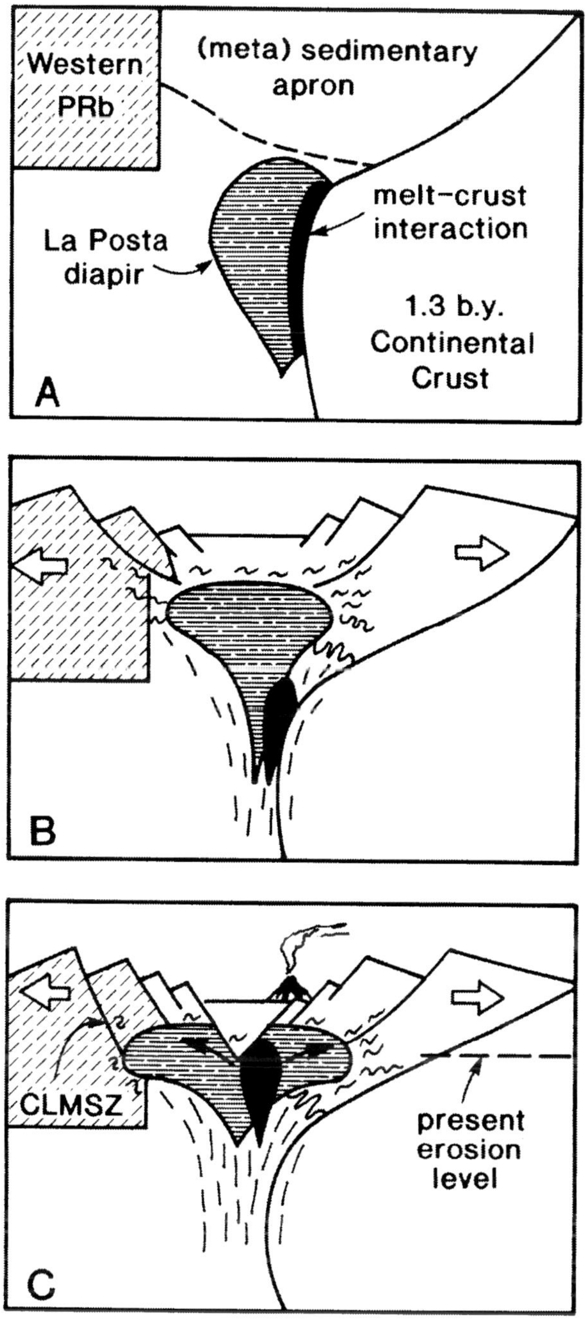

Figure 14. Schematic cross section showing development of the zoned La Posta pluton and associated structures. See text for explanation.

References Cited

Adams, M.A., Stratigraphy and petrography of the Santiago Peak Volcanics east of Rancho Santa Fe, California [M.S. Thesis]: San Diego, California, San Diego State University, 123 p., 1979.

Adams, M.A., and Walawender, M.J., Structure and stratigraphy of the Santiago Peak volcanics east of Rancho Santa Fe, California, in Geologic Studies in San Diego, P.L. Abbott, ed.: San Diego Association of Geologists, p. 104-117, 1982.

Allison, E.C., Middle Cretaceous gastropoda from Punta China, Baja California, Mexico: Journal of Paleontology, v. 29, P. 400-432, 1955.

Berggreen, R.G., and Walawender, M.J., Petrography and metamorphism of the Morena Reservoir roof pendant, southern California: California Division of Mines and Geology Special Report 129, p. 61-65, 1977.

Bushee, J., Holden, J., Geyer, B., and Gastil, R.G., Lead-alpha dates for some basement rocks of southwestern California: Geological Society of America Bulletin, v. 74, p. 803-806, 1963.

Clinkenbeard, J.P., The mineralogy, geochemistry, and geochronology of the La Posta pluton, San Diego and Imperial counties, California [M.S. Thesis]: San Diego, California, San Diego State University, 215 p., 1987.

Clinkenbeard, J.P., and Walawender, M.J., Mineralogy of the La Posta pluton: Implications for the origin of zoned plutons in the eastern Peninsular Ranges batholith, southern and Baja California: American Mineralogist, v. 74, p. 1258-1269, 1989.

Criscione, J.L., Davis, T.E., and Ehlig, P., The age of sedimentation/diagenesis for the Bedford Canyon Formation and the Santa Monica Formation in southern California - A Pb/Sr evaluation, in Howell, D.G. and McDougall, K.A., eds., Mesozoic paleogeography of the western United States: Pacific Section, SEPM, Pacific Coast Paleogeography Symposium 2, p. 385-396, 1978.

DePaolo, D.J., Sources of continental crust: Neodymium isotope evidence from the Sierra Nevada and Peninsular Ranges: Science, v. 209, p. 684-687, 1980.

DePaolo, D.J., A neodymium and strontium isotopic study of the Mesozoic calc-alkaline granitic batholiths of the Sierra Nevada and Peninsular Ranges, California: Journal of Geophysical Research, v. 86, p. 10470-10488, 1981.

Determan, M.E., Geology of the Metal Mountain district, In-Ko-Pah Mountains, San Diego County, California [M.S. Thesis]: San Diego, California, San Diego State University, 216 p., 1984.

Diercks, C., Structure and petrology of the southern portion of the Jacumba roof pendant [Senior Thesis]: San Diego, California, San Diego State University, v. 9, 37 p., 1981.

Dudley, P.H., Geology of a portion of the Perris Block, southern California: California Journal Mines and Geology, v. 31, pp 487-506, 1936

Everhart, D.L., Geology of the Cuyamaca Peak quadrangle, San Diego County, California, in Crystalline Rocks of Southwestern California: California Division of Mines Bulletin 159, pp 51-116, 1951.

Fife, D.L., Minch, J.A., and Crampton, P.J., Late Jurassic age of the Santiago Peak Volcanics, California: Geological Society of America Bulletin, v. 78, p. 229-304, 1967.

Gastil, R.G., Plutonic zones in the Peninsular Ranges of southern California and northern Baja California: Geology, v. 3, p. 361-363, 1975.

Gastil, R.G., and Miller, R.H., Prebatholithic paleogeography of peninsular California and adjacent Mexico, in Frizzell, V.A., Jr., ed., Geology of the Baja California Peninsula: Pacific Section, SEPM, v. 39, p. 9-16, 1984.

Gastil, R.G., Morgan, G.J., and Krummenacher, D., Mesozoic history of peninsular California and related areas east of of California, in Howell, D.G., and McDougall, K.A., eds., Mesozoic paleogeography of the western United States: Pacific Section, SEPM, Pacific Coast Paleogeography Symposium 2, p. 107-116, 1978.

Gastil, G., Diamond, J., Knaack, C., Walawender, M., Marshall, M., Boyles, C., and Chadwick, B., The problem of the magnetite/ilmenite boundary in southern and Baja California, in Anderson, J.L., ed., The nature and origin of Cordilleran magmatism: Geological Society of America Memoir 174, p. 19-32, 1990.

Germinario, M.P., The depositional and tectonic environments of the Julian Schist, Julian, California [M.S. Thesis]: San Diego, California, San Diego State University, 95 p., 1982.

Gorzolla, York R., Geochemistry and petrography of the Santiago Peak Volcanics, Santa Margarita and Santa Ana Mountains, southern California [M.S. Thesis]: San Diego, California, San Diego State University, 145 p., 1988.

Gromet, L.P., and Silver, L.T., REE variations across the Peninsular Ranges batholith; Implications for batholithic petrogenesis and crustal growth in magmatic arcs: Journal of Petrology, v. 28, p. 75-125, 1987.

Grove, M., Metamorphism and deformation in the Box Canyon area, eastern Peninsular Ranges, San Diego County, California [M.S. Thesis]: Los Angeles, California, University of California, 118 p., 1986.

Heaton, K.M., K-Ar age patterns in the Cibbets Flat pluton, Mount Laguna, California [M.S. Thesis]: San Diego, California, San Diego State University, 73 p., 1981.

Hoffman, D.R., Petrogenesis of the Lawson Peak orbicular gabbro [M.S. Thesis]: San Diego California, San Diego State University, 95 p.

Hoggat, W.C., and Todd, V.R., Geology of the Descanso quadrangle, San Diego County, California: USGS Open File Report 77-406, 1977.

Hudson, F.S., Geology of the Cuyamaca region of California, with special reference to the origin of nickeliferous pyrrhotite: University of California, Department of Geological Sciences Bulletin, v. 13, p. 175-252, 1922.

Hurlbut, C.S. Jr., Dark inclusions in a tonalite of southern California: American Mineralogist, V. 20, p. 609-630, 1935.

Jachens, R.C., Simpson, R.W., Grisiom, A., and Mariano, J., Plutonic belts in southern California defined by gravity and magnetic anomalies: Geological Society of America Abstracts with Programs, v. 18, p. 120, 1986.

Jones, D.A., and Miller, R.H., Jurassic fossils from the Santiago peak Volcanics, San Diego County, California, in Abbott, P.L., ed., Geologic studies in San Diego: San Diego Association of Geologists Publication, p. 93-103, 1982.

Kennedy, M.P., and Peterson, G.L., Geology of the eastern San Diego metropolitan area, California: California Division of Mines and Geology Bulletin 200, p. 43-56, 1975.

Krummenacher, D., Gastil, R.G., Bushee, J., and Doupont, J., K-Ar apparent ages, Peninsular Ranges batholith, southern California: Geologic Society of America Bulletin, v. 86, p. 760-768, 1975.

Larsen, E.S., Jr., Batholith and associated rocks of Corona, Elsinore and San Luis Rey quadrangles, southern California: Geological Society of America memoir 29, 182 p., 1948.

Lampe, C.M., Geology of the Granite Mtn. area; Implications of the extent and style of deformation along the southeast portion of the Elsinore Fault [M.S. Thesis]: San Diego, California, San Diego State University, 150 p., 1988.

Leeson, R.T., Jr., Fabric analysis of the Cuyamaca-Laguna Mountains shear zone [M.S. Thesis]: San Diego, California, San Diego State University, p. 136, 1989.

Leeson, R.T., Jr., Girty, G.H., Wardlaw, M.S., and Meier, D.B., Harper Creek gneiss, Peninsular Ranges, southern California: Geological Society of America Abstracts with Programs, v. 21, 105 p., 1989.

Lillis, P.G., Petrography geochemistry and structure of the Corte Madera pluton, San Diego County, California [M.S. Thesis], San Diego, California, San Diego State University 80 p., 1978.

Lombardi, M., The Alpine tonalite, image of a gabbroic source [M.S. Thesis], San Diego, California, San Diego State University, in prep.

Miller, W.J., A geologic section across the southern Peninsular Ranges of California: California Journal of Mines and Geology, v. 31, p. 115-142, 1925.

Naney, M.T. and Swanson, S.E., The effect of Fe and Mg on crystallization in granitic systems: American Mineralogist, v. 65 p. 639-653, 1980.

Nishimore, R.K., The petrology and geochemistry of the gabbros from the Peninsular Ranges batholith, California, and a model for their origin [Ph.D. Thesis]: San Diego, California, Scripps Institution of Oceanography 234 p., 1976.

Oliver, H.W., Interpretation of the gravity map of California and its continental region: California Division of Mines and Geology Bulletin 205, 52 p., 1980.

Parrish, K.E., Geology, Petrology, Geochemistry, and isotopic character of the Indian Hill pluton, Jacumba Mtns, California [M.S. Thesis]: San Diego, California, San Diego State University, 136 p., 1990.

Santillan, M., and Barrera, T., Los Posibilidades petroliferous en la costa occidental de la Baja California, enter los paralelos 30° y 32° de latitude norte: Instituto Geology Mexico Annles, v. 5, p. 1-37, 1930.

Siem, M.E., Comparisons of metamorphic grades across the Peninsular Ranges batholith, San Diego County, California [Senior Thesis]: San Diego, California, San Diego State University, 29 p., 1988.

Silver, L.T., Peninsular Ranges batholith: A case study in continental margin magmatic arc setting, characteristics and evolution: Geological Society of America Abstracts and Programs, v. 11, p. 517, 1979.

Silver, L.T., and Chappell, B.W., The Peninsular Ranges batholith; An insight into the evolution of the Cordilleran batholiths of southwestern North America: Transactions of the Royal Society of Edinburgh, Earth Sciences, v. 79, p. 105-121, 1988.

Silver, L.T., and Taylor, H.P., and Chappell, B., Some petrological, geochemical and geochronological observations of the Peninsular Ranges batholith near the international border of the U.S.A. and Mexico, in Abbott, P.L., and Todd, V.R., eds., Mesozoic Crystalline Rocks: San Diego, California, San Diego State University, Department of Geological Sciences, p. 83-110, 1979.

Smith, T.E., Huang, C.H., Walawender, M.J., Cheung, P. and Wheeler, C., The gabbroic rocks of the Peninsular Ranges batholith, southern California: Cumulate rocks associated with calc-alkaline basalts and andesites: Journal of Volcanology and Geothermal Research, v. 18, p. 249-278, 1983.

Tanaka, H., Smith, T.E., and Huang, C.H., The Santiago Peak Volcanics rocks of the Peninsular Ranges batholith, southern California, volcanic rocks associated with co-eval gabbros: Bulletin Volcanologique, v. 47, p. 153-171, 1984.

Taylor, H.P., and Silver, L.T., Oxygen isotope relationships in plutonic igneous rocks of the Peninsular Ranges batholith, southern and Baja California, in Zartman, R.E., ed., Short papers of the Fourth International Conference on geochronology, cosmochronology and isotope geology: U.S. Geological Survey Open File Report 78-701, p. 423-426, 1978.

Todd, V.R., Geologic map of the Cuyamaca Peak quadrangle, San Diego County, California: U.S. Geological Survey Open File Report 77-405, 1977.

Todd, V.R., Geologic map of the Mount Laguna quadrangle, San Diego County, California: U.S. Geological Survey Open File Report 79-862, 1979.

Todd, V.R., Geologic map of the Alpine Quad, San Diego County, California, USGS Open-File Report 80-82, 42 p., 1983.

Todd, V.R., Geological map of the Alpine Quad, San Diego County, California: U.S. Geological Survey Open File Report 83-781, 20 p., 1983.

Todd, V.R., and Shaw, S.E., Structural, metamorphic and intrusive framework of the Peninsular Ranges batholith in southern San Diego County, California, in Abbott, P.L., and Todd, V.R., eds., Mesozoic Crystalline rocks: San Diego, California, San Diego State University, Department of Geological Sciences, p.

177-232, 1979.

Todd, V.R., and Shaw, S.E., S-type granitoids and an I-S line in the Peninsular Ranges batholith, southern California: Geology, v. 13, p. 231-233, 1985.

Todd, V.R., Erskine, B.G., and Morton, D.M., Metamorphic and tectonic evolution of the northern Peninsular Ranges batholith, southern California, in Ernst, W.G., ed., Metamorphic and crustal evolution of the western United States (Rubey Volume VII): Englewood Cliffs, New Jersey, Prentice-Hall, p. 894-937, 1988.

Walawender, M.J., Petrology and emplacement of the Los Pinos pluton, southern California: Canadian Journal Earth Sciences, v. 13, p. 1288-1300, 1976.

Walawender, M.J., and Abbott, P.L., Quasi-sedimentary structures in the gabbroic complex of the Peninsular Ranges batholith, southern California, in Geologic Studies in San Diego, P.L. Abbott, ed: San Diego Association of Geologists, p. 148-158, 1982.

Walawender, M.J., and Smith, T.E., Geochemical and petrologic evolution of the basic plutons of the Peninsular Ranges batholith, southern California: Journal of Geology, v.88, p. 233-242, 1980.

Walawender, M.J., Hoppler H, Smith T.E., and Riddle, C., Trace element evidence for contamination in a gabbronorite-quartz diorite sequence in the Peninsular Ranges batholith: Journal Geology, v. 87, p. 87-97, 1979.

Walawender, M.J., Gastil, R.G., Clinkenbeard, J. P., McCormick, W.V., Eastman, B.G., Wernicke, R.S., Wardlaw, M.S., Gunn, S.H., and Smith, B.M., Origin and evolution of the zoned La Posta-type plutons, eastern Peninsular Ranges batholith, southern and Baja California, in Anderson, J.L., ed., The nature and origin of Cordilleran magmatism: Geological Society of America Memoir 174, p. 1-18, 1990.

Reprinted from *Geological Excursions in Southern California and Mexico*, Guidebook for the 1991 Annual Meeting, Geological Society of America, San Diego, California.

Landslides in The Peninsular Ranges, Southern California

MICHAEL W. HART

Consultant, 10365 Barrywood Way, San Diego, California 92131

INTRODUCTION

Many types of previously little-studied landslides including earthflows, wedge-slides, complex slides, and rock avalanches have occurred in the Peninsular Ranges of southern California. This fact is remarkable because, for the most part, the ranges are made up of crystalline rocks of the Peninsular Ranges Batholith with pre-batholithic and volcanic rocks underlying a much smaller area. In this study, four large landslides are described. Also presented are hypotheses accounting for their probable mechanisms of failure and runout.

The Peninsular Ranges are northwest-southeast oriented structural blocks extending 200km from the Transverse Ranges on the north to the Mexican border and another 1250km to the tip of Baja California, Mexico. They are bounded on the east by the Imperial Valley and the Gulf of California and on the west by a gently sloping coastal plain underlain principally by Tertiary and Late Cretaceous marine and fluvial sediments.

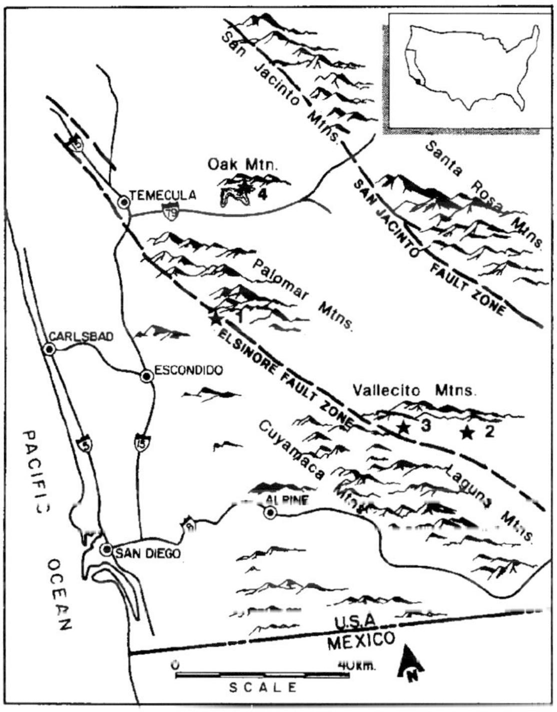

Figure 1. Location Map. (1) Tin Can Flat Landslide, (2) Vallecito Megabreccia, (3) Vallecito Landslide Complex, (4) Oak Mountain Landslide

Major northwest trending active faults separate many of the individual structural blocks within the Peninsular Ranges. The Elsinore and related faults form a complex fault system that extends from the Santa Ana Mountains southward into Mexico. The San Jacinto fault system extends from the Transverse Ranges on the north, crosses the Peninsular Ranges towards the south and extends southeasterly into the Colorado Desert and Mexico (Fig. 1). Additionally, recent studies by Eric Frost (personal communication; Schultejann, 1984; and Engel and Schultejann 1984) indicate that in addition to major right-lateral strike slip movement along the Elsinore and San Jacinto fault zones, Miocene detachment faulting has occurred as a result of east-west extension in the extreme easterly part of the Peninsular Ranges. These faults dip 5 to 45 degrees northeast and separate a lower core of gneissic Late Cretaceous granodiorite and Jurassic metasedimentary rocks from unconsolidated, unmetamorphosed megabreccia of Eocene to Miocene-age (Schultejann, 1984).

In many areas, the presence of faults and the highly fractured and jointed rocks associated with the wide fault zones have created conditions conducive to landsliding. The four landslides discussed in this paper; the Tin Can Flat Landslide, Vallecito Megabreccia, Vallecito Landslide Complex, and the Oak Mountain Landslide are all situated adjacent to faults that likely played an important role in the landsliding process.

THE TIN CAN FLAT LANDSLIDE

Introduction

The Tin Can Flat Landslide, named after a prominent topographic bench located near the head of the landslide, is the largest of many similar landslides that occur on the southern slopes of the Palomar Mountains in San Diego County, California (Figs. 1 and 2). These landslides were first recognized by the California Division of Mines and Geology and appeared on a geologic map of San Diego County compiled by Weber (1963). However, no details of the landslides were discussed in the text of his report.

The Palomar Mountains, and the contiguous Agua Tibia Mountains to the northwest, are an approximately 30km long, rectangular fault block bounded by the Elsinore Fault on the south and the Aguanga Fault on the north. This block is made up principally of Cretaceous plutonic rocks of the Peninsular Ranges Batholith. Pre-batholithic rocks, chiefly Julian Schist, locally form large roof pendants and near the Tin Can Flat Landslide make up the principal rock type at the head of the landslide. The lower part of the slide mass consists primarily of weakly-foliated tonalite.

Reprinted from; Geological excursions in southern California and Mexico, Walawender, M.J., and Hanan, B.B. eds., Guidebook for the 1991 annual meeting, Geological Soc. of America.pp.349-365.

Figure 2. The Tin Can Flat Landslide. Pauma Creek at left edge of photo; Jaybird Creek center; Harrison Cyn., right center; Sycamore Cyn, right edge of photo. Tin Can Flat lies at base of scarp at upper center. Landslide toe lies just beyond bottom edge of photo.

Slide Geometry and Morphology

The Tin Can Flat Landslide is approximately 3km long, 3km wide, and up to 150m thick. The southern distal edge of the landslide is partially buried by fanglomerates emanating primarily from Harrison Canyon (Fig. 2). The northern part of the distal edge of the slide, however, is well exposed along the banks of Pauma Creek and several of its minor tributaries. Erosion along one of these tributaries has exposed the slip surface and overlying breccia shear zones (Fig. 3, Pt. 1). The eastern limit of the landslide where it lies in contact with plutonic rocks, is not well-exposed. However, aerial photographs that predate the present orchards, show low lateral and distal pressure ridges that represent the limits of landsliding.

The arcuate headscarp above Tin Can Flat is well preserved and not heavily modified by erosion. This is in contrast to the highly eroded southern portion of the scarp at the head of Harrison and Sycamore Canyons.

The slide surface is characterized by topographic benches that step down toward the southwest, and probably represent the heads of small, secondary landslides. Three major canyons dissect the landslide; Jaybird Creek, Harrison, and Sycamore Canyons (Fig. 2). Harrison and Sycamore Canyons cut completely through the landslide and provide good exposures of its lithology, structure, basal rupture zone, and substrate.

Composition, Structure and Contact Relationships

The headscarp above Tin Can Flat is underlain by well-foliated pre-batholithic schist and gneiss. Foliation planes strike nearly parallel to the general trend of the scarp and dip 55° to 85° to the southwest. Gneissic rocks form the highly dissected scarp area south of Tin Can Flat.

Exposures along Nate Harrison Grade and in Harrison and Sycamore Canyons reveal the structure and composition of the landslide mass. In the upper part of Harrison Canyon below Tin Can Flat the steep, southwesterly dipping slip surface is exposed

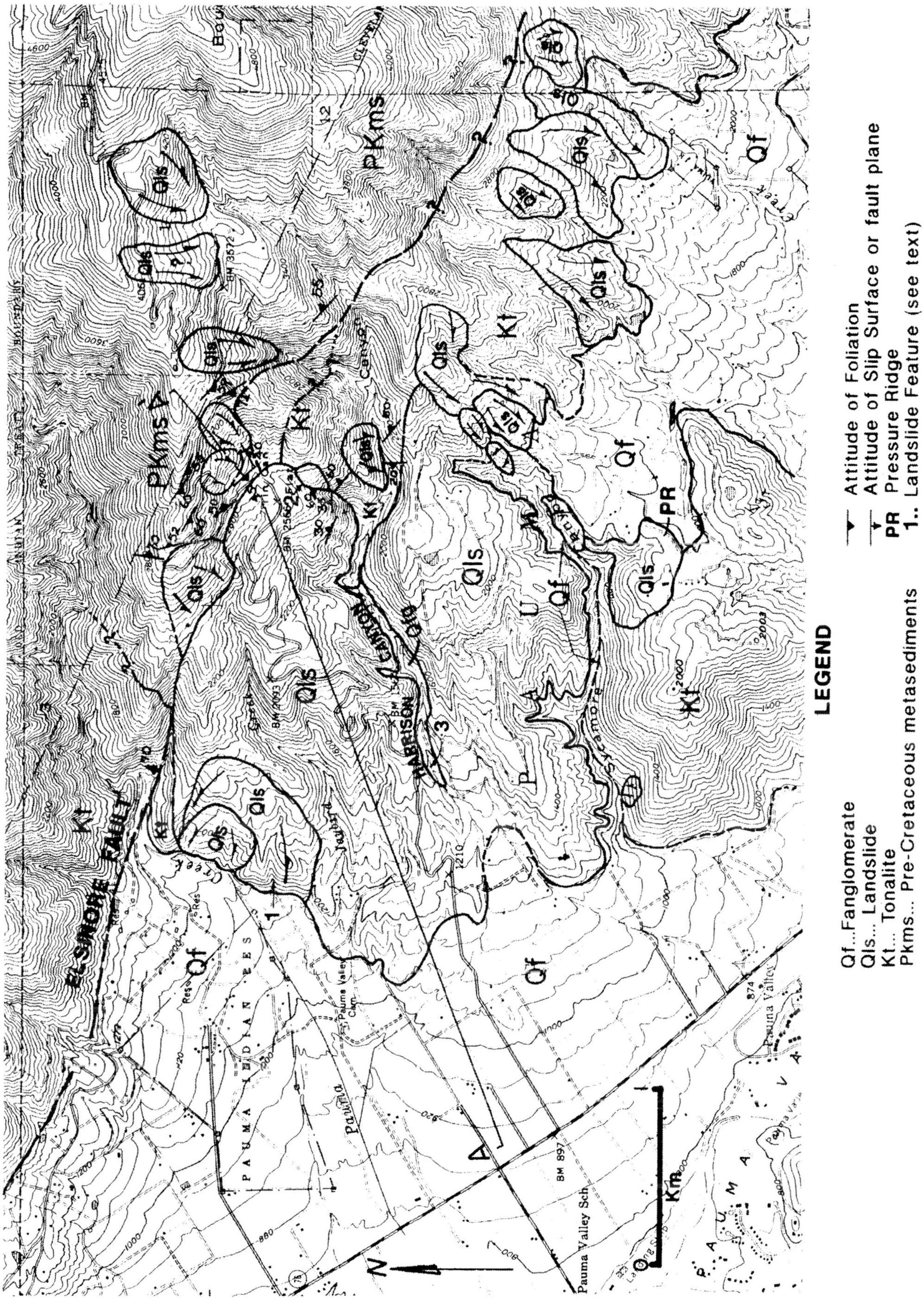

Figure 3. Geologic Map of Tin Can Flat Landslide

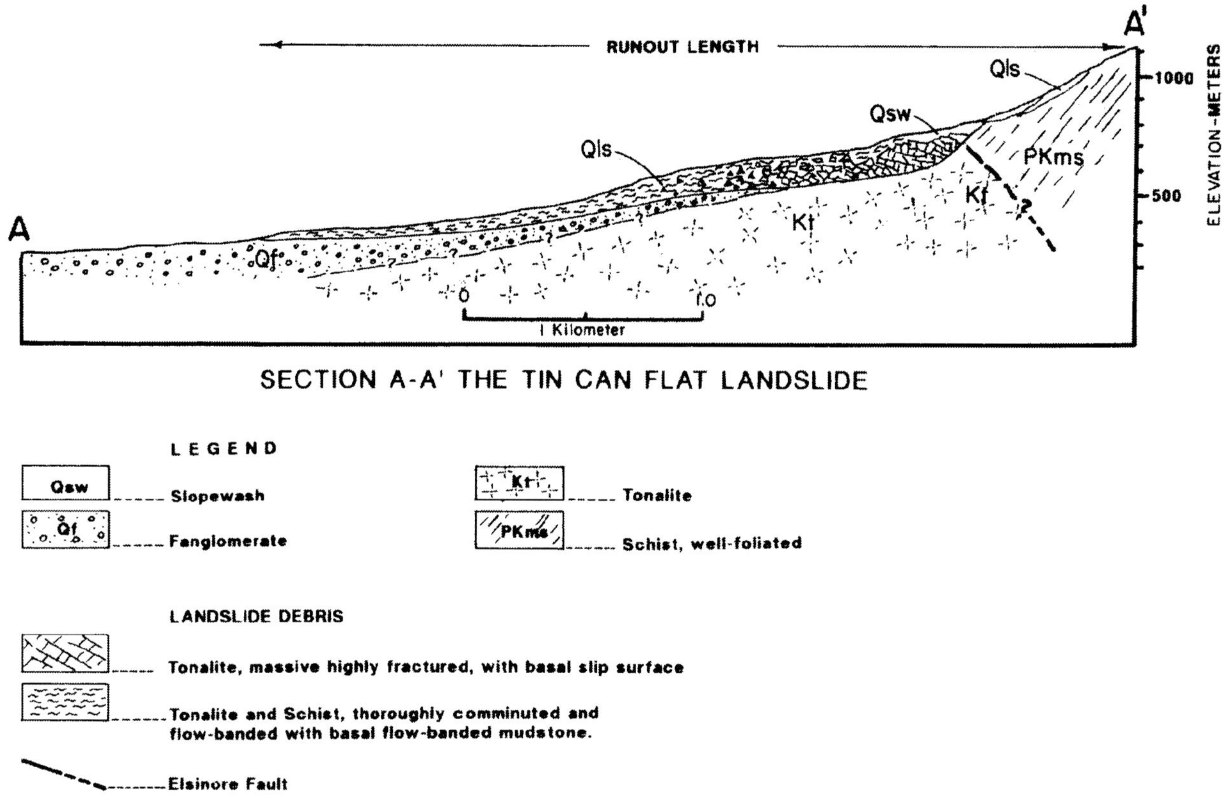

Figure 4. Section A-A', Tin Can Flat Landslide (see Fig. 3 for location of section)

near the base of a wide zone of sheared and crushed rock (Fig. 3, Pt. 2). Relatively unfractured, poorly-foliated, gneissic tonalite occurs below the shear zone and in hand-specimen appears identical in composition to the tonalite located above the shear zone.

Most of the distal slide mass is composed of thoroughly sheared and crushed granitic and metamorphic rocks that display flow-banding and slip surfaces within clayey gouge zones. The contact between the fractured but relatively intact-looking rock mass that makes up the eastern part of the slide and the completely comminuted western part is not exposed, but is probably gradational as depicted on Fig. 4.

The base of the western portion of the slide is well-exposed in the 15m to 20m high, vertical stream cuts of Harrison and Sycamore Canyons (Fig. 3, Pt. 3). Here the slide debris, consisting of thoroughly sheared or comminuted, flow-banded materials with fragments generally less than a few millimeters in size is in gradational contact with an approximately 1m thick, moderately well-cemented, crudely-stratified and deformed, dark brown to grey, flow-banded mudstone. The mudstone is, in turn, underlain by fanglomerate consisting of well-rounded, granitic and metamorphic cobbles and boulders in a well-sorted matrix of medium- to coarse-grained sand. Contact relationships between the fanglomerate and slide debris indicate a minimum runout onto a preexisting alluvial fan of approximately 1700m (Fig. 4). The contact between the slide breccia and mudstone is sharp but no slip surface is present.

In Harrison Canyon a clastic dike (Fig. 5) extends approximately 2m upward from the mudstone unit into the slide breccia. Similar clastic dikes have been reported in the Blackhawk Landslide by Johnson (1978) and in the Vallecito Megabreccia described later.

Mechanics of Failure and Runout

The mode of failure and mechanics of runout are of two of the most interesting aspects of landsliding. To determine the mechanics of failure, one must know the basal rupture surface and slide mass characteristics, geometry, and something of the landslide's erosional history. For example, the contrast between the well-preserved headscarp above the Tin Can Flat and the nearly obliterated portion of the headscarp to the southeast of Tin Can Flat suggests that there may have been at least two episodes of sliding separated by a sufficient amount of time to nearly erase the oldest headscarp. However, the stratigraphy of the slide debris on the west side of the canyon is identical to that on the east. Also, the basal rupture zone lies about the same elevation on the west and east sides of the canyon. It, thus, seems more likely that the landslide occurred as a single event, but that various components of the slide have been subject to differing rates of erosion.

Figure 5. Base of Tin Can Flat Landslide in Harrison Cyn. A = Comminuted landslide debris; B = Basal mudstone, arrow points to clastic dike approximately 1m in length; C = Fanglomerate.

The presence of a steep shear zone between fractured blocks of slide debris and in-place gneiss at the head of the slide and the broad topographic bench at Tin Can Flat suggest slumping as a mode of failure. Such failure was, therefore, probably initiated along foliation planes or along a possible continuation of the Elsinore Fault that has been mapped to the western edge of the landslide (Vaughan, 1987). The presence of flow-banded and comminuted rock in the western three-quarters of the slide and a long runout distance of 1700±m over fanglomerate suggests a rock-avalanche mode of failure. Most rock-avalanches, of which the author is aware, however, are completely detached from their source area and have travelled great distances relative to their drop height over nearly horizontal terrain. The ratio of height drop to length of runout (H/L) is a commonly used measure to determine if a landslide has excessive runout length and whether or not it can be classified as a "classic" rock-avalanche. Rock-avalanches generally have H/L ratios of less than 0.5 (Melosh, 1987). The Tin Can Flat Landslide has a H/L ratio of approximately 0.24, a ratio that could qualify it as a long runout rock-avalanche.

The above observations suggest a hybrid mechanism of failure; namely, slumping at the head of the slide followed by a rapid runout of the slide debris on a pressurized mud-slurry represented in outcrop today by the flow-banded mudstone in Harrison Canyon. The clastic dike at this location is evidence of the depressurization of the mud-slurry after the avalanche came to rest.

THE VALLECITO MEGABRECCIA (A POSSIBLE LONG RUNOUT LANDSLIDE)

Introduction

The Vallecito Megabreccia is located approximately 2.5km southeast of the Vallecito Mountains, a series of 1600m high, tonalite peaks (Todd, 1977) that form the northern boundary of the Vallecito-Carrizo Badlands of Anza-Borrego Desert State Park (Fig. 1). The megabreccia, consisting of tonalite blocks, occurs

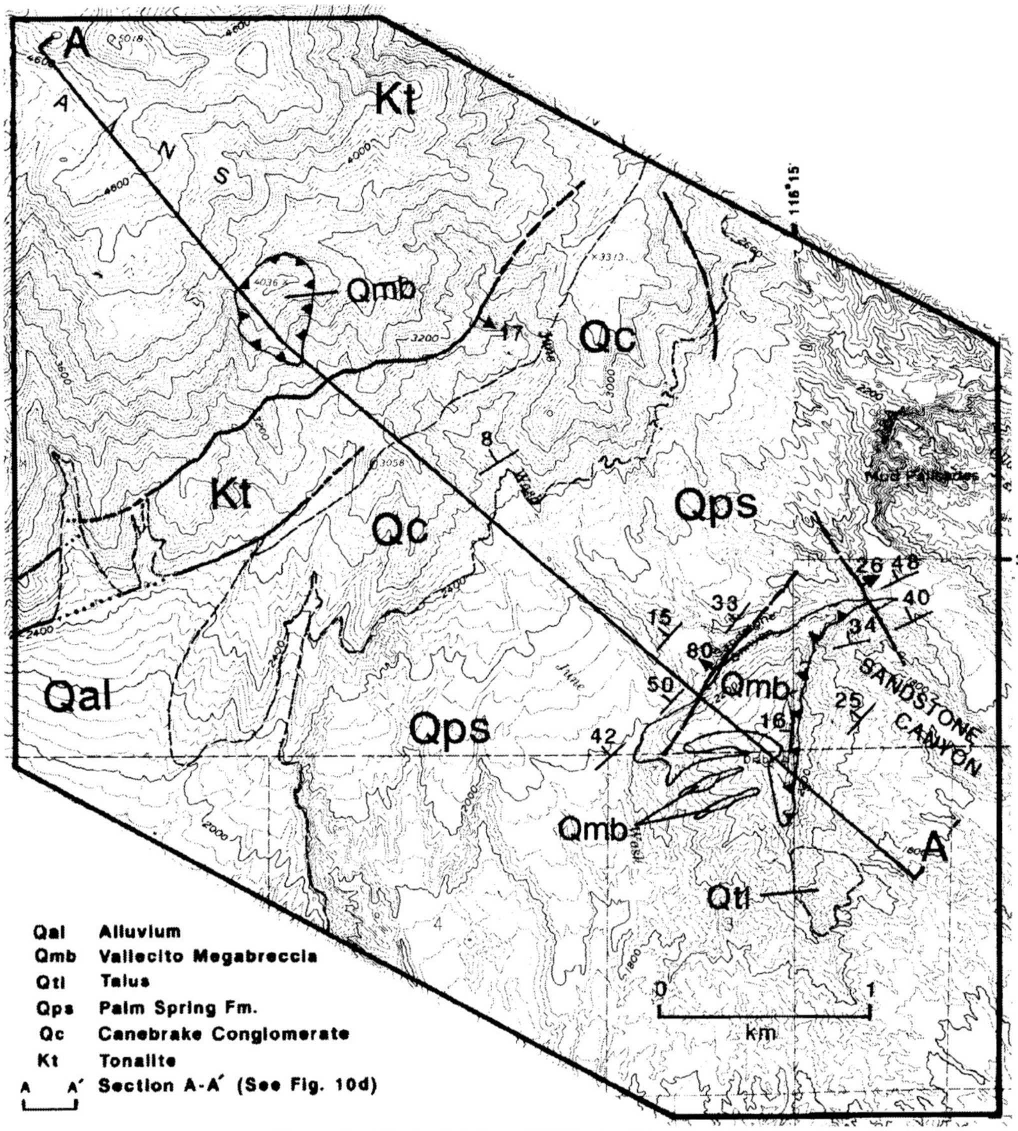

Figure 6. Geologic Map of Vallecito Megabreccia

within the lower member of the Palm Spring Formation, an approximately 3000m-thick sequence of Middle Pleistocene folded and faulted fluvial and lacustrine sediments (Woodard, 1963). In the Vallecito area these sediments lap onto the basement rocks along the northern and western edge of the basin. To the south and east, the Palm Spring sediments extend into the Imperial Valley and the Salton Trough. The Palm Spring Formation is overlain by and interstratified with the Pleistocene Canebrake Conglomerate, a massive, coarse, well-rounded boulder and cobble fanglomerate (Woodard, 1963).

Compared to other previously described megabreccia deposits (Kreiger, 1977; Yarnold and Lombard, 1989) interbedded in Cenozoic sedimentary rocks of the southwest, the Vallecito Megabreccia is unimpressive both in thickness and extent. However, its significance lies chiefly in that its structure and stratigraphy indicate a unique mode of failure first suggested by Johnson (1978) for the Blackhawk Slide located on the northern slopes of the San Bernardino Mountains.

Other megabreccias occur in lower Neogene sediments of the Fish Creek Mountains located approximately 15km to the east of the Vallecito Megabreccia (Robinson and Threet, 1974) and (Kerr, 1984). These landslide deposits are much thicker and have a greater areal extent than the Vallecito Megabreccia. The clast composition also differs markedly in that it consists of rounded granitic and metamorphic clasts supported throughout the section by a clayey matrix. Thus, their structure and stratigraphy suggest a much different mechanism of emplacement than that proposed for the Vallecito Megabreccia.

Slide Block Geometry

The Vallecito Megabreccia is well-exposed on the dip slope of a 160m-high cuesta near Sandstone Canyon, a major stream course that cuts through the breccia unit (Fig. 6). The breccia outcrop is presently 1.5km wide; however, a tonalite talus deposit exposed on the southern slope of the cuesta, beyond the current edge of the breccia unit, indicates that the megabreccia once extended at least 500m to 600m further to the southwest.

Figure 7. Crackle-breccia facies near upper Sandstone Cyn. Largest blocks are approximately 1m in greatest dimension.

The breccia averages 30m to 40m thick over most of outcrop length and thins over a distance of a few hundred meters to less than 1m at its northern edge. The nearly constant thickness and planar upper and lower contacts with fine-grained sediments of the Palm Spring Formation indicate than the Vallecito Megabreccia is essentially tabular-shaped body.

The breccia's presently exposed distal edge lies approximately 2km southeast of its inferred buried contact with tonalite basement rock of the Vallecito Mountains, the closest possible source of tonalite blocks that make up the breccia sheet. It is not known how much further the megabreccia may have travelled from its source area since erosion has removed any indication of its former presence to the southeast. Yarnold and Lombard (1989), however, developed a facies model for similar rock-avalanche deposits that presented characteristics of the proximal, medial, and distal edges of megabreccia deposits that are of use in determining where a particular outcrop might lie in relation to those positions on the slide. Their model for the distal zone indicated that deformation of the fine-grained sediments is generally intense immediately beneath the rock-avalanche/substrate interface, that the breccia sheet is relatively thick, and that there is a well-developed basal disrupted zone with abundant matrix-rich breccia and laterally extensive slip surfaces. Clastic dikes are reported to be characteristic of both the medial and distal zones. Because the Vallecito Megabreccia displays all the above characteristics except extensive deformation of the substrate, it is believed that the presently exposed distal edge of the deposit lies somewhere between the original medial and distal zones.

To approximate of the original volume of the rock-avalanche, the maximum observed dimensions previously discussed were reduced somewhat arbitrarily to account for the gradual thinning and narrowing of the breccia unit that most likely occurs toward its source area in the Vallecito Mountains. Assuming an average thickness of 20m, width of 1000m and length of 2500m, the volume of the Vallecito Megabreccia is estimated to have been $50x10^6m^3$.

Figure 8. Base of Vallecito Megabreccia, Sandstone Cyn.; A = matrix-rich facies; B = mudstone facies; C = sandstone of Palm Spring Fm. Arrow points to flame structures, clastic dike to left of hammer.

Megabreccia Structure and Stratigraphy

The Vallecito Megabreccia, as exposed in its most distal part, is essentially a two-layer system. Each layer is approximately 15m to 20m thick and is subdivided into two sublayers or facies members. The upper layer consists of a thin cap of boulders (upper facies) underlain by angular clasts separated by little or no matrix (crackle and jigsaw breccia facies of Yarnold and Lombard, 1989). The lower layer consists of an upper matrix-rich facies underlain by a thin basal mudstone facies.

The boulders exposed on the megabreccia surface are well-rounded and generally large, some attaining 5m to 8m in diameter. The rounding probably began immediately after the avalanche event by exfoliation and continued more recently after exhumation of the avalanche's surface by erosion. All internal clasts of both the upper and lower layers are angular and show no evidence of weathering or rounding.

In Sandstone Canyon, the crackle-breccia facies that makes up the bulk of the upper layer consists of shattered rock resembling undisturbed basement outcrops and angular pebble- to boulder-sized blocks fitting tightly together with little no matrix (Fig. 7).

The matrix-rich facies of the lower layer is separated from the upper clast-supported layer by a 5±m thick contorted zone of flow-banded breccia. There is a wide range of both clast size and percentage of matrix within this facies of the deposit. In general, the matrix makes up approximately 50 to 90 percent of the deposit, and the angular clasts vary in size from approximately 1m to a few centimeters. At least 80 percent of the clasts, however, are cobble- to pebble- sized. The matrix in the lower half of the

Figure 9. Clastic dikes in Sandstone Cyn. (arrow above rock hammer). A = matrix-rich facies; B = mudstone facies; C = Sandstone of Palm Spring Fm.

megabreccia is made up of rock flour and sand- to grit-sized material similar to that of the upper half present in the deposit.

The basal mudstone facies is approximately 0.5m thick and has a gradational contact with the overlying matrix-rich facies. The mudstone exhibits crude flow-banding that is generally parallel to bedding of the Palm Spring Formation. The mudstone also exhibits flame structures that indicate a southeast direction of transport (Fig. 8). Less than 10 percent of the mudstone volume consists of rounded pebbles and gravel derived from the underlying pebbly sandstone of the Palm Spring Formation and angular, cobble-sized clasts of tonalite.

There are four clastic dikes in Sandstone Canyon that extend upward from the top of the mudstone in a branching, lightning-bolt-like pattern 2.5m to 5m into the overlying breccia facies (Fig. 9). Similar features have been observed in the Blackhawk Slide (Johnson, 1978) and in other large rock-avalanches in the southwest, described by Yarnold and Lombard (1989) and Kreiger (1977).

Contact Relationships

The base of the megabreccia is well exposed for almost its entire outcrop width below the top of the cuesta and in Sandstone Canyon. Typically, one or more subparallel slip surfaces form the contact between the breccia and the underlying deformed basal mudstone. Southwest of Sandstone Canyon and near the top of the cuesta, the slip surface is exceptionally well-developed and exhibits striae that are oriented in a northwest-southeast direction. In Sandstone Canyon, the contact between the basal mudstone and Diablo member of the Palm Spring Formation is gradational. Well-developed bedding within the pebble conglomerate and medium-grained sandstone is essentially parallel with flow-banding in the mudstone and slip surfaces. Folding of the substrate sediments described by Yarnold and Lombard (1989) as being typical of the distal part of other large rock avalanches was not observed.

Origin and Mode of Emplacement

Todd (1977) first mapped the tonalite breccia as an allocthonous block lying in thrust fault contact with sandstones of the Palm Spring Formation. Todd also indicated that the deposits may be of landslide origin. The conformable nature of the contact between the breccia and underlying sandstone northeast of Sandstone Canyon, the generally matrix-supported nature of the lower part of the breccia unit, and its slab-like form and disconformable relationship to overlying sandstone of the Palm Spring Formation all indicate that the Vallecito Megabreccia was not emplaced by thrust faulting but rather by some type of land-sliding process. The nearest source of tonalite lies in the Vallecito Mountains approximately 2.5km to the northwest of the distal edge of the megabreccia unit. Steep, deeply incised canyons along the southeastern flank of the mountain front reveal an intensely jointed rock mass with closely-spaced joint planes dipping parallel to the mountain slope. Remnant megabreccia deposits exposed at an elevation of approximately 1160m on the southeast flank of the Vallecito Mountains are further evidence that the landsliding that produced the Vallecito megabreccia originated in this area.

Landslides that have had excessive travel distance or runout, such as the Blackhawk, Elm and Saidmarreh landslides (Voight and Pariseau, 1978; Hsu, 1975) as well as older Late Cenozoic landslides described by Yarnold and Lombard (1989) and Kreiger (1977) all have a characteristically low H/L ratio. These types of landslides have seemingly travelled far greater distances from the source areas than normal frictional resistance would allow. The Blackhawk Landslide, for instance, travelled over 7km from the base of the mountain source area (Shreve, 1968a; Johnson, 1978). Excessive travel distance requires very low shear resistance between the sliding materials and the base over which it is travelling. It is probable that the Vallecito Megabreccia travelled a distance (L) of at least 3.5km as measured from the upper edge of the landslide (megabreccia) remnant at an elevation of 1160m in the Vallecito Mountains (Fig. 6 and 10d). The total vertical drop (H) is more difficult to determine because of folding and faulting of the breccia unit and erosion that has removed evidence of the slide scarp and distal edge. However, it seems reasonable to assign the upper limit of the scarp to the top of a 1460m-high ridge that forms the eastern flank of the Vallecito Mountains. This ridge is approximately 250m higher than the elevation of the megabreccia remnant exposed south of the ridge (Fig. 6). If the Palm Spring Formation is unfolded, then the original elevation of the distal edge of the breccia unit may have terminated at an approximate elevation of 350m. H/L is then equal to approximately 0.32, a value comparable to that reported for long runout rock-avalanches such as the Elm Landslide of Switzerland (0.26), and the Frank, Alberta Landslide (0.28). Exceptionally long runout slides such as the Blackhawk, Saidmarreh, and El Capitan Landslides have H/L ratios of 0.13, 0.10 and 0.19, respectively (Kreiger 1977).

There are four primary mechanisms thought capable of explaining the apparently very low angles of internal friction that must account for the long travel distance of some rock avalanches. These are: (1) air-layer lubrication (Shreve, 1968a); (2) two-layer composite debris flow (Johnson, 1978); (3) dust suspension (Hsu, 1975); and (4) acoustic fluidization (Melosh, 1987). Hsu's theory, based on his study of the Elm Slide, does not seem to be the mechanism in this case because there is very little of the required rock dust or fines in the breccia except in its lower half. The acoustic fluidization hypotheses (Melosh, 1987) is difficult to test since no one knows exactly what diagnostic features to look for in a slide mass that travelled by such a mechanism. Melosh (1987) suggests that one-half the total shear would be concentrated in the lower 7.4 percent of the moving debris. Since the Vallecito Megabreccia is approximately 30m to 40m thick, the basal shear zone would be approximately 2m thick. Such a shear zone might consist of crushed or highly abraded relatively small rock fragments (in comparison to the overlying rock debris) with a matrix of rock flour generated by crushing and shearing during transport; however, no such zone is present.

Critically important to the air-lubrication hypothesis of Shreve (1968a) is the rate of air leakage through the slide breccia. Shreve calculated that the permeability of the Blackhawk Slide would be sufficiently low to support the slide mass (Shreve, 1968b). Erismann (1979), however, concluded that the permeability required to entrap the air is implausibly low. In other words, even though it may be reasonable for the slide to have a sufficiently low permeability during a relatively consolidated runout phase, the permeability of the mass at maximum dilation during air launch and landing would be too high to entrap air. If the relatively fine-grained marble breccias of the Blackhawk Landslide had too high a permeability to entrap air, then it seems probable that the much coarser-grained Vallecito Megabreccia would have had an even higher permeability. Thus, an air-layer mechanism of failure seems improbable. Based on the presence of fluid-like intrusion structures, clastic dikes, and limited exposures of a mudstone layer near the base the slide Johnson (1978) proposed a new, albeit much less dramatic, mechanism of failure for the Blackhawk Landslide. He suggested that the Blackhawk Slide traversed the alluvial apron north of the source area on Blackhawk Mountain as a two-layer composite debris flow in which an upper non-fluid rock mass rode "piggy-back" on a highly pressurized mud and rock slurry. Johnson concluded that the composite debris flow model could only be a crude working hypothesis since the basal section of the land-slide was too poorly exposed to make determination of the correct

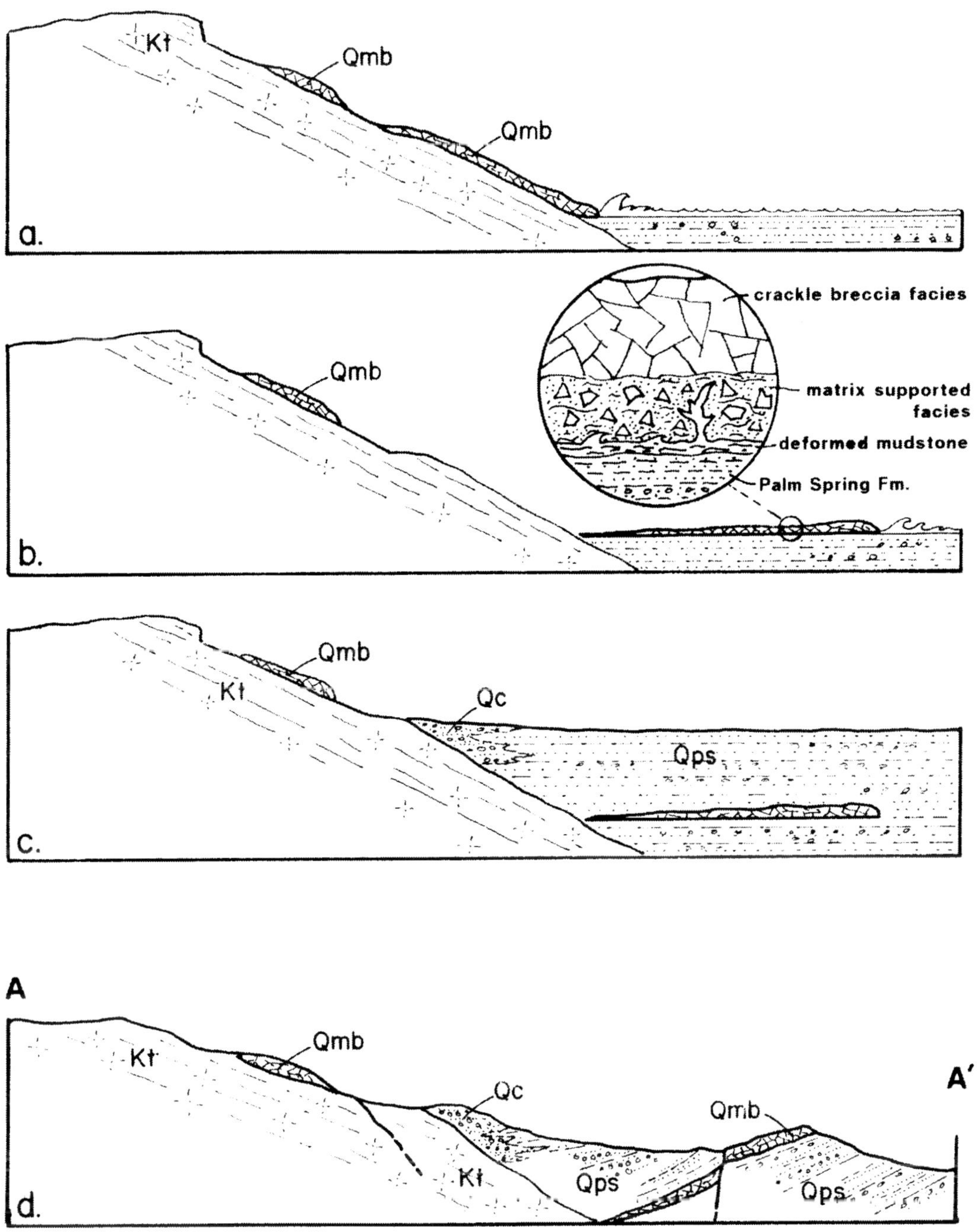

Figure 10. Probable sequence of landslide events: (a) Tonalite rock slab detaches from Vallecito Mtns. along well-developed joint system and begins to impact shallow lake; (b) megabreccia unit travels over shallow lake bed as two-layer system on pressurized mud slurry (see inset); (c) megabreccia buried by fluvial sediments of Palm Spring Fm.; (d) Section A-A', Fig. 6. Folding and faulting from late Pleistocene to present tilts Palm Spring beds to north and erosion exposes megabreccia;

Qmb = megabreccia, Qc = Canebrake Conglomerate,
Qps = Palm Spring Fm., Kt = tonalite.

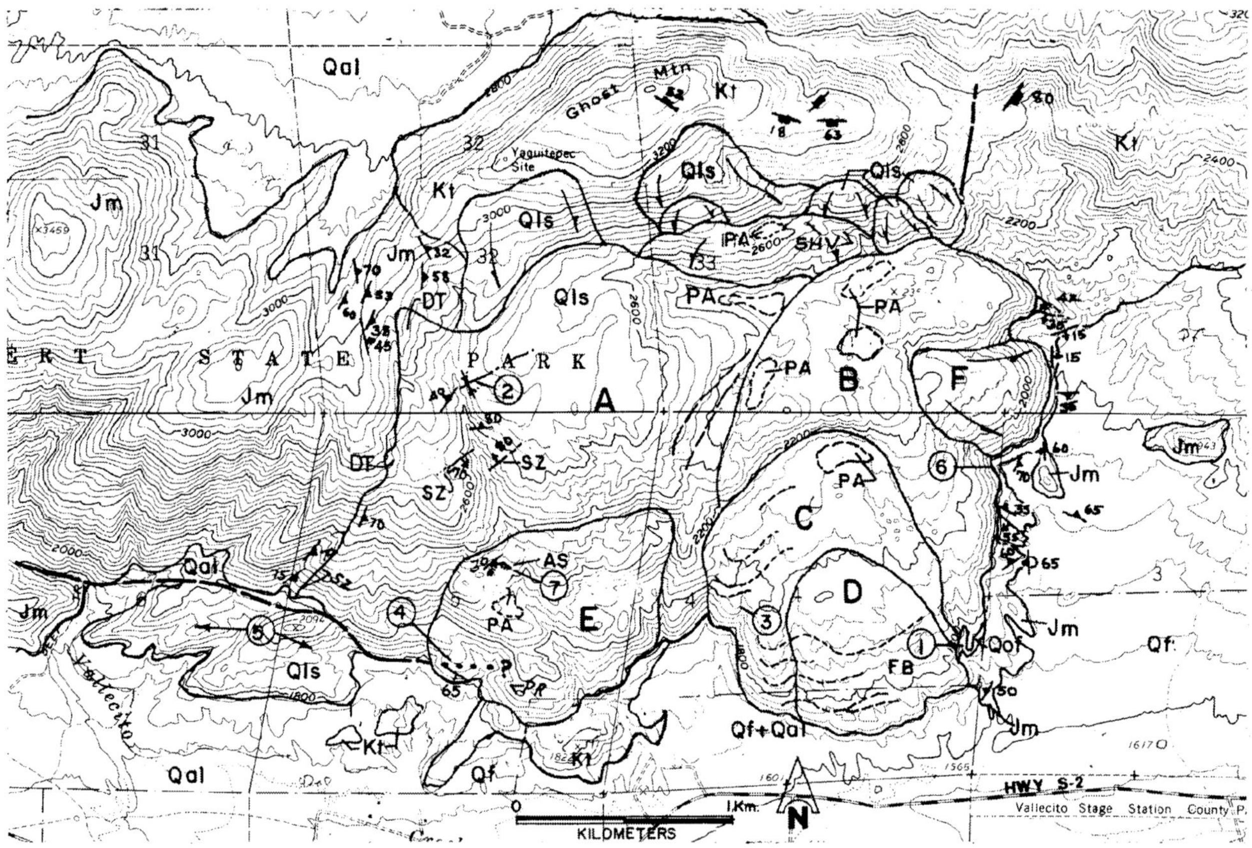

LEGEND

Qal....	Alluvium	A	Landslide designation 'see text)
Qf	Fanglomerate	①	Location of slide feature discussed in text
Qof	Older fanglomerate	FB	Flow banding
Qls	Landslide	PA	Ponded alluvium
Kt	Tonalite	DT	Debris train
Jm	Mixed granitic rock and metasediments	SZ	Shear zone
	Attitude of foliation	PR	Pressure ridge
	Active scarp	SHV	Side hill valley
	Attitude of shear surface	AS	Active scarp
	Attitude of bedding		
	Attitude of Joint		

Figure 11. Geologic Map of Vallecito Landslide Complex

mechanism. The occurrence of intrusion structures similar to those present at the base of the Blackhawk Slide and the presence of a flow-banded basal mudstone indicates that a once highly fluid mud-slurry was present beneath the Vallecito Megabreccia.

The foregoing observations suggest that the rock-avalanche began as a large slab failure along joint planes that dip parallel to the steep slopes forming the southeast side of Vallecito Mountains (Fig. 10a). The triggering mechanism for the initial failure is not known, however, the area lies between the active Elsinore Fault Based on the presence of fluid-like intrusion structures, clastic dikes, and limited exposures of a mudstone layer near the base of and San Jacinto Fault systems and major earthquakes along either system could have triggered the initial slope failure. The rapidly moving rock-avalanche (Figs. 10b through 10d) then impacted what must have been a nearly horizontal flood plain occupied by a braided stream system or swamp (Woodward, 1963). The sudden loading of the saturated sediments beneath the rock-avalanche created high pore-pressures and sufficiently low shear strength to allow a minimum 2km runout over the valley floor. The upper part of this two-layer composite debris flow consisted of a nearly dry, non-fluid, rock rubble mass whereas the lower part was a saturated, highly fluid mud and rock slurry. The mass slowed as water escaped and pore pressures lowered. During the final few feet of movement, slip surfaces formed in the rapidly consolidating mud. Clastic dikes, feeding from the depressurizing mud, extended upward into the breccia during the final few moments before or after the slide came to a halt, since they are not significantly deformed.

Of great importance to this sequence of hypothetical events is required saturation of the sediments impacted by the rock-avalanche. Woodard (1963) described the coarser sediments of the Diablo member of the Palm Spring Formation as massive

Figure 12. Vallecito Landslide Complex, Ghost Mtn. top center; slide designations A-E same as Fig. 11

sheet-flood deposits laid down by sediment-choked streams. The siltstones and mudstone deposits within the Diablo member were accumulated in low-lying marshy or swampy areas which were periodically inundated by northerly transgressions of the Gulf of California. Conditions such as these would have been ideal for providing the saturated sediments required for the two-layer composite debris-flow model.

VALLECITO LANDSLIDE COMPLEX

Introduction

The Vallecito Landslide Complex is located in the Anza-Borrego Desert north of County Route S-2 and west of the Vallecito Stage Station County Park in San Diego County, California (Fig. 1). The slide complex consists of at least six individual slide masses of different ages that occupy approximately 5.5km^2 (Figs. 11 and 12). The primary rocks represented in the slides and source areas are quartz diorite, and pre-batholithic schist. Late Pleistocene alluvium and fanglomerate underlie portions of the toe of the landslide complex. Middle Pleistocene fluvial sediments of the Palm Spring Formation lie in contact with the basement rock south and east of the slide complex. The Elsinore Fault has been mapped by Pinault (1984) along the southern edge of the landslide complex and has, in places, offset landslide debris.

Geometry and Geomorphology

The Vallecito Landslide Complex is approximately 3km wide and 2.4km long (Fig. 11). The northern limits of the Complex are clearly expressed on aerial photographs by tonal differences between undisturbed rock and landslide debris. The western edge of the slide is not well exposed; however, there is a well-defined lithologic contact between coarse-grained porphyritic tonalite and biotite gneiss along the western portion of the mapped area that probably represents the limit of landsliding. In this same area there is also a several-meters-thick shear zone coinciding with the lithologic contact that may represent the western slide limit. An alternative interpretation for the shear zone is that it may only represent a small surficial slide or possibly a relatively shallow debris-flow deposit.

The height of the scarp measured from the crest of Ghost Mountain to the topographic bench at the head of Slide A, probably the oldest of the slide masses (Fig. 11), is approximately 130m. Another minor scarp resulting from the second largest slide event (Slide B) is located along the eastern limit of Slide A and is approximately 90m high. Superimposed on Slide C is a "secondary" landslide (Slide D) with a subtle scarp 10m to 15m high. The surface of Slides C and D is hummocky and strewn with large, rounded to subangular boulders, some of which are 10m or more in dimension. Slides D and E are probably secondary landslides that formed sometime after the parent slide events.

Several small, unlabeled slides, occur on the backscarp of the slide complex below the crest of Ghost Mountain. West of these small slides, the remnant of a "debris train," designated DT on Fig. 11, forms a flat-iron on the scarp.

The west side of Slides C and D projects over 900m onto the floor of Vallecito Valley as measured from the face of Slide A. The east side of the Slide C/D complex, however, extends less than 600m onto the valley floor past the most southerly limit of Slide B from which it originated. These distances do not necessarily indicate the distance travelled by the landslides because of the irregularities that may have been present in the shape of the toe of the parent slide mass. A better measure of the minimum travel distance was obtained at the toe of Slide B where it has over-ridden alluvium (Fig. 11, Pt. 1) for a distance of approximately 60m.

Composition, Structure and Contact Relationships

Glimpses of the composition and structure of the several landslides making up the Vallecito Landslide Complex occur along numerous drainages occurring within the slides and in relatively good exposures near the toe of Slides B and C. A deep canyon near the western edge of Slide A has exposed fractured and relatively undisturbed-looking tonalite in the lower half and progressively more disturbed and sheared tonalite, schist, and gneiss in the upper half of the slide. Several excellent exposures of 2m to 3m wide shear zones occur just below the upper bench in Slide A at elevations between 2400 and 2600 feet. Movement along one of these steep southerly dipping shear zones has resulted in brittle drag-folding of adjacent metasedimentary rocks (Fig. 11, Pt. 2).

The slide debris is well exposed in a drainage that cuts through Slide C (Fig. 11, Pt. 3). In this locality, the slide debris consists of mostly large ($1\pm$m), angular blocks of fresh tonalite that fit closely together with little or no interstitial matrix. Elsewhere, however, clasts varying in size from a few centimeters to several meters or more are set in a matrix of pulverized rock.

The best exposures of the basal rupture zone occur on the easterly side of the slide complex. Several poorly exposed basal shear zones are located just west of Slide C a few meters above the alluviated valley floor and in a minor drainage at the toe of Slide D. At Pt. 4, (Fig. 11) near the toe of Slide A, there is an interesting exposure of what Todd (1977) mapped as Middle Pleistocene Palm Spring Formation. Here, the Palm Spring Formation is steeply tilted and displays several minor faults. This tilting and faulting may have been caused by either the bulldozing effect of the landslide or possibly by activity along strands of the Elsinore Fault mapped by Todd (1977) and Pinault (1984). Pinault (1984) interpreted the very unusual outcrop of tonalite at Pt. 5 (Fig. 11) as part of the toe of Slide A that has been displaced northward along the Elsinore Fault. The evidence for this interpretation is strong because the rocks at this locality could not have originated by sliding from an uphill area that is mostly underlain by schist and gneiss.

Several excellent exposures of the basal rupture zone of Slide B occur along the eastern edge of the slide complex. The best exposures are located just south of Slide F (Fig. 11, Pt. 6). The rupture zone at this locality lies horizontally on granitic residuum and consists of a upper 10m to 15m thick matrix-rich breccia containing angular metamorphic and granitic clasts less than .5m in size and a lower 30cm thick black clay gouge containing numerous slickensided slip surfaces.

It is within this clay gouge that most of the frictional sliding occurred. X-ray analysis of the clays by Richard Berry at San Diego State University revealed that they consist primarily of expandable smectite with quartz making up approximately 50 percent of the sample. Also present in nearly equal amounts were feldspar, illite, calcite, and a trace of manganese oxide. In order for the slide to have occurred, there must have been a preexisting zone of low shear resistance such as that represented by the clay. How did the nearly horizontally-dipping shear zone originate? It is hypothesized that landsliding may have occurred on a low-angle detachment fault since such faults have been reported in nearby areas.

Thin-sections of the clay in the shear zone, however, revealed nothing diagnostic of a deep-seated origin for the shear zone. On the contrary, a relatively shallow origin of the shear zone is suggested by the presence of brittlely-fractured quartz grains that make up 50 to 70 percent of the samples. No plastically deformed or highly strained mineral grains that might be indicative of deep-seated tectonism were observed. In fact, under the microscope, the claystone appears as a debris-flow deposit in microcosm consisting of minute, highly angular quartz clasts set in a matrix of clay and quartz flour. Because the deformation in the shear zone occurred at a relatively shallow depth, no definite conclusions can be drawn as to whether clays originated as a detachment fault.

Age of Sliding

There are several lines of evidence that indicate differing ages for the several landslides making up the Vallecito Landslide Complex. Relative ages can be determined from the degree of dissection, surface roughness, and degree of scarp degradation. Based on the above criteria, the landslides in order of decreasing age are designated Slides A through F.

The numeric age of Slide A can be determined approximately by observing to what degree, if any, the toe of slide is offset by the Elsinore Fault and dividing the offset by the estimated yearly slip rate. Pinault (1984), for example, observed that the toe of Slide A is offset by approximately 1km. He also determined that the slip rate for the Elsinore Fault in the Coyote Mountains, approximately 25km to the south, is 3.4 to 5.5mm/year (minimum). If it is assumed that the slip rate in Vallecito Valley is essentially the same as in the Coyote Mountains, then Slide A

Figure 13. Oak Mountain Landslide Complex. Dashed line is base of slide complex (see Fig. 1 for location).

is a maximum of 180,000 to 290,000 years old. Slides C, D, and E are apparently not displaced by faulting and, hence, may be only a few thousand or even a few hundred years old. In one small area (Fig. 11, Pt. 7) within Slide E, there is a fresh 1m to 2m high vertical scarp that displaces present-day alluvium in a sidehill valley. This feature indicates that at least portions of Slide E are still active. Movement on this most recently active part of the landslide was most-likely triggered by seismic activity since it is unlikely that groundwater is present above the slip surface.

A late Pleistocene age of sliding for Slide B is suggested at Pt. 1 (Fig. 11) where a fanglomerate, whose upper surface is approximately 25m above the present active fan surface to the east, has been buried by the landslide.

Mechanics of Failure

Flow-banding observed on the surface of Slides C and D suggests that they had some characteristics of a slow-moving, highly viscous mass. The arcuate headscarps on all slides indicates slumping, at least at the heads of the slides. Horizontal slip surfaces observed at the toe of the slide complex suggest translational movement. The presence of a matrix-rich flow-banded breccia at the base of Slide B suggests a two-layer composite debris-flow mechanism of failure similar to that previously described for the Tin Can Flat Landslide and Vallecito Megabreccia. The mechanism of failure then could be described as extremely complex and in need of more study in order to fully understand.

Oak Mountain Landslide Complex

Introduction

The Oak Mountain landslide complex is located in southern Riverside County near Vail Lake (Fig. 1 and 13). This group of superimposed and coalescing slides is approximately 4.8km wide, 800m long, and occurred in coarsely crystalline gabbro.

Although shallow landslides and flows in gabbroic residuum are common in the Peninsular Ranges (Hart, and Farkas, 1983), massive, deep-seated landslides involving large volumes of relatively unweathered gabbroic rock are rare. The Oak Mountain Landslide Complex represents one of the few such deep-seated slides in gabbroic rocks that is sufficiently exposed to permit study of its structure and mechanism of failure.

The toe of the landslide complex is well-exposed at an elevation of 500m$\pm$ and may be observed in several roadcuts approximately 400m north of Vail Lake. The upper limits of the slide complex are approximately defined by a series of nearly horizontal topographic benches that lie at elevations of 630m to 700m. Many smaller and younger slides, some of which are active, occur along the toe of the main slide mass. The oldest and largest of these secondary slides probably originally extended much further to the south, and perhaps at one time temporarily blocked Temecula Creek in the area below Vail Dam.

Composition and Structure

The landslides consist of loose to moderately well-cemented breccia containing clasts varying from a few millimeters to approximately 4m in longest dimension and set in a light grey, silty to clayey matrix of rock flour. The basal shear zone contains multiple planar rupture surfaces that dip several degrees to the south. Exposures of the toe of the landslide complex reveal a complex relationship with the underlying Temecula Arkose, a Pliocene fluvial and lacustrine sandstone and claystone unit that overlies the gabbro beyond the limits of landsliding. Exposures of the Temecula Arkose that now underlie the landslide locally reveal multiple imbricated, shallow northward-dipping slip surfaces similar to those produced by large-scale thrust faults. The primary shear zone lies at the base of the gabbro breccia that is in one locality approximately 10m above the lowest series of slip surfaces observed in the Temecula Arkose. This relationship suggests a complex mode of failure in which movement of the slide mass produced sufficient drag forces below the primary rupture surface to cause additional shearing at greater depth in the underlying claystones.

Mechanism of Failure

The triggering mechanism for the two largest landslides comprising this complex group is not known; however, the region is located in an area of high seismicity between the active Elsinore Fault to the south and the San Jacinto Fault to the east. The Lancaster Fault has been mapped to the eastern limits of the slide complex. Movement along this fault probably resulted in extensive fracturing in the gabbro that eventually contributed to the sliding process.

Several outcrops that expose the head of the slide complex suggest that the largest and oldest slides began as large rock-slides failing along fault planes just below the crest of Oak Mountain. The sliding rock mass then impacted gently southerly-dipping slopes of Temecula Valley. The existence of debris-flow-like sediments now exposed at the base of the slide just east of Vail Dam suggest that the substrate was wet or saturated at the time it was overridden by the slide mass. The sudden impact of the overriding slide could have created sufficiently high pore-pressures in the wet sediments to temporarily reduce shear strengths and allow a longer run-out than one would expect based on the nature of the rock involved in sliding.

Summary And Conclusions

Landsliding is an important geomorphic process in the crystalline and metamorphic rocks making up the Peninsular Ranges of southern California. While there are examples of most major types of landslides including earth-flows, wedge-slides, slumps, complex slides, and rock-avalanches, the complex landslides (combinations of slumps and translational slides) are most common. While examples of rock-avalanches are rare, the Vallecito Megabreccia, an erosional remnant of a large rock-avalanche, is preserved as an interbed in the Middle Pleistocene Palm Spring Formation.

The complex landslides are generally very large. The Tin Can Flat Landslide and the Oak Mountain and Vallecito Landslide complexes, for example, have areas of 5 to 6km^2. Debris-flow-like sediments that are present to some extent beneath all landslides studied suggest that water was a factor in the mechanism of failure. Water was a major factor in the failure mechanisms of The Tin Can Flat Landslide and the Vallecito Megabreccia. Both of these landslides occurred as two-layer systems in which an upper non-fluid rock mass rode "piggy back" on a pressurized mud-slurry. Another common feature of all four landslides discussed in this study is the presence of faulting. The landslides have either utilized fault planes along portions of their failure surfaces or the faults have set the stage for sliding by extensively fracturing the host rock.

In most cases, the relative ages of sliding can only be estimated by the degree if dissection; however, the toe of the one slide within the Vallecito Landslide Complex has been displaced approximately 1km by the Elsinore Fault. A maximum age of sliding, based on the faults's estimated slip rate of 3.4 to 5.5 mm/yr, is 180,000 to 290,000 years.

Acknowledgements

The author is grateful to R. Shlemon for his critical review of this paper and to R. Scaramella, D. Streiff, and W. Elliott for their helpful suggestions in the field.

References

Erismann, T. H., Mechanisms of large landslides: *Rock Mechanics*, v. 12, pp. 15-46, 1979.

Hart, M. W. and A. E. Farkas, Reconnaissance of landslides and debris flows in the Cuyamaca Peak Gabbro, San Diego County, California, in *Abs. & Program, Assn. of Engineering Geologists 26th Annual Meeting*, p. 71, 1983.

Hsu, K. J., Catastrophic debris streams (sturzstroms) generated by rockfalls: *Geological Society of America Bulletin*, v. 86, pp. 129-140, 1975.

Johnson, B., Blackhawk landslide, California, U.S.A., in Voight, B., ed., *Rockslides and Avalanches, v.1; Natural Phenomena;* Amsterdam, Elsevier, pp. 481-504, 1978.

Kerr, D. R., Early Neogene continental sedimentation in the Vallecito and Fish Creek Mountains, western Salton Trough, California, *Sedimentary Geology*, v. 38, pp. 217-246, 1984.

Kreiger, M. H., Large landslides, composed of megabreccia, interbedded in Miocene basin deposits, southeastern Arizona, *U. S. Geological Survey Professional paper*, 1008, 25p, 1977.

Melosh, H.J., The mechanics of large rock avalanches, in Costa, J.E., and Wieczorek, G.F., eds., *Debris Flows/ Avalanches; Process, Recognition, and Mitigation, Reviews in Engineering Geology v. VII, Geological Society of America*, pp. 41-49, 1987.

Pinault, T. C., Structure, tectonics geomorphology, and neotectonics of the Elsinore Fault between Banner Canyon and the Coyote Mountains, southern California, *unpublished M. S. Thesis*, San Diego State University, 1984.

Robinson, J. W. and R. L. Threet, Geology of the Split Mountain area, Anza-Borrego Desert State Park, Hart, M. W. and R. V. Dowlen eds. in *Recent Geologic and Hydrologic Studies, Eastern San Diego County and Adjacent areas*, San Diego Assn. Geologist Field Trip Guidebook, p. 47-56, 1974.

Shreve, R.L., The Blackhawk landslide, *Geological Society of America special paper*, 108, p. 47, 1968a.

Shreve, R.L., Leakage and fluidization in air layer lubricated avalanches, *Geological Society of America Bulletin*, v. 79, pp. 653-658, 1968b.

Todd, V.R., Geologic map of Agua Caliente Quadrangle, San Diego County, California, *U. S. Geological Survey Open File Report*, 77-742, 17 p., 1977.

Vaughan, P. R., Alluvial stratigraphy and neotectonics along the Elsinore Fault at Agua Tibia Mountain, California, *unpublished M. S. Thesis*, San Diego State University, 1987

Voight, B., and Pariseau, W.G., Rockslides and avalanches: an introduction, in Voight B., ed., *Rockslides and Avalanches, v. 1; Natural Phenomena*: Amsterdam, Elsevier, pp. 1-67, 1978.

Weber, H. W., Jr., Geology and mineral resources of San Diego County, California, *County Rpt. 3, Calif. Div. Mines and Geology*, 305 p., 1963.

Woodard, G.D., The Cenozoic succession of the west Colorado desert, San Diego and Imperial Counties, southern California, *PhD. Thesis*, University of California, Berkeley, 173 p., 1963.

Yarnold, J. C., and Lombard, J.P., A facies model for large rock-avalanche deposits found in dry climates, in Colburn, Ivan P., Abbott, P., and Minch, J., eds., *Conglomerates in basin analysis: A symposium dedicated to A. O. Woodford*: Pacific Section S.E.P.M., v. 62, pp. 9-31, 1989.

Reprinted from *Geological Excursions in Southern California and Mexico*, Guidebook for the 1991 Annual Meeting, Geological Society of America, San Diego, California.